Greater Kenyir Landscapes

Mohd Tajuddin Abdullah • Aqilah Mohammad
Mohamed Nor Zalipah • Muhamad Safiih Lola
Editors

Greater Kenyir Landscapes

Social Development and Environmental Sustainability: From Ridge to Reef

Springer

Editors
Mohd Tajuddin Abdullah
Institute of Tropical Biodiversity and Sustainable Development
Universiti Malaysia Terengganu
Kuala Nerus, Terengganu, Malaysia

School of Marine and Environmental Sciences
Universiti Malaysia Terengganu
Kuala Nerus, Terengganu, Malaysia

Mohamed Nor Zalipah
School of Marine and Environmental Science
Universiti Malaysia Terengganu
Kuala Nerus, Terengganu, Malaysia

Aqilah Mohammad
School of Marine and Environmental Science
Universiti Malaysia Terengganu
Kuala Nerus, Terengganu, Malaysia

Muhamad Safiih Lola
School of Informatics and Applied Mathematics
Universiti Malaysia Terengganu
Kuala Nerus, Terengganu, Malaysia

Institute of Tropical Biodiversity and Sustainable Development
Universiti Malaysia Terengganu
Kuala Nerus, Terengganu, Malaysia

ISBN 978-3-030-06395-5 ISBN 978-3-319-92264-5 (eBook)
https://doi.org/10.1007/978-3-319-92264-5

Softcover re-print of the Hardcover 1st edition 2019

This Springer imprint is published by the registered company Springer Nature Switzerland AG
The registered company address is: Gewerbestrasse 11, 6330 Cham, Switzerland

Preface and Acknowledgements

The Greater Kenyir landscapes are varied from Mount Gagau in the pristine Taman Negara (National Park) to the man-made Tasik Kenyir, Sungai Terengganu valley, coastal zone and the Terengganu offshore islands. The vast area is subjected to constant exploitation for economic development that calls for the need for research to document its physical features, ecosystems and the communities that are constantly adapting to the environment. Many researchers from the Universiti Malaysia Terengganu have taken the challenge to observe and record the natural and disturbed environments and aspects of the human livelihood and management. The physical scientists describe the climatic condition of the Tasik Kenyir, physical characteristics of the Setiu Wetland and polycyclic aromatic hydrocarbon in the National Park. The biologists describe the wide-ranging resources from bryophytes, fungi, ginger, woody plants, invertebrates and vertebrates. The final section of the book deals with modeler who explore the eco-tourism potentials of Tasik Kenyir and social scientists who examine the livelihood of the local communities along Sungai Terengganu and the domestic water issues and challenges.

This book will provide the basic knowledge on the Greater Kenyir as well as the gap for deep research by future scientists who are interested in the environmental issues that are facing the tropical rainforest region. As we progress for economic prosperity, humans have greater appetite for land, while the primary forest is shrinking and rivers are getting murky. We also aspire to be a First World Nation by 2020. As a benchmark of a developed nation, our children should be able to swim in a pristine Sungai Terengganu and hear the glorious sounds of birds and gibbons in the nearby forests.

The editors and authors are very grateful to numerous research grants and support provided by the Ministry of Education and the Universiti Malaysia Terengganu that are acknowledged in the chapters, and the untiring staff members of Springer Publication for the support and engagements with the book project editors.

Universiti Malaysia Terengganu
Kuala Nerus, Terengganu, Malaysia
28 March 2018

Mohd Tajuddin Abdullah
Aqilah Mohammad
Mohamed Nor Zalipah
Muhamad Safiih Lola

Contents

Contributors

University's Non-Academic Staffs and Postgraduate Students

The remarkable contributors of this book are the administrative and support staffs including the administrative editors Amirah Azizah Zakaria and Amera Hanna Mohd Tajuddin, staffers of the Springer Publisher, postgraduate students from the Institute of Tropical Biodiversity, School of Marine and Environmental Sciences, School of Social and Economic Development, Institute of Oceanography and Environment, School of Fundamental Science, School of Maritime Business and Management, School of Informatics and Applied Mathematics, Research Publication Section and Corporate Communication Centre from Universiti Malaysia Terengganu who have supported in the overall researches and compilations of articles in this book.

Photo Credit

Additionally, the astonishing photographs in this book were taken by our professional photographer, better known as Dome.

Greater Kenyir: Landscapes, Social Development and Environmental Sustainability

From Ridge to Reef

This book offers an uncommon glimpse of the vibrant tropical rainforest biome and the physical features in the state of Terengganu situated on the east coast of Peninsular Malaysia. Greater Kenyir is a vast landscape of natural topographies intermixed with man-made structures. Combining those landscapes from the natural heritage of Taman Negara (National Park), well-preserved ridge of Mount Gagau to the pristine riparian and lowland dipterocarp forest, man-made structure of Kenyir Lake (Tasik Kenyir), industrial agricultural oil palm crop, local fruit orchard, villages, townships, wetlands, coastal forest and to the off-coast islands that provided the dynamism of Terengganu's ecosystems. All over these areas there are numerous cultural and biological diversity for the maintenance of livelihood of many Malay communities and indigenous people.

The tropical rainforest is functioning countless ecological and environmental services from the regulation of climate and temperatures, pollination for the healthy maintenance of the forest trees, as specialised niches for hyperdiversity of fauna and flora, as carbon sink and also to provide clean water supply that feed many rivers and man-made lake of Kenyir. During the north-east monsoon seasons that usually produce high precipitation to sustain the Kenyir reservoir capacity to produce continuous cheaper hydroelectric power to feed for industry and domestic use in Terengganu as well as to other industrial areas on the western parts of Malaysia.

Construction of Sultan Mahmud Dam in 1984 (Photograph by Ab Wahab)

High rate of development is manifested with the fragmentation of habitats and wildlife populations in the Greater Kenyir. Our dedicated scientists spent thousands of hours observing and documenting those biological and physical interactions affecting the ecosystems in the landscapes. Anthropogenic activities produce particulate matters that infiltrate from the terrestrial areas into the aquatic environment.

Kenyir Dam area at present time

The dynamic of social, economic and ecological interactions resulted in the resilience of the rural community to maintain sustainable livelihood. The river communities in the Greater Kenyir are relatively poor small-scale farmers earning about USD230 per month, while the farmer-trader group earn about USD460. For the rural communities, they are able to tackle the challenge of low income in the face of uncertainties of environmental and economic changes.

This book will provide basic information for new researchers and undergraduate students as reference point for the Greater Kenyir. It is hoped that it will stimulate future debate on the sustainability of our development that is decreasing the resources in the Kenyir landscape.

Cleaning site processes of Kenyir in 1979 (Photograph by Ab Wahab)

View of Tasik Kenyir in 1984 (Photograph by Ab Wahab)

Lower Sungai Terengganu below the Kenyir Dam

About the Editors

Mohd Tajuddin Abdullah is a Professor at the School of Marine and Environmental Sciences, Universiti Malaysia Terengganu. He received his PhD in Zoology from the University of Queensland, Australia; MSc and BSc from the West Virginia University, USA; and Diploma in Forestry from the Institut Teknologi MARA, Malaysia. He was elected and inducted as a Fellow in the Academy of Sciences Malaysia on 8 June 2013 and DIMP from the Royal Highness the Sultan of Pahang. His research discipline is in biology with specialisation in zoology, molecular ecology, biodiversity and protected area management. In the 1970s to 1990s, he studied on the endangered Sumatran rhinos. In the late 1990s, he studied in Borneo on the wildlife diversity, distribution, population, ecology and species conservation and their natural habitats. From 2000 to 2014, he used molecular methods to study on the ecology, biogeography and evolution of mammals in Borneo and Peninsular Malaysia. From 2011 he received grants to study on zoonoses, genome and metagenomic of primates. From 2014 onwards he worked in the Greater Kenyir, on the canopy wildlife distribution patterns in the dipterocarp forest, wetlands and islands on the east coast of Malaysia.

Aqilah Mohammad is a Senior Lecturer who teaches Mycology at the School of Marine and Environmental Sciences, Universiti Malaysia Terengganu. She received her PhD in Biological Sciences from the Royal Holloway, University of London, United Kingdom, and MSc degree in Conservation Biology from Universiti Kebangsaan Malaysia, Bangi, Malaysia. Underlying themes of her research include fungal ecology, biodiversity, identification, fungal interaction, fungal fruiting phenology and ethnomycology. Her current research mainly focuses on the utilisation of wild mushrooms for food and medicine by indigenous communities in the east coast of Peninsular Malaysia. She is also working on an identification of edible bolete fungus found growing nearby Melaleuca trees in Setiu Wetlands, Terengganu.

Mohamed Nor Zalipah is a Senior Lecturer at the School of Marine and Environmental Sciences, Universiti Malaysia Terengganu, Malaysia. She received her Bachelor degree in Zoology from Universiti Sains Malaysia, Penang, Malaysia, and MSc degree in Environmental Biology from the same university. For her MSc degree, she studied the edge effect on the non-volant small mammal's community in Perak, Malaysia. She worked on bat pollination of mangroves called *Sonneratia* in Malaysia and obtained her PhD degree from the University of Bristol, UK. Her current work is mainly on plant-animal interactions, particularly in pollination and seed dispersal.

Muhamad Safiih Lola is an Associate Professor at the School of Informatics and Applied Mathematics, Universiti Malaysia Terengganu.He received his PhD in Econometrics Modelling from the Universiti Sains Malaysia, Penang, Malaysia; MSc in Applied Statistics from the Universiti Putra Malaysia, Serdang, Malaysia; and BEc (Honours) in International Trade and Business from Universiti Utara Malaysia, Kedah, Malaysia. He started his career as a statistician and mathematics lecturer in 2001 and has over 15 years of experience in applied statistics, econometrics and hybrid modelling with specialisation in forecasting, eco-tourism, total economic value and loyalty programme. He has obtained research grants on projects involving fuzzy parametric and fuzzy nonparametric sample selection models, studies on cost of living particularly on loyalty and membership programmes and also studies in dynamic modelling.

Part I
Physical Environment

The Weather and Climate of Tropical Tasik Kenyir, Terengganu

Samsuri Abdullah and Marzuki Ismail

Abstract This study investigates the trend of weather and climate in Tasik Kenyir, Terengganu State based on seasonal monsoons utilizing nearest meteorological station of Kuala Terengganu. Data from 1985 to April 2016 was used to describe the physical environment of Kenyir Lake in terms of rainfall amount (mm), relative humidity (%), ambient temperature (°C), MSL pressure (hPa), and wind speed (m/s) and wind direction (degrees). The data was first divided by different monsoon seasons faced by Kuala Terengganu; Southwest Monsoon (SWM), Northeast Monsoon (NEM) and Inter Monsoon (IM). There exists significant different ($p < 0.05$) during these three monsoon seasons for rainfall amount, ambient temperature, and MSL pressure. Conversely, relative humidity shows no significant different between the monsoon seasons. Rainfall amount, relative humidity and MSL pressure noted high intensity during NEM with an increment of 250.1 mm, 0.3%, and 2.2 hPa from SWM and by 189.2 mm, 0.2%, and 1.3 hPa from IM, respectively. Meanwhile, there was a decrease of ambient temperature during NEM by 1.1 °C from SEM and 0.8 °C from IM. Prevailing wind direction was noticed from NE direction with mean value of 3.5 m/s. In conclusion, regardless of any monsoon seasons, there is no significant change in relative humidity variation, but rainfall amount, ambient temperature and MSL pressure shows significant change for each and every monsoon. The understanding of indigenous community on the changing of monsoon seasons might help them in early preparation for foods and other materials for their own survival and festive activities.

Keywords Meteorological parameter · Monsoon seasons · Tasik Kenyir · Terengganu

S. Abdullah · M. Ismail (✉)
School of Marine and Environmental Sciences, Universiti Malaysia Terengganu, Kuala Nerus, Terengganu, Malaysia
e-mail: marzuki@umt.edu.my

M. T. Abdullah et al. (eds.), *Greater Kenyir Landscapes*,
https://doi.org/10.1007/978-3-319-92264-5_1

Introduction

Malaysia is characterized as having climate of hot and humid throughout of the year, since it is situated near the equator (Ismail et al. 2015a, b). The weather and climate is generally defined through several meteorological parameters. The most important meteorological parameter is rainfall intensity or rainfall amount. The rainfall plays an important role in the distribution of heavy or light rain over the areas. Not to be neglected, some of the other meteorological parameters are also important in defining the weather of an area. Those are ambient temperature, relative humidity and MSL pressure. Wind speed and wind direction are the two key meteorological parameters in determining the prevailing wind of an area.

Tasik Kenyir is situated in Terengganu State of Peninsular Malaysia. Terengganu State is located in East Coast of Peninsular Malaysia which was popularized with the known beaches and islands. The east coast of Peninsular Malaysia experienced several monsoons seasons, southwest monsoon (SWM), northeast monsoon (NEM) and inter monsoon (IM) (Abdullah et al. 2017). The main characteristic of SWM is dry seasons and NEM is wet seasons. In defining the weather and climate of Tasik Kenyir, the analysis was based on these three monsoon seasons. This study highlights the trend of selected meteorological parameters and determines the significant change of each parameter throughout different monsoon seasons. The changes of monsoon seasons is believe in affecting indigenous populations in terms of restriction in accessing a safe and nutritious food, which including traditional foods that is considered very important for cultural practices. The changes of seasons also trigger to the threatening of cultural identities of indigenous people as such plants and animals used in traditional practices or scared ceremonies become less available. By understanding of the trend of monsoon seasons, early steps can be taken by this community for early preparation for their foods and festive activities.

In conjunction describing the weather and climate of Tasik Kenyir, the nearest meteorological station was used in acquisition of data. Kuala Terengganu Meteorological Station was selected in representing the weather and climate of Tasik Kenyir. It is precisely located at coordinate: 5°23′07.4″N; 103°06′37.0″E as shown in Fig. 1. The distance between the Kenyir Lake (5°00′N; 102° 48′E) and Kuala Terengganu Meteorological Station was 66.94 km.

The parameters taken into consideration were rainfall amount (mm), mean sea level pressure (MSLP) (hPa), relative humidity (%), temperature (°C), wind speed (m/s), and wind direction (degrees), all are monthly basis data. The data covers period from year 1985 to 2016 (32 years). The completeness of data in an analysis is important in terms of maintaining the data reliability (Suhaila et al. 2010). In this study, no imputation of missing data was performed as the data was complete for that period. Climate of Terengganu is best described by different monsoon seasons. The data management was performed in dividing the data into three main monsoon seasons. The southwest monsoon (SWM) occurs from May to August, northeast monsoon occurs from November to February and inter-monsoon occurs during September to October and March to April (Daryabor et al. 2014; Suhaila et al.

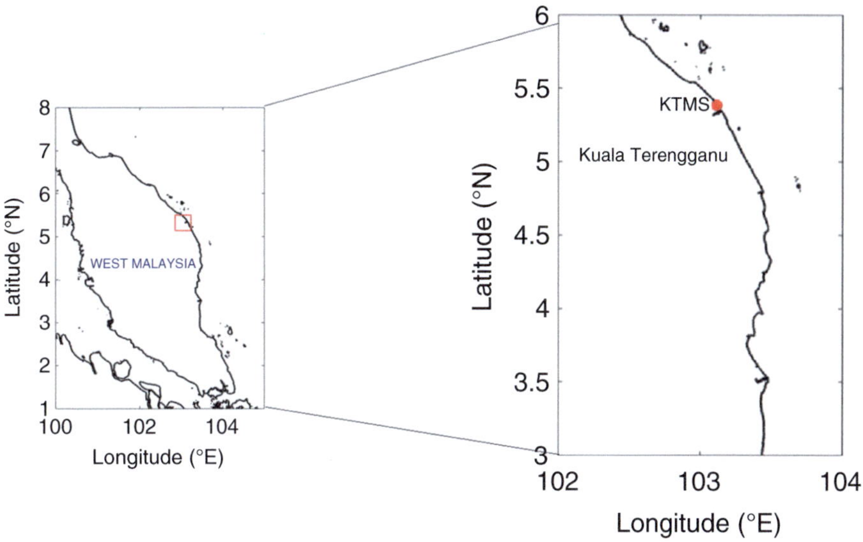

Fig. 1 Location of Kuala Terengganu meteorological station (KTMS)

2010). Therefore three main data sets were extracted which having different monsoon seasons (except for wind speed and wind direction) and statistical analysis was performed on these data sets.

The trend of weather and climate of Tasik Kenyir was determined by constructing the boxplot. Boxplot is describe statistically and act as visual aid which the data was represented by distribution of rectangle containing several information in the data set such as minimum, maximum, first quartile, third quartile, median and mean values (Ismail et al. 2016). The statistically significant different among the monsoon seasons for each parameters considered was tested by applying Analysis of Variance (ANOVA) whereby the evaluated of P-value of less than (<0.05) concludes that there exist significant different among monsoon seasons and vice versa (Ismail et al. 2015a, b). The analysis was taken at 95% confidence interval. Microsoft Excel Spreadsheet 2013 was used for tabulation and analysis of data.

Patterns of Meteorological Parameters During Different Monsoon Seasons

The trend of rainfall amount, temperature, relative humidity and MSL pressure during different monsoon seasons was described in Fig. 2. The trend of rainfall amount on average was same with MSL pressure which the highest was denoted during NEM with rainfall amount of 370.0 mm, and MSL pressure of 1011.4 hPa. Temperature shows that on average, NEM has lowest temperature with 26.5 °C. It indicates that as the increasing rainfall amount, will result in decreasing of

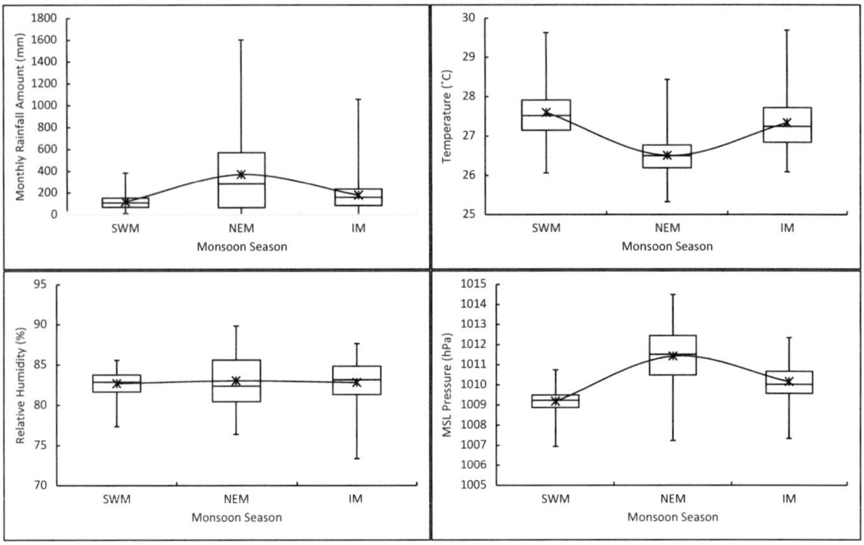

Fig. 2 Trend of rainfall amount, temperature, relative humidity and MSL pressure during different monsoon seasons

temperature. The trend of relative humidity is slightly constant during the three monsoon seasons with range of 82.8–83.0%. Descriptively, the highest mean for rainfall amount was 370.0 mm, where it increased by 191.4 mm from IM and 252.2 mm from SWM. Generally, rainfall amount at east coast of Peninsular Malaysia during NEM is higher as it receives heavy rainfall. Eastern part of Peninsular Malaysia is not blocked by Titiwangsa Range, and therefore heavy rain was bring by northeasterly winds towards east coast, consequently has wettest area during NEM (Akhir and Chuen 2011; Suhaila et al. 2010). NEM recorded lowest mean value of temperature with 26.5 °C, where it decreased by 0.8 °C during IM and 1.1 °C during SWM. It is noted SWM is the drier period for the whole Malaysian Peninsular. There is no slight change of mean values in relative humidity during the monsoon seasons with 82.7%, 83.0%, and 82.8% for SWM, NEM, and IM, respectively. The highest mean for MSL pressure was 1011.4 hPa, where it increased by 1.2 hPa from IM and 2.2 hPa from SWM. The analysis of wind speed and wind direction was based on the data January 1970 to April 2016. Prevailing wind direction was noticed from NE direction with mean value of 3.5 m/s (Fig. 3). The other summary of wind direction and mean values of wins speed were; N (3.3 m/s), E (2.8 m/s), SE (2.3 m/s), S (1.5 m/s), SW (1.5 m/s), W (1.5 m/s), and NW (2.4 m/s).

The differences of each selected weather parameters is further confirmed with the ANOVA. ANOVA was performed in determining the significant different of each parameter during different monsoon seasons. Results show that rainfall amount, temperature and MSL pressure has statistically significant different ($P < 0.05$) with the P-value of 1.37×10^{-16}, 5.04×10^{-39} and 8.06×10^{-51}, respectively

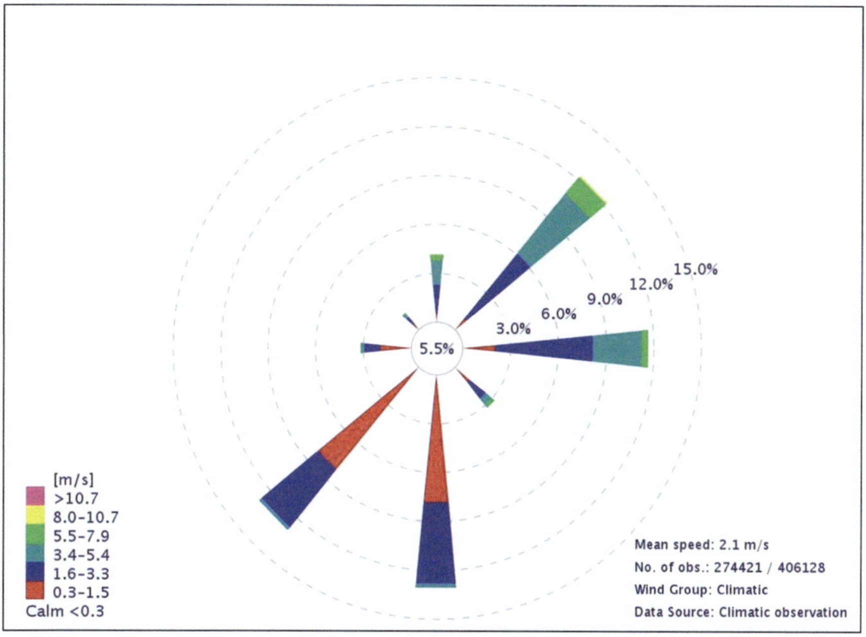

Fig. 3 Wind rose for Kuala Terengganu meteorological station, between 1.1.1970 and 30.4.2016

during the three monsoon seasons, conversely relative humidity has no significant different ($P > 0.05$) with 0.5918 during the different monsoon seasons.

Conclusion

The data of monthly basis from year 1985 to April 2016 was acquired form MMD of Kuala Terengganu Meteorological Station representing Tasik Kenyir, Terengganu. Parameters used were rainfall amount, relative humidity, ambient temperature, MSL pressure, and wind speed and wind direction. The analysis was conducted based on the different monsoon seasons in Peninsular Malaysia namely; SWM, NEM and IM. Results revealed that the trend of rainfall amount and MSL pressure is slightly same with highest mean values during NEM, conversely, ambient temperature has lowest mean values during NEM. Prevailing wind direction was noticed from NE direction with mean value of 3.5 m/s. ANOVA shows that there exist statistically significant different of rainfall amount, ambient temperature and MSL pressure during different monsoon season, while there is no significant different for relative humidity over the monsoon seasons. The comprehension of indigenous group on the different seasons may help them in early readiness for sustenance and different materials for their own particular survival.

Acknowledgements This study was funded by Universiti Malaysia Terengganu Scholarship Scheme (BUMT) to Author 1. The authors also would like to thank the Malaysian Meteorological Department (MMD) for the meteorological data.

References

Abdullah, S., Ismail, M., & Fong, S. Y. (2017). Multiple Linear Regression (MLR) models for long term PM_{10} forecasting during different monsoon seasons. *Journal of Sustainability and Management, 12*(1), 60–69.

Akhir, M. F. M., & Chuen, Y. J. (2011). Seasonal variation of water characteristics during inter-monsoon along the east coast of Johor. *Journal of Sustainability and Management, 6*(2), 206–214.

Daryabor, F., Tangang, F., & Juneng, L. (2014). Simulation of southwest monsoon current circulation and temperature in the east coast of Peninsular Malaysia. *Sains Malaysiana, 43*(3), 389–398.

Ismail, M., Abdullah, S., & Fong, S. Y. (2015a). Trend and status of particulate matter (PM_{10}) concentration at three major cities in east coast of Peninsular Malaysia. *Research Journal of Chemical and Environmental Sciences, 3*(5), 25–31.

Ismail, M., Suroto, A., & Abdullah, S. (2015b). Response of Malaysian local rice cultivars induced by elevated ozone stress. *EnvironmentAsia, 8*(1), 86–93.

Ismail, M., Abdullah, S., Fong, S. Y., & Ghazali, N. A. (2016). A ten year investigation on ozone and it precursors at Kemaman, Terengganu, Malaysia. *EnvironmentAsia, 9*(1), 1–8.

Suhaila, J., Deni, S. M., Zin, W. Z. W., & Jemain, A. A. (2010). Trends in Peninsular Malaysia rainfall data during the southwest monsoon and northeast monsoon seasons: 1975–2004. *Sains Malaysiana, 39*(4), 533–542.

A Preliminary Assessment of Water Quality Status in Tasik Kenyir, Malaysia

Suhaimi Suratman, Edmand Bedurus, Suzana Misbah, and Norhayati Mohd Tahir

Abstract The preliminary study of water quality which involved the measurement of physical, chemical and microbiological parameters was conducted in Tasik Kenyir, Malaysia from August until October 2010. Results showed that the water quality of the surface water was in 'clean' status based on Malaysian Water Quality Index. However, the bottom water was in 'slightly polluted'. According to National Water Quality Standard, all stations was in Class I which is suitable for conservation of natural environment, water supply with practically no treatment and fishery activity for very sensitive aquatic species.

Keywords Water quality index · National water quality standard · Sustainable water · Tasik Kenyir

---

S. Suratman (✉)
Institute of Oceanography and Environment, Universiti Malaysia Terengganu, Kuala Nerus, Terengganu, Malaysia

Institute of Tropical Biodiversity and Sustainable Development, Universiti Malaysia Terengganu, Kuala Nerus, Terengganu, Malaysia

School of Marine and Environmental Sciences, Universiti Malaysia Terengganu, Kuala Nerus, Terengganu, Malaysia
e-mail: miman@umt.edu.my

E. Bedurus
Institute of Oceanography and Environment, Universiti Malaysia Terengganu, Kuala Nerus, Terengganu, Malaysia

S. Misbah
School of Fundamental Sciences, Universiti Malaysia Terengganu, Kuala Nerus, Terengganu, Malaysia

N. Mohd Tahir
Institute of Oceanography and Environment (INOS), Universiti Malaysia Terengganu, Kuala Nerus, Terengganu, Malaysia

School of Marine and Environmental Sciences, Universiti Malaysia Terengganu, Kuala Nerus, Terengganu, Malaysia

M. T. Abdullah et al. (eds.), *Greater Kenyir Landscapes*,
https://doi.org/10.1007/978-3-319-92264-5_2

Introduction

Numerous studies have been conducted in the early years since the establishment of Tasik Kenyir to determine water quality in relation to the effects of sustainable aquaculture, fish diversity and fish stock assessment, zooplankton dynamics and production (Jamaludin and Machiels 1999; Yusoff and Ambak 1999; Verhallen and Verhagen 1994). The increase of visitor numbers to Tasik Kenyir and the rapid development in the area might increase the possibility of deforestation and sewage discharge into the lake. These factors might then decrease the water quality of the lake. Therefore, the present study was carried out to establish the current status of the water quality by measuring the physical, chemical and microbiological (***Escherichia coli*** and total coliform) parameters. The Water Quality Index (WQI) was also calculated according to Department of Environment-WQI (DOE-WQI) as tabulated in Table 1 (Department of Environment 2011). In addition, data obtained was compared to National Water Quality Standard (NWQS) classification to understand the beneficial uses of the water body (Table 2) (Department of Environment 2011).

Table 1 DOE-WQI calculation formula (Department of Environment 2011)

Subindex DO (SIDO) (% saturation)	$x \leq 8$	SIDO = 0
	$8< x <92$	SIDO = 100
	$x \geq 92$	$SIDO = -0.395 + 0.03x^2-0.0002x^3$
Subindex BOD (SIBOD) (mg/L)	$x \leq 5$	$SIBOD = 100.4-4.23x$
	$x >5$	$SIBOD = 108e^{-0.055x} - 0.1x$
Subindex COD (SICOD) (mg/L)	$x \leq 20$	$SICOD = -1.33x + 99.1$
	$x >20$	$SICOD = 103e^{-0.0157x} - 0.04x$
Subindex AN (SIAN) (mg/L)	$x \leq 0.3$	$SIAN = 100.5 - 105x$
	$0.3< x <4$	$SIAN = 94e^{-0.573x} - 5 \lvert x - 2 \rvert$
	$x \geq 4$	SIAN = 0
Subindex TSS (SITSS) (mg/L)	$x \leq 100$	$SITSS = 97.5e^{-0.00676x} + 0.05x$
	$100< x <1000$	$SITSS = 71e^{-0.0061x} - 0.015x$
	$x \geq 1000$	SITSS = 0
Subindex pH (SIpH)	$x <5.5$	$SIpH = 17.2-17.2x + 5.02x^2$
	$5.5\leq x <7$	$SIpH = -242 + 95.5x - 6.67x^2$
	$7\leq x <8.75$	$SIpH = -181 + 82.4x - 6.05x^2$
	$x \geq 8.75$	$SIpH = 536 - 77x + 2.76x^2$

DOE-WQI = (0.22*SIDO) + (0.19*SIBOD) + (0.16*SICOD) + (0.15*SIAN) + (0.16*SITSS) + (0.12*SipH)

DOE-WQI = 0–59 (Polluted); 60–80 (Slightly Polluted); 81–100 (Clean)

Table 2 National water quality standards for Malaysia (Department of Environment 2011)

Parameter	Unit	Class					
		I	IIA	IIB	III	IV	V
Ammoniacal Nitrogen	mg/L	0.1	0.3	0.3	0.9	2.7	>2.7
BOD	mg/L	1	3	3	6	12	>12
COD	mg/L	10	25	25	50	100	>100
DO	mg/L	7	5–7	5–7	3–5	<3	<1
pH	–	6.5–8.5	6–9	6–9	5–9	5–9	–
TSS	mg/L	25	50	50	150	300	300
Temperature	°C	–	Normal +2 °C	–	Normal +2 °C	–	–
Total Coliform	count/100 mL	100	5000	5000	50,000	50,000	>50,000

Class	Uses
Class I	Conservation of natural environment
	Water supply I – Practically no treatment necessary
	Fishery I – Very sensitive aquatic species
Class IIA	Water supply II – Conventional treatment required
	Fishery II – Sensitive aquatic species
Class IIB	Recreational use with body contact
Class III	Water supply III – Extensive treatment required
	Fishery III – Common of economic value and tolerant species; livestock drinking
Class IV	Irrigation
Class V	None of the above

Three sampling surveys were conducted from August until October 2010. Generally, there were two major areas involved, the eastern part of Tasik Kenyir (ELK, stations K1–K15) and Terengganu National Park (TNP, stations N1–N17) which is located in western part of Tasik Kenyir (Fig. 1). In ELK, the water samples were collected from both surface (~1 m depth) and bottom waters (i.e. 30 m depth), along two main transects, which were Transect one (station K1–K10) and Transect two (station K11–K15) using a Van Dorn sampler. In TNP, scattered sampling stations were distributed throughout the region and only surface waters (~1 m) were collected due to shallowness of the water column. In addition, samples for microbiological parameters were only carried out at TNP. Water samples were filtered through 0.45 μm pore size cellulose acetate membrane filters and decanted into 1 L

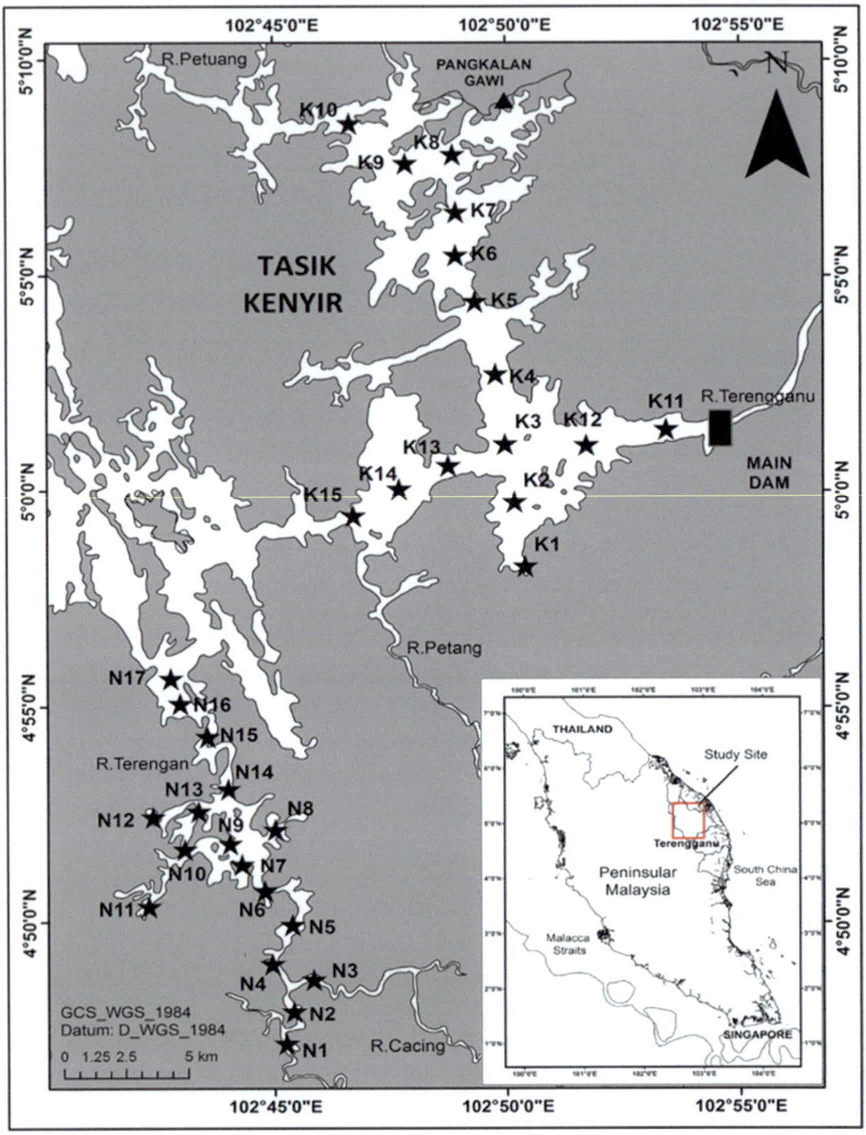

Fig. 1 Sampling stations in Tasik Kenyir

polyethylene bottles for storage prior to dissolved ammonia determination. Surface water samples for microbiological analysis were subsampled directly into 500 mL sterilized Scott bottles. All water samples were stored in an ice chest and transported to the laboratory where all samples were analysed on the same day.

Temperature, pH and dissolved oxygen (DO) were measured in-situ using YSI 6600 multiparameter data logger. Calibration of the instrument was made 24 h before the sampling as per the manufacturer's recommendation. Ammonia was determined colorimetrically according to Grasshoff et al. (1983) with total suspended solids (TSS), biochemical oxygen demand (BOD) and chemical oxygen demand (COD) were based on standard methods (American Public Health Association, 2005). The membrane filtration technique used for the detection and quantification of *E. coli* and total coliform according to the standard methods (American Public Health Association 2005). The number of coliforms were reported on a colony-forming unit per 100 mL basis (CFU/100 mL). Six parameters i.e. pH, DO, BOD, COD, TSS and ammonia were used to calculate the DOE-WQI (Department of Environment 2011). Significant difference for each parameter between different stations and depths were determined by using Analysis of Variance (ANOVA) test.

Temperature

The temperature for surface water of ELK was in the range of 29.9–30.8 °C (mean: 30.4 ± 0.3 °C) (Fig. 2a, b). The range of temperature for surface water in TNP was wider, with values between 23.7 and 31.7 °C (mean: 29.6 ± 2.5 °C). The temperature of bottom water in ELK stations are not reported due to loss of data. Low values recorded at few stations upstream of TNP were expected to be due to sampling being carried out under the trees which shaded the water column.

ANOVA test showed there no significance difference between sampling stations ($p > 0.05$). Temperature values were in Class IIA and above based on NWQS classification. In general, the temperature range was normal for tropical Malaysia rainforest climate especially in lake system (Jawan and Sumin 2012; Wan Mohd Afiq and Md Pauzi 2012; Othman et al. 2009).

pH

In ELK, the pH value varied from 7.0 to 7.7 (mean: 7.3 ± 0.2) for surface water (Fig. 2). Meanwhile, the pH value for the bottom water was between 3.9 and 7.2 (mean: 5.4 ± 1.1). For TNP area, the pH value was ranging from 6.6 to 8.0 (mean: 7.5 ± 0.4). The two-way ANOVA test showed no significant difference between the stations ($p > 0.05$) for both ELK and TNP. However, there was significant difference between the surface and bottom water ($p < 0.05$) for ELK. Based on NWQS classification, most of the surface water was in Class I for both ELK and TNP. In contrast, most of bottom water fell into Class IV.

The pH values for surface and bottom waters were similar at station K1 and K8, due to the shallow of the water (<10 m). Generally, the pH value was in neutral range (6.6–8.0) for surface water while the bottom water (for depth ≥30 m) was acidic (pH ≤6.5). This was probably due to the decomposition process of organic matter (dead trees) at the bottom part of the lake that caused the lower pH of the bottom water. The decomposition produced dissolved CO_2 gases in which turn into the weak acid known as carbonic acid (H_2CO_3) (Moran and Stottrup 2011).

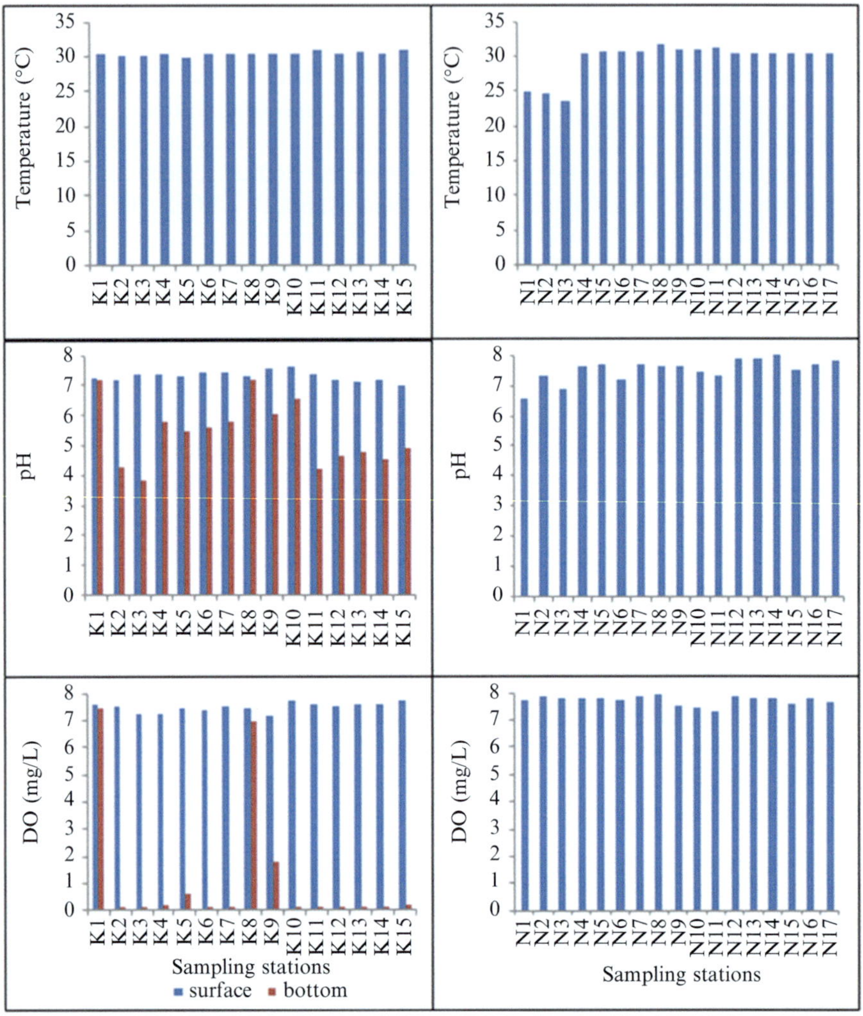

Fig. 2 (**a**) Physical and chemical characteristics of Tasik Kenyir during present study. (**b**) Physical and chemical characteristics of Tasik Kenyir during present study

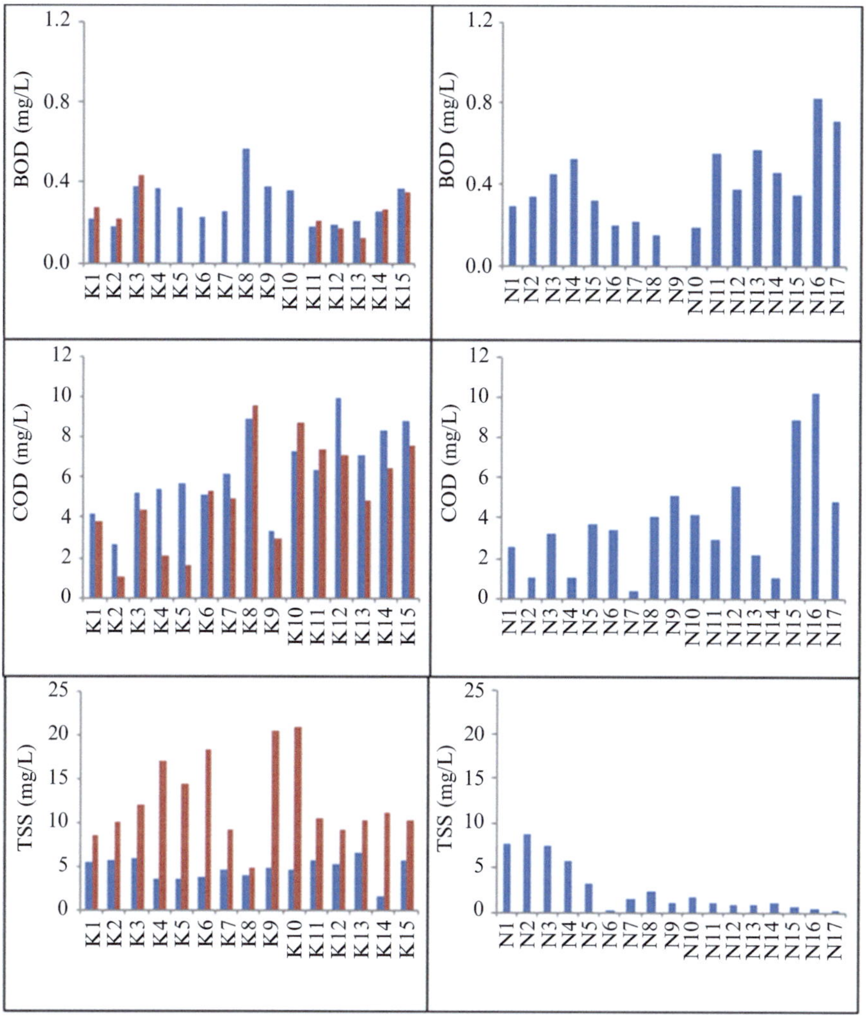

Fig. 2 (continued)

Dissolved Oxygen

Figure 2 shows the DO concentration for surface water in ELK was in the range of 7.2–7.8 mg/L (mean: 7.5 ± 0.2 mg/L) while the bottom water was 0.1–7.5 mg/L (mean: 1.2 ± 2.5 mg/L). A two-way ANOVA test showed no significant difference between the stations ($p > 0.05$). However, the difference was significant between surface and bottom water ($p < 0.05$). Meanwhile, in TNP, the concentration of DO was in the range of 7.3–7.9 mg/L (mean: 7.7 ± 0.2). Similar to ELK for surface water, the values showed no significant difference between stations ($p > 0.05$). According to NWQS, all the surface water was classified as Class I for both ELK and TNP area. Meanwhile, most of the bottom water in ELK fell into Classes IV and V.

In this study, K1 (7.6 mg/L) and K8 (7.5 mg/L) had higher DO value for bottom water because the depth was less than 10 m. Most of the stations had DO value <1 mg/L for bottom water, due to oxygen depletion, caused by the decomposition of submerged vegetation, leaf litter and phytoplankton. There is no obvious trend of high and low DO values between stations, although stations located near the human activities (jetty and restaurants) such as Pangkalan Gawi (K8) may has low DO value due to waste discharge which contained organic compounds directly into the lake. This is contrast to those observed at other lake systems such as Chini Lake, Pahang (Othman et al. 2009), oxbow lakes, Sabah (Jawan and Sumin 2012) and Titiwangsa Lake, Selangor (Said et al. 2012) which recorded low DO content nearby human activities. Waste which contained organic compounds will decrease the DO content in the water column as a result of degradation process by microbial activity (Chapman 1992). It is suggested that due to great volume of the Tasik Kenyir relative to degradation process, the water column of the lake did not show the lower DO often associated with waste and associated nutrient. Thus, the DO content in the water column is still high.

Biochemical and Chemical Oxygen Demand

The BOD concentration for surface water in ELK varies from 0.18 to 0.57 mg/L with a mean concentration of 0.30 ± 0.11 mg/L (Fig. 2). In addition, the bottom water recorded almost similar values to the surface water ranging from 0.12 to 0.44 mg/L (mean: 0.26 ± 0.10 mg/L). No data was available for bottom water in stations K4–K10 due to sample loss. According to two-way ANOVA test, the results showed no significant difference between the stations ($p > 0.05$) and depth profiles ($p > 0.05$). On the other hand, in TNP, the BOD value ranged from 0.15 to 0.71 mg/L (mean: 0.41 ± 0.19 mg/L). There was no significant difference between stations ($p > 0.05$) based on ANOVA test for TNP. For both ELK and TNP area, BOD values were in Class I of the NWQS classification.

In ELK, the lowest and highest values of COD for surface water were 2.70 mg/L and 9.98 mg/L, respectively (Fig. 2). For bottom water in the same area, the lowest value was 1.08 mg/L and the highest value was 9.54 mg/L. The mean values of COD for surface and bottom water were 6.31 ± 2.11 mg/L and 5.19 ± 2.60 mg/L, respectively. The two-way ANOVA test revealed significant difference ($p < 0.05$) between all the stations, and between surface and bottom waters. According to NWQS, all the stations in ELK were in Class I for both surface and bottom waters. Meanwhile, in TNP, the value of COD varied from 0.4 to 10.2 mg/L (mean: 3.80 ± 2.65 mg/L). The values of COD in TNP showed a significant difference between stations ($p < 0.05$) according to ANOVA test. Based on NWQS classification, all stations in TNP were classified as Class I.

BOD is a measurement of amount of DO used by bacteria to oxidise the biodegradable organic compounds present in water. Whereas COD is the amount of oxygen used to oxidise all organic compounds which include biodegradable and non-biodegradable. Hence, as observed in this study, the COD value is always greater than BOD value. Calculations were made in order to estimate to the ratio of biodegradable and non-biodegradable organic compounds in Tasik Kenyir for surface water (Suratman et al. 2015; Kumar et al. 2010). Most of the stations in ELK recorded higher non-biodegradable compounds i.e. 9–53 times much higher compared to biodegradable compounds. In contrast, much lower values were found in TNP i.e. 2–27 times higher between non-biodegradable and biodegradable compounds. These results suggested that most of the organic compounds in Tasik Kenyir is non-biodegradable especially at ELK.

Total Suspended Solids

In ELK, the highest value of TSS for surface water was 65 mg/L and the lowest was 1.5 mg/L (Fig. 2a, b). For bottom water, the highest TSS concentration was 20.9 mg/L and the lowest was 8.6 mg/L. The mean concentration of TSS for surface and bottom water was 4.7 ± 1.3 mg/L and 12.5 ± 4.7 mg/L, respectively. From the result, the bottom water showed higher TSS values compared to surface water. As the dead trees and leaf litter may fall into the lake and then decay underwater, suspended organic particles were released and contribute to high TSS concentration in the bottom water. A two-way ANOVA test showed no significant difference between the stations ($p > 0.05$). However, there was significant difference demonstrated between the surface and bottom water ($p < 0.05$). According to NWQS, all surface and bottom waters into Class I.

Meanwhile, in TNP, the highest and lowest concentrations of TSS were 8.8 mg/L and 0.2 mg/L. The mean concentration of TSS was 2.7 ± 2.9 mg/L. No significant different of TSS between stations ($p > 0.05$) based on the ANOVA test. Due to low TSS value, surface water in TNP area was classified as Class I, according to NWQS. The result showed a drastic reduction of TSS in TNP area, which was a protected and reserved area. The high TSS value in ELK was probably due to the higher amount of precipitation from boats activities and land clearing into the ELK, which was more densely visited by tourists and fishing enthusiasts.

Ammonia

For ELK, the range of ammonia concentrations recorded for surface and bottom waters was 20.3–77.2 μg/L N (mean: 50.95 ± 18.42 μg/L N) and 37.8–88.4 μg/L N (mean: 63.8 ± 16.2 μg/L N), respectively (Fig. 3). Based on two-way ANOVA test, the ammonia concentration showed significant difference between the sampling stations ($p < 0.05$) as well as between the surface and bottom water ($p < 0.05$). According to NWQS, the concentration for all sampling stations in surface and

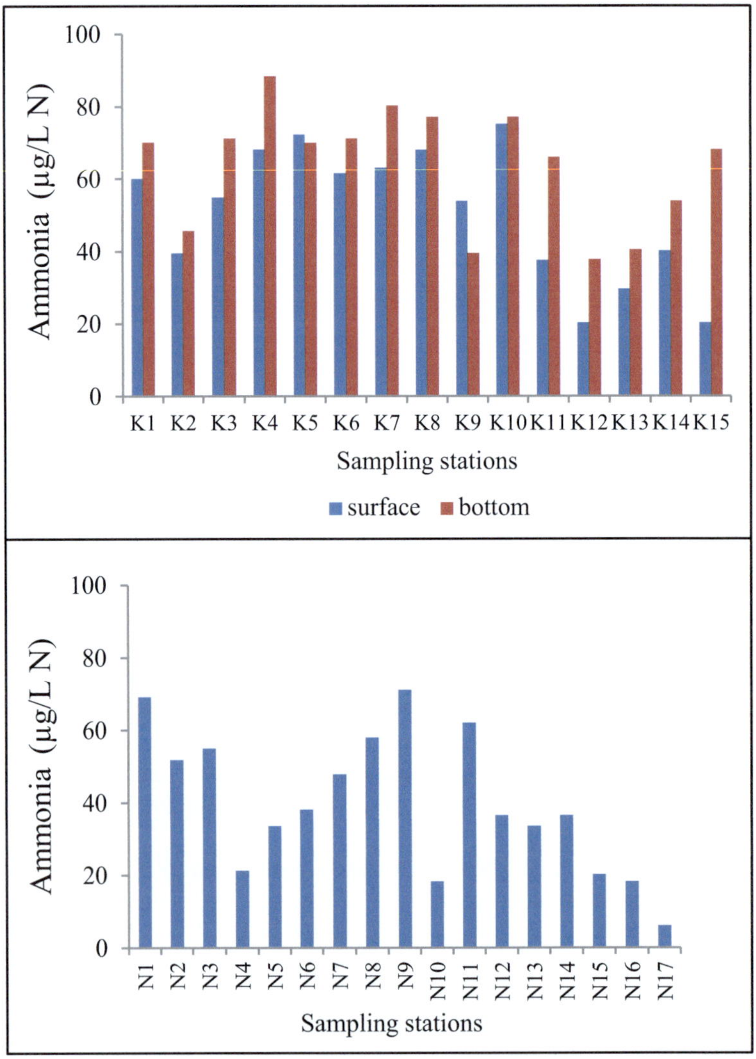

Fig. 3 Ammonia concentration of Tasik Kenyir during present study

bottom water were classified as Class I. In addition, the concentration of ammonia in TNP varied from 6.1 to 71.1 μg/L N (mean: 39.8 ± 19.3 μg/L N). There was a significant different between stations ($p < 0.05$) according to ANOVA test. Similar to ELK, all stations in TNP was in Class I based on NWQS classification.

In ELK, most of bottom water recorded higher concentrations of ammonia compared to surface water. Most probably the decomposition of organic matter from dead trees, leaf litter and phytoplankton in the bottom waters of the lake by heterotropic bacteria that had increased the concentration of this nutrient. Previous study at Bakun Reservoir, Sarawak also recorded the same trend of lower and higher concentrations of ammonia at surface and bottom waters, respectively due to remineri-sation of organic matter to ammonia (Ling et al. 2016a). In addition, a previous study showed that nutrients, especially ammonia, present in deep water will not be utilised by the phytoplankton as light could not reach the bottom of the lake, with these nutrients 'trapped' in this hypolimnion due to thermal stratification in ELK (Yusoff et al. 1994). In addition, our results agree with Yusoff and Ambak (1999), in which they found that lower nutrients level were present in the epilimnion layer (upper layer of the lake) compared to the metalimnetic and hypolimnetic areas (lower layers of the lake) during the period of strong stratification.

The concentrations of ammonia were compared with some selected previous studies in Malaysia area. In general, the ammonia concentrations in the current study covered a lower range of concentration as relative to Chini Lake, Pahang (0–597 μg/L N; Shuhaimi-Othman et al. 2007), Bakun Reservoir, Sarawak (20–1340 μg/L N; Ling et al. 2016a) and Batang Ai Reservoir, Sarawak (17–567 μg/L N; Ling et al. 2016b). The lower concentration in Tasik Kenyir was probably due to less anthropogenic activities in this lake. In contrast, wastes from human and aquaculture activities contributed to high ammonia concentration in Chini Lake, Bakun and Batang Ai Reservoirs.

Microbiological Parameters

Figure 4 shows the concentrations of *E. coli* and total coliform in TNP. No measurement was made at ELK and some stations (N4, N9, N10, N14, N15, and N16) in TNP. The range of concentrations of *E. coli* and total coliform were undetected-126 counts/100 mL (38 ± 41 counts/100 mL) and 68–408 counts/100 mL (139 ± 66 counts/100 mL), respectively. Based on one-way ANOVA test, there was a significant difference between sampling stations ($p < 0.05$) for these two parameters. The concentrations of total coliform in most of the stations were in Class I (≤100 counts/100 mL) according to NWQS classification, although some stations recorded higher values. No comparison could be made for *E. coli* as it is still not listed in this classification

Total coliform and *E. coli* are used as microbiological indicators of water quality. Total coliform count has been used as a tool to test the bacterial contamination in water. The bacteria are not necessarily pathogenic, but indicates the presence of

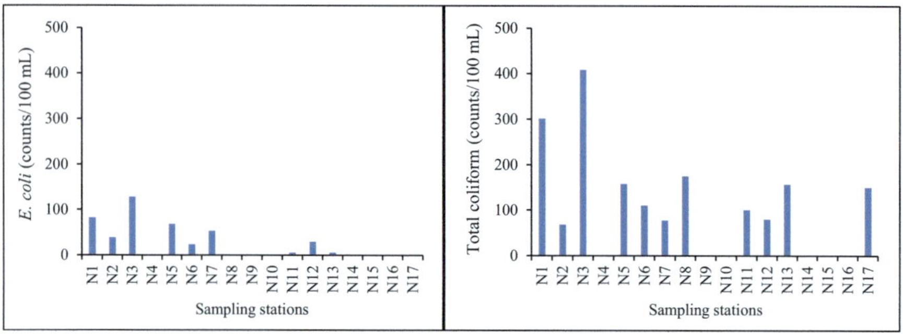

Fig. 4 Microbiological concentration of Tasik Kenyir during present study

disease-causing organisms. It is a large collection of different kinds of bacteria and can be found in the environment (plant and soil material), digestive systems and feces of all warm-blooded animals including humans. *E. coli* is a species of fecal coliform bacteria (i.e. sub-group of total coliform) are harmless and capable of causing many diseases. They exist specifically in the intestines and feces of humans and animals. Due to their origin they are more specific to what compared to general total coliform group of bacteria, and *E. coli* are therefore considered a more accurate indication of presence of animal or human feces than the total coliform.

Generally, *E. coli* were detected at most of the stations (i.e. N1, N2, N3, N5, N6, N7 and N12) in Tasik Kenyir with relatively higher concentration of *E. coli* recorded at station N3 (Cacing River) (126 counts/100 mL). The presence of *E. coli* in this present study was from animal or human feces. Tasik Kenyir is one of tourist attraction especially for fishing activities. Most of them rent the houseboats which do not have the proper facilities of disposal of sewage and waste water. All were disposed directly into the lake water. However, the *E. coli* found in Tasik Kenyir is far below the recorded value for Chini Lake, Pahang with a maximum value of 52×10^4 counts/100 mL (Hamzah and Hattasrul 2009). This was due to the use of traditional methods of sewage disposal by the indigenous people where the untreated sewage enters directly into Chini Lake.

Water Quality Index

The DOE-WQI of Malaysia guideline was used to determine the water pollution status in Tasik Kenyir. The DOE-WQI scale classifies the water quality as 'clean' (81–100%), 'slightly polluted' (60–80%) and 'polluted' (0–59%) (Department of Environment 2011). In ELK, the WQI values were varied from 94.8% to 97.5% (mean: 96.2 ± 0.7%) for surface water while 65.2–96.4% (mean: 70.9 ± 10.4%) for bottom water. The results suggest that the surface water of ELK has a 'clean' status for all stations whereas for bottom water is classified as 'slightly polluted'. Low

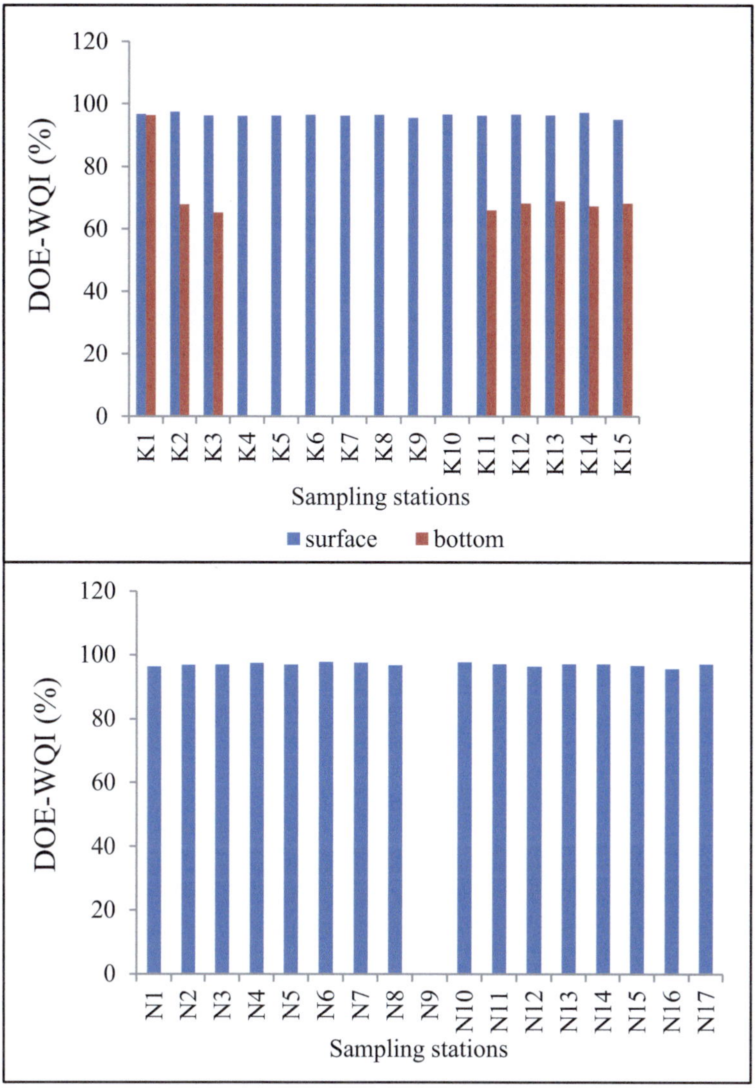

Fig. 5 DOE-WQI of Tasik Kenyir during present study

values of DOE-WQI for bottom water was due to deterioration of the water quality parameters especially the pH and DO. In TNP, the WQI value for the surface water throughout the area was lies within 95.6–97.9% (mean: 96.9 ± 0.6%) (Fig. 5). Therefore, the surface water for each station in TNP can be classified as 'clean' status.

Conclusion

The results from this preliminary study have shown that most of the stations in Tasik Kenyir (ELK and TNP) can be classified as ‘clean’ status based on the DOE-WQI with respect to surface water. However, ‘slightly polluted’ status was recorded for bottom water which was due to deterioration of pH and DO values. Based on the NWQS classification, all stations for surface water in Tasik Kenyir were classified as Class I which is suitable for conservation of natural environment, water supply with practically no treatment and fishery activity for very sensitive aquatic species. Data from this study can be used by other researchers as a baseline comparison for their study in this area. Although this study has shown that Tasik Kenyir is still in good condition, our findings is only based on the short term sampling and this may therefore vary over other times of the year when sampling did not take place and also interannually. We therefore recommend that long term water quality monitoring is carried out at Tasik Kenyir on water quality to help ensure the water quality and conservation of this status is maintained and the associated mechanisms better understood for this important conservation site.

Acknowledgements Thanks to the Department of Chemical Sciences, Universiti Malaysia Terengganu for sponsoring this study. This study was partially supported by Ministry of Education, Malaysia (Fundamental Research Grant Scheme Vote No. 59313).

References

American Public Health Association. (2005). *Standard methods for examination of water and wastewater* (21st ed.). Washington, DC: APHA-AWWA-WEF.

Chapman, D. (1992). *Water quality assessment* (2nd ed.). New York: World health Organization.

Department of Environment. (2011). *Malaysia environmental quality report 2010*. Kuala Lumpur: Department of Environment, Ministry of Natural Resources and Environment Malaysia.

Grasshoff, K., Ehrhardt, M., & Kremling, K. (1983). *Methods of seawater analysis* (2nd ed.). Weinheim: Verlag Chemie.

Hamzah, A., & Hattasrul, Y. (2009). Water quality and microbiological study at Tasik Chini, Pahang. In I. Mushrifah, O. Muhammad Shuhaimi, A. R. Sahibin, H. Khatijah, & A. Nur Amelia (Eds.), *Sumber asli Tasik Chini: Ekspedisi Saintifik* (pp. 65–88). Bangi: Universiti Kebangsaan Malaysia.

Jamaludin, I. B., & Machiels, M. A. M. (1999). Zooplankton dynamics and production in two Malaysian reservoirs. In W. L. T. van Densen & M. J. Morris (Eds.), *Fish and fisheries of lakes and reservoirs in Southeast Asia and Africa* (pp. 59–76). Otley: Westbury Publishing.

Jawan, A., & Sumin, V. (2012). The effect of land used on the water quality of oxbow lakes in Sabah. *The Malaysian Journal of Analytical Sciences, 16*, 273–276.

Kumar, A., Dhall, P., & Kumar, R. (2010). Redefining BOD: COD ratio of pulp mill industrial wastewaters in BOD analysis by formulating a specific microbial seed. *International Biodeterioration and Biodegradation, 64*, 197–202.

Ling, T. Y., Nyanti, L., Muan, T., Grinang, J., Sim, S.-F., & Mujahid, A. (2016a). Physicochemical parameters of Bakun Reservoir in Belaga, Sarawak, Malaysia, 13 months after reaching full supply level. *Sains Malaysiana, 45*, 157–166.

Ling, T. Y., Nyanti, L., Nurul-Safinaz, M. K., Sim, S. F., & Grinang, J. (2016b). Comparisons of water quality near cage culture sites in Batang Ai Reservoir, Sarawak, Malaysia. *Iranica Journal of Energy and Environment, 7*, 169–176.
Moran, D., & Stottrup, J. G. (2011). The effect of carbon dioxide on growth of juvenile Atlantic cod *Gadus morhua* L. *Atlantic Toxicology, 102*, 24–30.
Othman, M. S., Lim, E. C., & Gasim, M. B. (2009). Status kualiti air tasik dan sungai. In I. Mushrifah, O. Muhammad Shuhaimi, A. R. Sahibin, H. Khatijah, & A. Nur Amelia (Eds.), *Sumber asli Tasik Chini: Ekspedisi Saintifik* (pp. 33–42). Bangi: Universiti Kebangsaan Malaysia.
Said, K. S. A., Othman, M. S., & Ahmad, A. K. (2012). The evaluation of water quality and metal concentrations of Titiwangsa Lake, Selangor, Peninsular Malaysia. *Pakistan Journal of Biological Sciences, 15*, 459–468.
Shuhaimi-Othman, M., Lim, E. C., & Mushrifah, I. (2007). Water quality changes in Chini Lake, Pahang, West Malaysia. *Environmental Monitoring and Assessment, 131*, 279–292.
Suratman, S., Mohd Sailan, M. I., Hee, Y. Y., Bedurus, E. A., & Latif, M. T. (2015). A preliminary study of water quality index in Terengganu River basin, southern South China Sea (Malaysia). *Sains Malaysiana, 44*, 67–73.
Verhallen, A., & Verhagen, J. (1994). Hydrodynamics and nutrient availability in Kenyir reservoir, Malaysia, presented at the ASEAN-Australia Development Cooperation Program Component 2 Workshop, Wageningen, The Netherlands, 5–9 December 1994. Jakarta: AADCP II.
Wan Mohd Afiq, W. M. K., & Md Pauzi, A. (2012). Seasonal influence on water quality status of Temenggor Lake, Perak. *The Malaysian Journal of Analytical Sciences, 16*, 163–171.
Yusoff, F. M., & Ambak, M. A. (1999). Trends and fluctuations in the environmental characteristics of surface waters in Kenyir Reservoir, Malaysia. In W. L. T. van Densen & M. J. Morris (Eds.), *Fish and fisheries of lakes and reservoirs in Southeast Asia and Africa* (pp. 59–76). Otley: Westbury Publishing.
Yusoff, F. M., Zaidi, M. Z., & Ambak, M. A. (1994). Fisheries and environmental management of Lake Kenyir, Malaysia, presented at the Regional Symposium on Sustainable Devlopment of Inland Fisheries under Environmental Constraints, Bangkok, Thailand, 19–21 October 1994. Rome: Food and Agriculture Organization of the United Nations.

Comparative Study of Physico-Chemical Analyses of Different Water Resources in Setiu Wetland, Terengganu

Mohd Nizam Lani, Nurul Fatin Malek Rivan, Adilah Ismail, Wan Bayani Wan Omar, Zaiton Hassan, and Zarizal Suhaili

Abstract Physico-chemical properties of water are usually reflecting the functioning of an aquatic ecosystem like Setiu Wetland, Terengganu. Therefore, this study was conducted to determine the physico-chemical properties and chemical nature of their water resources (trench, rivers, drainage and tap water) and their impact on sustainable water resource in Setiu Wetland, Terengganu. A total of 12 water samples were collected aseptically from in this area and the field work was conducted from August to September 2015. In-situ measurements of physico-chemical properties of the water samples were recorded using YSI Professional Plus Parameter (XYLEM, USA) with multi-sensor probe. The analyses of physico-chemical of water include of water temperature, barometric pressure, dissolved oxygen, specific conductance, conductance, redox potential, total suspended solids, salinity and pH. In comparison to previous study conducted in 2008, most of the parameters showed that water quality at Setiu Wetland, Terengganu remained Class 1 based on National Water Quality Standards for Malaysia. In conclusion, the various water resources in Setiu Wetland was in very good condition indicating that the habitats in Setiu Wetlands have conditions suitable to support a biotic community. Further study should be carried out in monitoring water qualities of the water resources at Malaysia to increase safety of water consumption.

M. N. Lani (✉) · N. F. Malek Rivan · A. Ismail
School of Food Science and Technology, Universiti Malaysia Terengganu, Kuala Nerus, Terengganu, Malaysia
e-mail: nizamlani@umt.edu.my

W. B. Wan Omar
School of Marine and Environmental Science, Universiti Malaysia Terengganu, Kuala Nerus, Terengganu, Malaysia

Z. Hassan
Faculty of Science and Technology, Universiti Sains Islam Malaysia, Bandar Baru Nilai, Negeri Sembilan, Malaysia

Z. Suhaili
Faculty of Bio-resources and Food Industry, Universiti Sultan Zainal Abidin, Tembila Campus, Besut, Terengganu, Malaysia

M. T. Abdullah et al. (eds.), *Greater Kenyir Landscapes*,
https://doi.org/10.1007/978-3-319-92264-5_3

Keywords Aquatic ecosystem · Physico-chemical properties · Setiu Wetland · Sustainable water resources

Introduction

Water is one of the most important components of life and life without it, is impossible. However, due to increase of environmental pollution, the quality of water is decreasing continuously (Khan et al. 2012). The world's rapidly growing human population and its associated consumption of water resources affect the whole planet (Vitousek et al. 1997). These inland water resources are particularly under threat by a multitude of impacts originating mainly from human activities (Dokulil 2013). River, trench and drainage are open and easily connected to water disposal from domestic (kitchen garbage and human excreta) and agricultural/aquaculture wastes (pesticides, manure, animal excreta, rainwater runoff and soil). However, tap water should be free from pathogens and pollution as it is systematically treated before reaching to consumers.

Physico-chemical properties of water can influence the levels of the primary productivity, trophic structure and total biomass of aquatic food web (Wetzel 2001). The physico-chemical properties of water play significant roles in determining the quality of aquatic ecosystem and stability to support life forms in aquatic environment. The balanced interaction between biotic and abiotic components in the aquatic environment would be the main factor of sustainable aquatic ecosystem. With the influence of external factors, the combination between internal and external factors will determine the structure of ecosystem and the aquatic species may develop certain pattern of abundance, biomass, seasonality and stratification. Any change in the abiotic environment will have an impact on the biotic life.

Setiu Wetland is one of the least developed area compared to other Wetlands in the state of Terengganu. It is located in the north of Kuala Terengganu, Malaysia with only 10.49% area of the State of Terengganu Darul Iman (Setiu Wetland Council 2013). Since it is the least developed and rich of nature, the pollution level should be low. However, the loss of mangrove forest due to the actively aquaculture activities such as brackish water cage culture, pond culture and oyster farming, have led to a series of serious flooding in the Setiu Wetland, Terengganu when it comes to monsoon season. The rapid growth of these activities had caused dwindling fish stock due to disease outbreaks and discharge of particulate organic materials to the water resources (Eng et al. 1989; Tovar et al. 2000; Suratman et al. 2014). Flooding has been known to affect the water quality as contaminants and pollutants including sewage and manure are washed into surface water bodies (Merolla 2011).

Physico-chemical properties of water gets varied season wise and in addition, anthropogenic activities such as agriculture, urbanization and domestic sewage are among factors that result in the deterioration of water quality (Verma et al. 2012). Pesticides and fertilizers applied to lawns and crops can accumulate and migrate to the water resources thus affecting the physico-chemical and microbial quality of water (Palamuleni and Akoth 2015). Temperature, turbidity and dissolved oxygen are some of the important factors that play a vital role for the growth of living organisms in the water body (Qureshimatva et al. 2015).

A total of 12 water samples from the Setiu Wetlands were collected from 4 locations; which were 3 samples from trench, 3 samples from rivers, 3 samples from drain and the another 3 samples were from tap water using 500 ml sterile blue cap bottles. The samples collection were carried out from August to September 2015. The physico-chemical test of water was done by in-situ measurement using YSI Professional Plus Multiparameter (XYLEM, USA) with multi-sensor probe consist of pH, temperature, salinity, dissolved oxygen, electrical conductivity, specific conductance, resistivity, redox potential, barometric pressure and total soluble solids. The calibration of YSI Professional Plus, Multiparameter was conducted in the laboratory before field sampling and once again after sampling progress work was done at room temperature (25 ± 2 °C).

To the best of our knowledge, it is very limited data on the physico-chemical properties of water resources in Setiu Wetlands that may give misleading information on the safety level of water resources for the use of industry and domestic. This study compare physico-chemical properties of water from previous study conducted in the year of 2008 (Suratman et al. 2014) in order to evaluate physico-chemical water quality from various resources in Setiu Wetland over a time period of several years due to various anthropogenic activities in this area. The values obtained were then compared to available Malaysia guidelines, Malaysian Marine Water Quality Criteria and Standards (MMWQCS) (DOE 2017).

Water Temperature of River, Trench, Drainage and Tap Water in Setiu

Water temperature is an important parameter for aquatic environment, where it is governed by physical, chemical and biochemical properties (Andrew 2012). Most aquatic organisms such as fish, insects, zooplankton and phytoplankton are cold-blooded as they are unable to internally regulate their body temperature (Jain et al. 2013). These aquatic organisms have chosen temperature ranges. Changes in temperature affect aquatic life as it determines which organisms will thrive and which will diminish in numbers and size. Figure 1 shows variability of water temperature (°C) from four different locations which were Setiu River, trench, drainage and tap waters in Setiu, Terengganu.

From the Fig. 1, tap water from restaurant had the highest mean value of 28.15 ± 4.17 °C, while the lowest was Setiu river sample from midstream with 26.00 ± 3.68 °C of mean value. There was no significant different ($P > 0.05$) between all the water resources at Setiu, Terengganu. All the water from all locations had only minimal differences in the temperature within the range of 26–28 °C, which not significant different with temperature range of 28.8–31.1 °C reported by Suratman et al. (2014). The range of this temperature obtained in this study is the

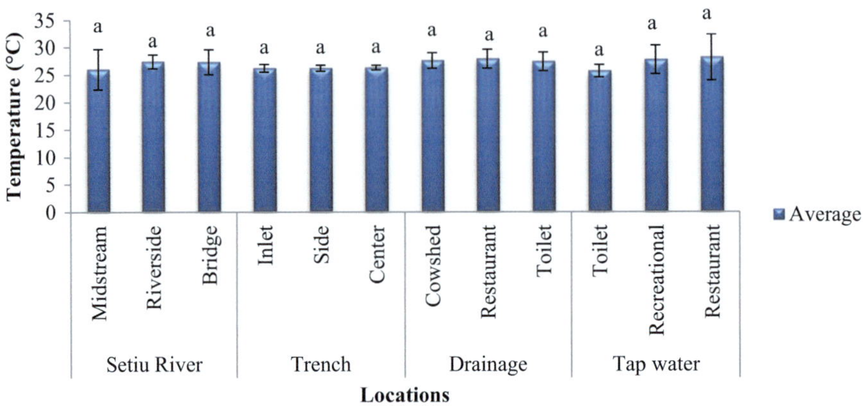

Fig. 1 Water temperature (°C) of water samples from different locations at Setiu Wetland. Mean value with the same letter in the different water resources from different locations did not significantly different at $P > 0.05$

typical temperature of the tropical coastal waters (Alongi et al. 2009). Thermal changes always occur as a result of discharge of municipal or industrial effluents in inland water (Jain et al. 2013). As there was no significant change of thermal change, it strongly indicated no industrial effluent was discharged in the water resources in Setiu. In short, the temperature of water resources in Setiu are within the safe limit for the aquatic life and typical for tropical coastal water.

Barometric Pressure of River, Trench, Drainage and Tap Water in Setiu

From the Fig. 2, the highest barometric pressure was observed at Setiu River with 759.25 ± 0.64 mmHg mean value and the lowest was at the trench with mean value of 756.90 ± 1.56 mmHg. The Setiu river at site 3 and the trench at site 1 and 2 was did not significantly different ($P > 0.05$) with the others location of water resources. However, the trench at site 3 was significantly different at $P < 0.05$ to the other locations of water resources.

The range of barometric pressure at Columbia River was between 728 and 731 mmHg (United States, Department of the Interior 2006). The large difference of barometric pressure between these two places was due to the different climatic factors. In Malaysia, the water temperature was higher (27–34 °C) than the Columbia River (17.2–27.1 °C). Temperature effects on the structure of liquid water, density, viscosity, refractive index anomalies, isotope effects and pressure (Cho et al. 2002). The higher of water pressure, it will reduce the level of dissolved oxygen that can affect the aquatic life.

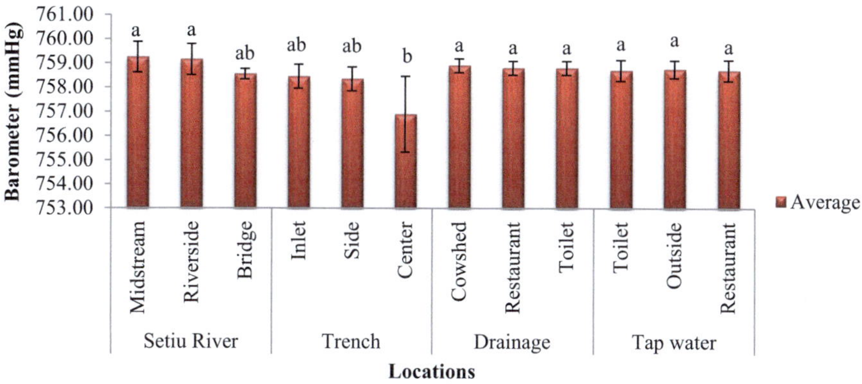

Fig. 2 Barometric pressure (mmHg) of water samples from different locations at Setiu Wetland. Mean value with the same letter in the different water resources from different locations did not significantly different at $P < 0.05$

Dissolved Oxygen of Water Samples (mg/L) of River, Trench, Drainage and Tap Water in Setiu

From Fig. 3, the highest concentration of dissolved oxygen was recorded at drainage from cowshed with 5.04 ± 0.35 mg/L, while the lowest concentration at Setiu River with 1.74 ± 0.13 mg/L. All the water samples from Setiu river and trench did not significantly different at $P < 0.05$. Meanwhile, the cowshed drain was significantly different ($P < 0.05$) with all samples from Setiu river and trench, but did not significantly different ($P > 0.05$) with the samples from other drains and tap waters. The water samples from drainage from restaurant and toilet, as well as all samples from tap waters was significantly different ($P < 0.05$) with Setiu river at midstream but did not significantly different at $P < 0.05$ with the other samples.

The previous study shows that the dissolved oxygen concentration at Setiu Wetland was within 2.21–6.52 mg/L (Suratman et al. 2014). This study showed the present value of dissolved oxygen is lower than the previous study reported by Suratman et al. (2014). Since the temperature and pressure at Setiu River was high, the solubility of oxygen in the water was reduced giving low concentration of dissolved oxygen. The data from the study showed that the highest level of DO was located at the freshwater area with low salinity level. The range of water resources at Setiu Wetland, Terengganu showed that the water was categorized as Class III since it was within the range of 3–5 mg/L (DOE 2017). The amount of dissolved oxygen of Chandlodia Lake water samples ranged between 3.13 and 5.89 mg/L were higher than the dissolved oxygen in water resources in the Setiu. The higher value of dissolved oxygen indicates good aquatic life.

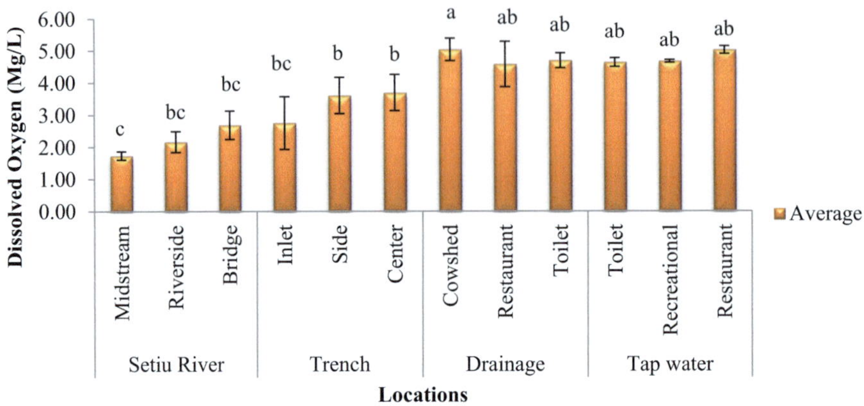

Fig. 3 Dissolved oxygen (mg/L) of water samples from different locations at Setiu Wetland. Mean value with the same letter in the different water resources from different locations did not significantly different at $P < 0.05$

Specific Conductance (µS/cm) of Water Samples of River, Trench, Drainage and Tap Water in Setiu

From Fig. 4, the specific conductance of water was in the range of 5.10 µS/cm and 146.90 µS/cm where the lowest and highest values are from recreational tap water and cowshed drain. The cowshed and toilet drain were significantly different at $P < 0.05$ with other water samples from other locations. All sites from the trench and toilet drain did not significantly different ($P > 0.05$) with the Setiu river at riverside and under the bridge and all samples from tap waters. Meanwhile, all the samples from tap waters were significantly different (P < 0.05) with all samples from Setiu river, trench and drainage.

High specific conductance indicates that high dissolved solids concentration in the water that may affect the aquatic life and suitability of water for domestic, industrial and agricultural uses.

Electrical Conductivity (µS/cm) of River, Trench, Drainage and Tap Water in Setiu

From the Fig. 5, the highest conductance (µS/cm) was observed at drainage location. The toilet drain had highest mean value compared to the other drain with 154.05 ± 8.56 µS/cm. Meanwhile, the lowest mean value of conductance was outside tap water with 5.40 ± 0.42 µS/cm. All water samples from Setiu river was significantly different ($P < 0.05$) with all samples from trench, drainage, and toilet and outside tap water but did not significantly different ($P > 0.05$) with restaurant tap water. The cowshed drain was significantly different at $P < 0.05$ with all other samples from different sites and locations.

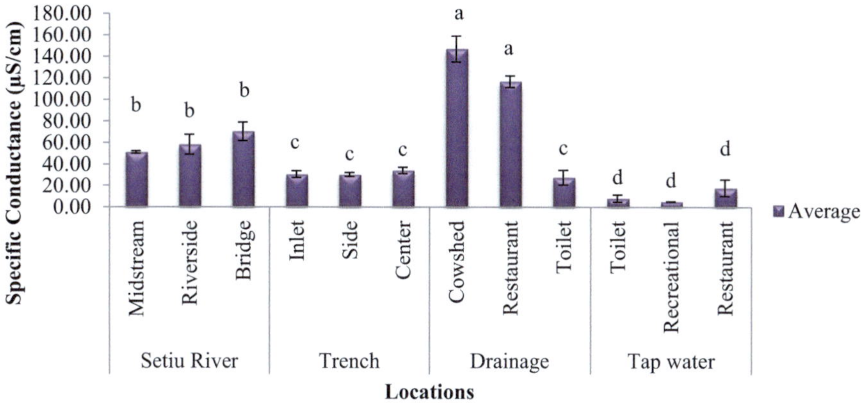

Fig. 4 Specific conductance (µS/cm) of water samples from different locations at Setiu Wetland. Mean value with the different letter in the different water resources from different locations was significantly different at $P < 0.05$

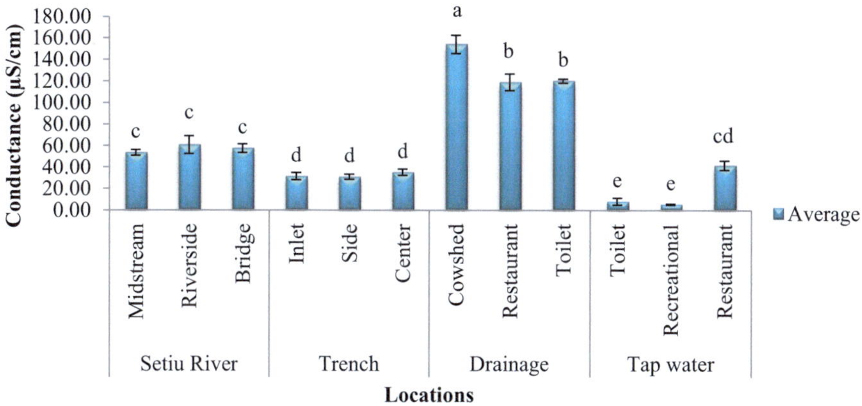

Fig. 5 Conductivity (µS/cm) of water samples from different locations at Setiu Wetland. Mean value with the different letter in the different water resources from different locations was significantly different at $P < 0.05$

Water capability to transmit electric current is known as electrical conductivity and serves as tool to assess the purity of water (Athimoolam et al. 2006). This ability depends on factors such as the presence of ions, their total concentration, mobility, valence, relative concentrations and temperature of measurements (Shinde et al. 2011). Based on National Water Quality Standards for Malaysia, the electrical conductivity for Class 1 of water quality was 1000 µS/cm (DOE 2017). The results confirmed the condition of water bodies in Setiu Wetland, Terengganu shows a good level of conductivity since it is within the safe range that can support variety of aquatic life.

Redox Potential (mV) of River, Trench, Drainage and Tap Water in Setiu

Figure 6 shows that the range of redox potential in different sites of water resources at Setiu, Terengganu. The highest value of redox potential was 132.20 mV from outside tap water while the lowest is 26.85 mV from centre of the trench. All the samples from Setiu river, trench, and drainage did not significantly different ($P > 0.05$) except for the tap waters.

In healthy waters, the redox potential values should be in the range of 300 and 500 millivolts (Li et al. 2014) to provide healthy environment for the aquatic life. With these values, the sanitizer in the water is active enough to decompose all harmful organisms as well as other contaminants and dead tissues almost instantaneously (Lowry and Dickman 2013). The low redox potential values of water resources in Setiu Wetland, Terengganu have been proven that these water did not provide healthy environment for the aquatic life.

Total Suspended Solids (mg/L) of River, Trench, Drainage and Tap Water in Setiu

In Fig. 7, the range of total suspended solids was 3.25–58.2 mg/L with the lowest and the highest from outside tap water and restaurant drain. All water samples collected from four locations were significantly different at $P < 0.05$. The total suspended solid in water resources at Setiu Wetland studied in 2008 was ranged from 32 to 56 mg/L (Suratman et al. 2014). The highest desirable limit for total suspended solid was 20 mg/L and maximum permissible limit was 150 mg/L.

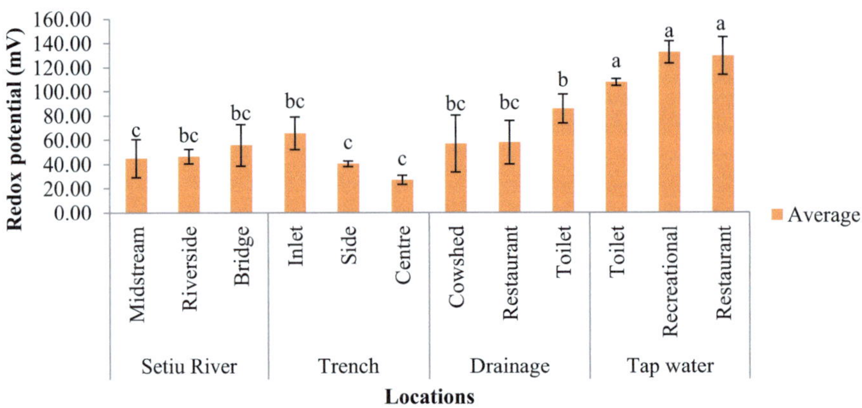

Fig. 6 Redox potential (mV) of water samples from different locations at Setiu Wetland. Mean value with the different letter in the different water resources from different locations at Setiu, Terengganu was significantly different at $P < 0.05$

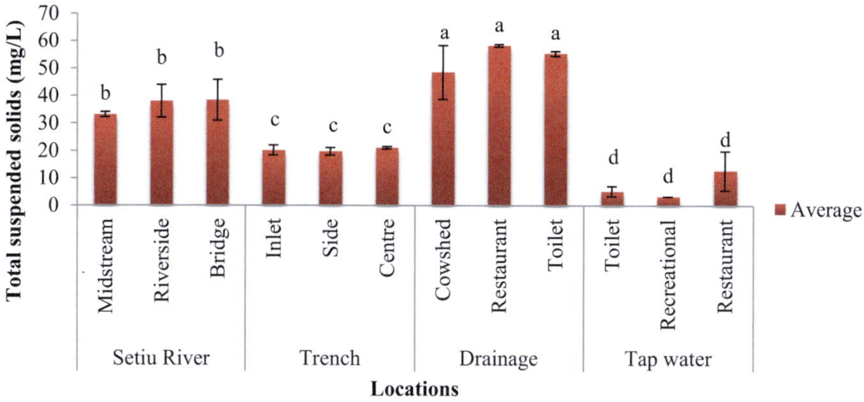

Fig. 7 Total suspended solids (mg/L) of water samples from different locations at Setiu Wetland. Mean value with the different letter in the different water resources from different locations was significantly different at $P < 0.05$

According to National Water Quality Standards for Malaysia (DOE 2017), the total suspended solids within the range of 25–50 mg/L were considered as Class 2 while 50–150 mg/L were for Class III of water quality. Therefore, the water samples from Setiu river and trench was in Class II while drainage in Class III where extensive treatment was required. However, the total suspended solid in tap waters was low since there was less organic and inorganic matter that was suspended in the water and considered as Class I of water quality since its value for total suspended solids was below then 25 mg/ L where there is no treatment necessary. In the study reported by Andrew (2012), the total soluble solids of Ogun River in Nigeria were higher than international standards due to various activities such as trading, construction and automobile industries that were carried out at Lafenwa City, Nigeria.

Salinity (ppt) of River, Trench, Drainage and Tap Water in Setiu

Based on Fig. 8, the salinity of the water resources from Setiu, Terengganu was in the range of 0.0–0.06 ppt with the highest and lowest at drainage and tap waters, respectively. All four locations were significantly different at $P < 0.05$ but did not significantly different ($P > 0.05$) between the sites. Suratman et al. (2014)), the salinity of water resources in Setiu Wetland was in the range of 12.0–33.5 ppt. The highest salinity in Setiu Wetland was located in the inshore area about 500 m outside the wetland that represents the seawater condition and there are very little anthropogenic activities within surrounding area while the freshwater area had the lowest salinity.

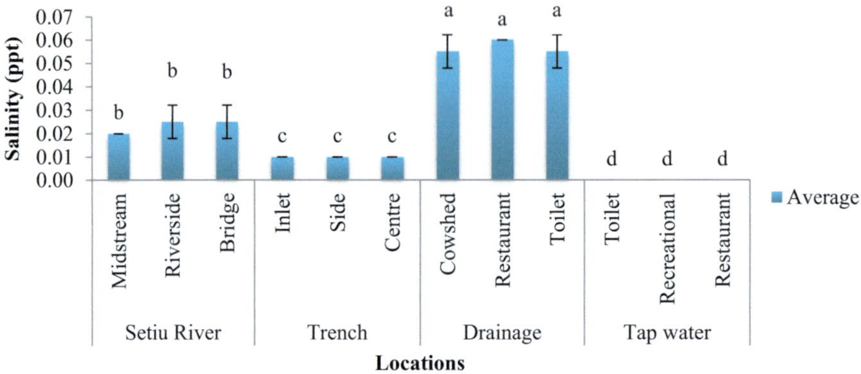

Fig. 8 Salinity (ppt) of water samples from different locations at Setiu Wetland. Mean value with the different letter in the different water resources from different locations was significantly different at $P < 0.05$

The excessive amount of salt in water can affect the agriculture activity, drinking water supplies and ecosystem health. Moreover, high concentration of salt also reduced the ground cover that lead to soil erosion. Consequently, it will increase sediment and making it unsuitable for the consumption of human and animal. However, the water quality standards for Malaysia stated that the salinity for Class 1 of water quality was 0.05 ppt (DOE 2017). The salinity of the water resources at Setiu Wetland, Terengganu is in Class I, where it is safe limit for water resources and needs to be maintained in order to save the ecosystem.

pH of River, Trench, Drainage and Tap Water in Setiu

From Fig. 9, the range of pH in the area of Setiu was 6.75–8.84 with lower and higher values recorded at Setiu River and tap water from toilet, respectively. As far as pH is concerned, the results did not significantly different ($P < 0.05$) among the sampling sites. The previous study showed that the range of pH was from 6.9 to 8.6 (Suratman et al. 2014). In comparison to previous study, there was not much different on pH changes as it was typical pH range for tropical water. According to National Water Quality Standards for Malaysia, the ranges of pH for water resources are within 6.5–8.5 for the Class 1 of water quality (DOE 2017). The water resources in the Setiu Wetland, Terengganu are considered safe and good quality since the ranges for the pH are within the limit.

The distribution of surface water quality of Setiu in the present study and the previous study conducted in 2008 (Suratman et al. 2014) as in Table 1.

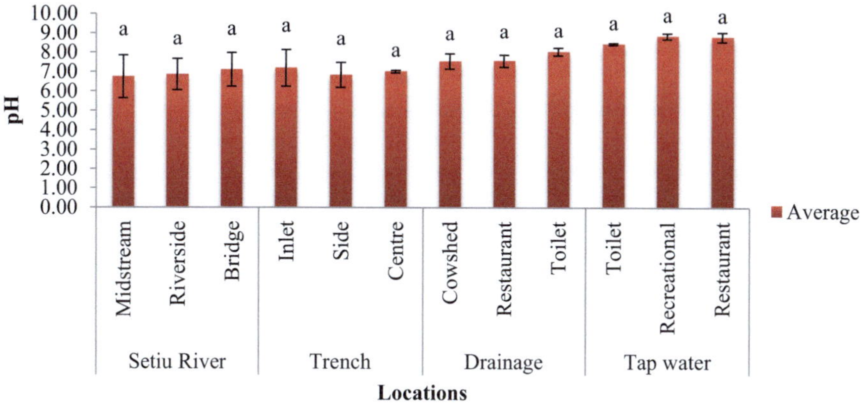

Fig. 9 pH of water samples from different locations at Setiu, Terengganu. Mean value with the different letter in the different water resources from different locations at Setiu, Terengganu did not significantly different at $P < 0.05$

Table 1 Comparison of water quality in Setiu

Physico-chemical properties	Previous study conducted in 2008 (Suratman et al. 2014)	Present study conducted in 2015
Temperature (°C)	28.8–31.1	25.0–30.0
Barometric pressure (mmHg)	728–731	756.9–759.25
Dissolved oxygen (mg/L)	2.21–6.52	3.25–58.2
Specific conductivity (μS/cm)	500–800	5.10–146.90
Conductance (μS/cm)	50–1500	5.40–154.05
Redox potential (mV)	300–500	26.85–132.20
Total suspended solid (mg/L)	32–56	3.25–58.2
Salinity (ppt)	12.0–33.5	0.00–0.06
pH	6.9–8.6	6.75–8.84

Conclusion

Based on all of the parameters of physico-chemical properties, it can be concluded that the water resources at Setiu Wetland, Terengganu was categorized in Class I based on National Water Quality Standards for Malaysia. The importance of this study relies on the measurements given for physico-chemical properties of water resources at Setiu Wetland that determine the level of degradation of environmental quality and ecosystem. The significant of this finding was the ability to compare the quality of water during the 7-years gap and see how much significant changes have occurred due to anthropogenic activities. When it relates to the food safety, the quality of water is very important to be ensured it is safe for public health at each step of the food chain from production to consumption. From this study, these water resources were pristine and safe for consumption. However, continuous monitoring of water quality should be enforced by policy makers to ensure the sustainable water resources are maintained for benefits of community, aquatic ecosystem and environment.

Acknowledgements The study was funded by the School of Food Science and Technology, Universiti Malaysia Terengganu (UMT).

References

Alongi, D. M., McKinnon, A. D., Brinkman, R., Trott, L. A., Undu, M. C., Muawanah, & Rachmansyah. (2009). The fate of organic matter derived from small-scale cage aquaculture in coastal waters of Sulawesi and Sumatra, Indonesia. *Aquaculture, 295*, 60–75.

Andrew, O. A. (2012). Physicochemical analysis of Ogun River (water samples) within two locations (Akin-Olugbade and Lafenwa) in Abeokuta, Ogun State, Nigeria. *IOSR Journal of Applied Chemistry, 1*(4), 24–27.

Athimoolam, M., Ramu, A., & Kannan, N. (2006). Water quality assessment from Uthamapalayam municipality in Theni District, Tamil Nadu, India. *Pollution Research, 25*, 163–166.

Cho, C. H., Urquidi, J., Singh, S., Park, S. C., & Robinson, G. W. (2002). Pressure effect on the density of water. *Journal of Physical Chemistry, 106*, 7557–7561.

Department of Environment (DOE). (2017). Malaysian Marine Water Quality Criteria and Standard. Retrieved from: https://www.doe.gov.my/portalv1/info-umum/piawaian-dan-kriteria-kualiti-air-marin-malaysia/301.

Dokulil, M. T. (2013). Impact of climate warming on European inland waters. *Inland Waters, 4*, 27–40.

Eng, C. T., Paw, J. N., & Guarin, F. Y. (1989). The environmental impact of aquaculture and the effects of pollution on coastal aquaculture development in Southeast Asia. *Marine Pollution Bulletin, 20*, 335–343.

Jain, S., Sharma, G., & Mathur, Y. P. (2013). Effects of temperature variations on fish in lakes. *International Journal of Engineering Research and Technology, 2*(10), 2516–2523.

Khan, M. Y., Shabeer, M., Raja, I. A., & Wani, N. A. (2012). Physico-chemical analysis of River Jhelum (Kashmir). *Global Journal of Science Frontier Research Interdiciplinary, 12*, 1–3.

Li, X., Blancheton, J.-P., Liu, Y., Triplet, S., & Michaud, L. (2014). Effect of oxidation-reduction potential on performance of European sea bass (*Dicentrarchus labrax*) in recirculating aquaculture systems. *Aquaculture International, 22*(4), 1263–1282.

Lowry, R. W., & Dickman, D. (2013). The ABC's of ORP: Clearing up some of the mystery of oxidation-reduction potential. Retrieved from http://www.danmonitoring.com/wp-content/uploads/2009/09/the-abc-of-orp.pdf.

Merolla, S. (2011). *The effect of floods and high rainfall on the water quality in selected sub-areas of the upper vaal catchmen* (M.Sc. Thesis). Faculty of Science, University of Johannesburg, South Africa.

Palamuleni, L., & Akoth, M. (2015). Physico-chemical and microbial analysis of selected borehole water in Mahikeng, South Africa. *International Journal of Environmental Research and Public Health, 12*, 8619–8630.

Qureshimatva, U. M., Maurya, R. R., Gamit, S. B., Patel, R. D., & Solanki, H. A. (2015). Determination of physico-chemical parameters and water quality index (Wqi) of Chandlodia Lake, Ahmedabad, Gujarat, India. *Journal of Environmental & Analytical Toxicology, 5*(4), 1–6.

Setiu Wetland Council. (2013). Background of Setiu. Official Portal Setiu Wetland Council. Retrieved from: http://mds.terengganu.gov.my/en/latar-belakang-setiu.

Shinde, S. E., Pathan, S. A., Raut, K. S., & Sonawane, D. L. (2011). Studies on the physico-chemical parameters and correlation coefficient of Harsool-savangi Dam, District Aurangabad, India. *Middle-East Journal of Scientific Research, 8*, 544–554.

Suratman, S., Hussein, A. N. A. R., Latif, M. T., & Weston, K. (2014). Reassessment of physico-chemical water quality in Setiu Wetland, Malaysia. *Sains Malaysiana, 43*(8), 1127–1131.

Tovar, A., Moreno, C., Manuel-Vez, M. P., & Garcia-Vargas, M. (2000). Environmental implications of intensive marine aquaculture in earthen ponds. *Marine Pollution Bulletin, 40*, 981–988.

United States, Department of the Interior. (2006). Water Resources Data. Retrieved from http://pubs.usgs.gov/wdr/2005/wdr-wi-05-1/WDR-WI-05-1_part1.pdf.

Verma, P. U., Purohit, A. R., & Patel, N. J. (2012). Pollution status of Chandlodia Lake located in Ahmedabad Gujarat. *International Journal of Engineering Research and Applications, 2*, 1600–1610.

Vitousek, P. M., Mooney, H. A., Lubchenco, J., & Melillo, J. M. (1997). Human domination of Earth's ecosystems. *Science, 277*, 494–499.

Wetzel, R. G. (2001). *Limnology: Lake and river ecosystems* (3rd ed.). San Diego: Academic.

Part II
Biological Environment

A Baseline Assesment on the Distributions and Sources of Aliphatic and Polycyclic Aromatic Hydrocarbons in Abiotic Environment of Terengganu National Park

Norhayati Mohd Tahir, Tan Hock Seng, Muhammad Fais Fadzil, Suhaimi Suratman, and Marinah Ariffin

Abstract A baseline study to assess the distributions of aliphatic and polycyclic aromatic hydrocarbons in abiotic environment consisting of soil, sediment and atmospheric particulate samples was carried in Terengganu National Park. Aliphatic hydrocarbons were found in all three media studied whilst polycyclic aromatic hydrocarbon (PAH) compounds were detected, only in the atmospheric particulate samples, at trace levels. The 24 hr-PM_{10} and Total Suspended Particulate (TSP) concentrations is still below the recommended Malaysian Air Quality Guidelines for the respective parameter. Analyses of the results suggest that terrestrial plant waxes is the main source of aliphatic hydrocarbons in the study area whereas sources for PAHs in the particulate samples are most likely derived from combustion sources. The detection of PAHs in the aerosol samples also suggest possible anthropogenic input of particulate matter to the study area; however the levels are still well below the values reported for other locations in Malaysia.

Keywords Particulate matter · Soils and sediment · Biogenic sources · Biomass burning · Tasik Kenyir · Sustainable management

N. Mohd Tahir (✉)
Institute of Oceanography and Environment (INOS), Universiti Malaysia Terengganu, Kuala Nerus, Terengganu, Malaysia

School of Marine and Environmental Sciences, Universiti Malaysia Terengganu, Kuala Nerus, Terengganu, Malaysia
e-mail: hayati@umt.edu.my

T. H. Seng · S. Suratman
Institute of Oceanography and Environment (INOS), Universiti Malaysia, Kuala Nerus, Terengganu, Malaysia

M. F. Fadzil · M. Ariffin
School of Marine and Environmental Sciences, Universiti Malaysia, Kuala Nerus, Terengganu, Malaysia

M. T. Abdullah et al. (eds.), *Greater Kenyir Landscapes*,
https://doi.org/10.1007/978-3-319-92264-5_4

Introduction

Hydrocarbons found in forest soil or lakes sediment are derived from biogenic (or natural) and/or anthropogenic sources. Biogenic sources of organic matter in forest soil are mostly contributed by decomposed organic materials such as plant residues (e.g. wood, leaves and roots) and plant resins. Hydrocarbons in water or sediment, on the other hand, may come from aquatic plants and organisms (e.g. algae and phytoplankton, bacteria, fungi, protozoa, rooted and floating aquatic macrophytes, and etc.) as well as surface soil run-off into the adjacent water bodies. Vegetation by far is the main biogenic source of hydrocarbons in the atmosphere, particularly the volatile compounds such as isoprene and monoterpenes as well as epicuticular plant waxes containing light to heavy molecular weight n-alkanes.

The main anthropogenic source of hydrocarbons into the atmosphere is through usage of fossil fuels such as in motor vehicle and industrial use and biomass burning. Two main classes of hydrocarbons widely use to ascertain sources, transport and extent of contamination are the aliphatic hydrocarbons, specifically n-alkanes, and polycyclic aromatic hydrocarbons. Polycyclic aromatic hydrocarbons (PAHs) represent a class of organic compounds that are comprised of two or more fused aromatic rings and they are primarily formed by incomplete combustion or pyrolysis of organic matter (Wang et al. 2013) hence their presence in the environment is often associated with possible anthropogenic sources. These PAHs are believed to be carcinogenic and/or mutagenic (Ravindra et al. 2008). Their primary route to the pristine environment is from long range atmospheric transport of aerosol particles associated with PAH compounds that were transported by trade winds from their source of origin (Wang et al. 2013; Simoniet 2002). In addition, the local source of PAHs may also contribute to their presence in the abiotic environment. Examples of local sources that could contribute to the presence of PAHs in Terengganu National Park includes camp fires, open burnings of biomass materials from nearby Tasik Kenyir campsites, surrounding villages and the use of fossil fuel to power the fishing and recreational boats in Tasik Kenyir. To date, no study has been carried out to establish the distribution of hydrocarbon compounds in Terengganu National Park and the surrounding areas. With increasing developments surrounding Tasik Kenyir and other related human activities, it is timely that a study be conducted to determine the levels and distribution of aliphatic and polycyclic aromatic hydrocarbons in the area. Using molecular markers relevant to these compounds, sources that contribute to the presence of aliphatic and polycyclic aromatic hydrocarbons in Terengganu National Park could be assessed. Thus the data generated could serve as a baseline information for future reference.

Sampling was carried out during a two day expedition to Tanjung Mentong, Terengganu National Park organized by members of the Environmental Research Group, Research and Innovations Affairs of Universiti Malaysia Terengganu (UMT) on October 15–16, 2010. Tanjung Mentong is located at the southern end of Tasik Kenyir and it serves as gateway into the Terengganu's national park. The facilities at Tanjung Mentong are under the jurisdiction of the Department of Wildlife and National Park of Terengganu.

The sampling of particulate matter (PM) was carried out at an open area, uphill of Tanjung Mentong Hill (TMH) which was surrounded by lake and forest. Total Suspended Particulate (TSP) and particulate matter less than 10 μ (PM_{10}) was collected on a pre-cleaned glass fiber filter using two individual high volume air sampler. The PM samples were collected for a period of 24 h. Soil and sediment sampling were carried out at selected areas of Terengganu National Park. A total of four sites were established for the soil sampling, viz. Tanjung Mentong Hill (TMH), Sungai Cacing (SC), Sungai Chenana (SCHE) and Sungai Terengan (ST). Sediment sampling was carried at only one site, Sungai Terenggan (ST). Soil, sediment and aerosol samples were solvent extracted with a mixture of dichloromethane and methanol (3,1, v:v) using ultrasonic agitation technique. Quantification of aliphatic hydrocarbons and polycyclic aromatic hydrocarbons compounds were carried out by using Gas Chromatography-Mass Spectrometry (GCMS), whilst identification of the compounds was done by comparing with aliphatic hydrocarbons and polycyclic aromatic hydrocarbons external standards. Quality control for all laboratory analysis was done by spiking internal standards and the results showed good recovery in the range of 80–117%.

Aerosols Particulate Mass

The 24 h average TSP and PM_{10} concentration recorded at TMH was 68.94 μg m^{-3} and 40.14 μg m^{-3}, respectively. The values recorded were well below the Recommended Malaysian Air Quality Guideline for TSP (260 μgm^{-3}) and PM_{10} (150 μgm^{-3}). The TSP value recorded at TMH is within the range reported at various locations within Kuala Terengganu city (range: 9.33–162 μgm^{-3}; mean: 17.2–148 μgm^{-3}) (Mohd Tahir et al. 2009). Likewise the PM_{10} concentration found at TMH is within the values reported for the Politeknik Sultan Omar station in Kuala Terengganu but higher than the mean value observed at the same station (5.21–63.48 μgm^{-3}; mean 24.71 μgm^{-3}) (Mohd Tahir et al. 2013). It must be emphasized that the TSP and PM_{10} values reported for TMH station is solely based on a single monitoring of the air quality at the site thus could not be said to represent a true picture of the air quality over Terengganu National Park and Tasik Kenyir as a whole, nonetheless these values provide a snapshot picture of the air quality during the study period. In contrast TSP study reported for Kuala Terengganu was based on spatial monitoring over a three month monitoring period (Mohd Tahir et al. 2009) whilst the PM_{10} study was based on a 17 months period of monitoring at a single location in the heart of Kuala Terengganu city center (August 2006–December 2007).

Aliphatic Hydrocarbons

Total identified resolved aliphatic hydrocarbons (TIRAH) detected in soil and sediment samples are shown in Table 1 whilst Fig. 1 shows the distribution of individual aliphatic hydrocarbon present in the samples studied. Main aliphatic hydrocarbons

Table 1 Total identified resolved aliphatic hydrocarbons (TIRAH) and diagnostic ratios for soils and sediment samples

Type	Surface Soil				Surface sediment
Location	TMH	SC	SCHE	ST	ST
C_{max}	C_{31}	C_{31}	C_{31}	C_{31}	C_{31}
TIRAH	28.2	8.72	10.1	3.52	0.68
$CPI_{(12–36)}$	7.39	7.09	7.58	7.81	6.93
$CPI_{(24–36)}$	7.52	7.27	7.77	7.93	7.20
% WNA	72.6	73.3	73.9	76.5	75.2
% ALK $_{terr}$	81.2	83.8	83.6	85.8	83.6

TMH Tanjung Mentong Hill, *SC* Sungai Cacing, *SCHE* Sungai Chenda, *ST* Sungai Terengan, C_{max} Maximum carbon number detected in n-alkanes homologue series, *TIRAH* Total identified resolved aliphatic hydrocarbon, *CPI* Carbon Preference Index, *%WNA* Percentage of wax n-alkane, %ALK terr, Percentage of terrestrial n-alkane

identified in this study are the n-alkanes in the carbon range of C_{20} and above. Soil of TMH exhibited higher TIRAH concentration compared to other soil samples with value of 28.2 μgg^{-1}, followed by Sungai Chenana (SCHE), Sungai Cacing (SC) and Sungai Terengan (ST) with 10.1 μgg^{-1}, 8.72 μgg^{-1} and 3.52 μgg^{-1}, respectively. Much lower value of TIRAH concentration was found in sediment of Sungai Terengan (0.68 μgg^{-1}). The trend observed is expected as river only receives input of organic matter (hence aliphatic hydrocarbons) through run-offs of land-based organic matter that are accumulated in soils of surrounding land and with relatively fast flowing water at the upstream end of the river, the potential of organic materials to sediment out of the water column is also low.

Carbon preferential index (CPI) value is a measure of the carbon number predominance between the odd and even n-alkanes and carbon maximum ($C_{max)}$ is the carbon number of the most abundant n-alkane detected. These two indices are useful as an indicator of relative source input for aliphatic n-alkanes (Mazurek and Simoneit 1984) and are often used to differentiate between biogenic and petrogenic n-alkanes source. Plant wax n-alkanes usually show C_{max} in the range of C_{25}–C_{33} depending on the plant species and its locality (Simoneit and Mazurek 1982; Abas and Simoneit 1998). Meanwhile, petrogenic n-alkane sources normally exhibit CPI value lower or equal to one with no carbon predominance whilst biogenic sources show CPI values larger than one with odd carbon predominance (Simoniet 2002). The $CPI_{12–36}$ value calculated in soil of ST, SC, TMH and SC is 7.91, 7.09, 7.39 and 7.09, respectively and in Sungai Terengan sediment is 6.93 with all sites exhibiting a C_{max} of C_{31}. Furthermore, $CPI_{24–36}$ for the high molecular weight n-alkanes (C_{24}–C_{36}) calculated also gave values similar to that of $CPI_{12–36}$ for all soil and sediment samples. All soil and sediment samples show the distribution of n-alkanes in the range of C_{12}–C_{36} with dominance of carbon number above 20, in particular C_{29}–C_{33} range. These high CPI values and n-alkane dominance in the range of C_{29}–C_{33} clearly indicate the input of plant wax n-alkanes into the soil and sediment of the selected sites.

In addition, the diagnostic ratio of percentage wax n-alkanes (%WNA) and percentage of terrestrial n-alkanes (% ALK_{terr}) are normally used to determine the con-

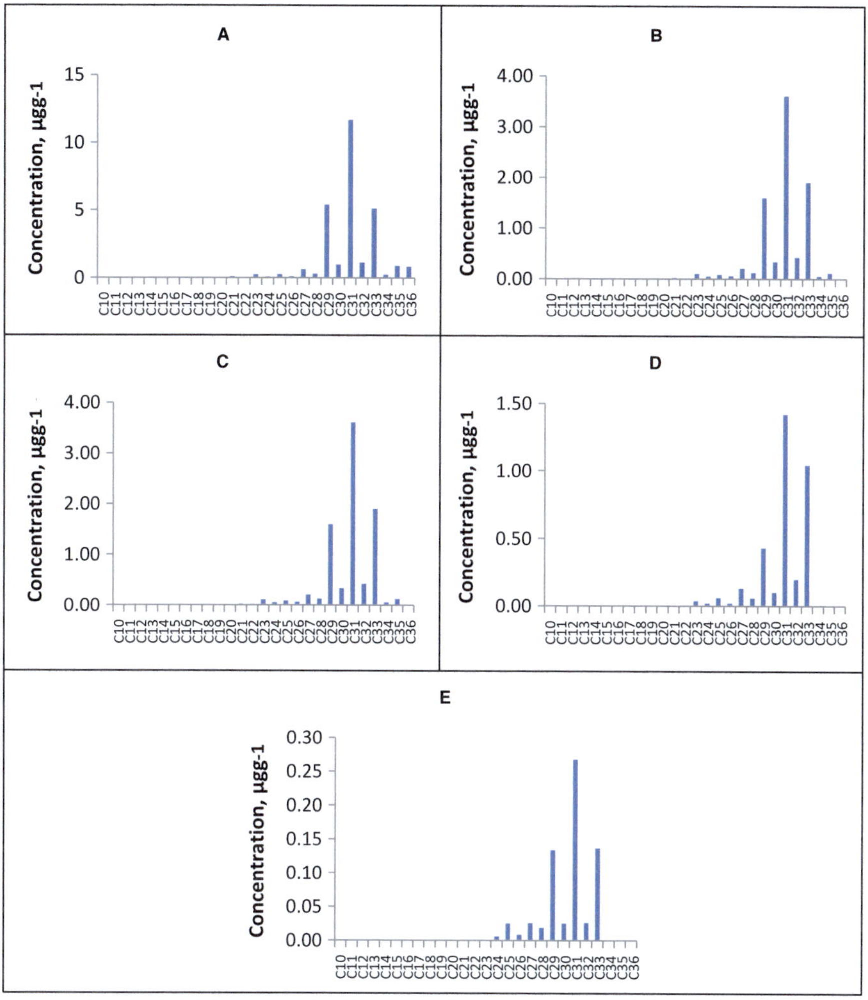

Fig. 1 Distribution of alipahtic hydrocarbon in soils and sediment of Terengganu National Park. (**a**) Tanjung Mentong Hill soil, (**b**) Sungai Chenana soil, (**c**) Sungai Cacing soil, (**d**) Sungai Terengan soil, (**e**) Sediment of Sungai Terengan

tribution of n-alkane from plant wax and terrestrial plant input. High percentage of %WNA and $\%ALK_{terr}$ provide further evidence in support of plant wax input into the soils and sediment of the study area where all studied samples gave percentage values of %WNA and $\%AlK_{terr}$ in the range of 72.6–76.5% (mean: 74.3%) and 81.2–85.8% (mean: 83.6%), respectively. Thus from the n-alkane results obtained, it could be concluded that the aliphatic hydrocarbons in Terengganu National Park soils is mainly derived from terrestrial plant input, most likely, from the plant litter fall, plant resins or leaves abrasion which introduce plant waxes into the soil. In the case of sediment, leaves abrasion could introduce plant waxes directly into the aquatic environment that could settled at the bottom of the river. On the other hand, surface runoff of organic detritus in soils could leach out hydrocarbons into the nearby river and slowly accumulates in the sediment. Input of hydrocarbons from aquatic organisms is not significant owing to absence of low molecular weight n-alkanes ($n \leq C_{20}$) in the n-alkane distribution obtained from the sediment (Fig. 1e).

The distribution of individual aliphatic hydrocarbon obtained in TSP and PM_{10} samples are shown in Fig. 2 whilst the aliphatic indices calculated using selected diagnostic parameters are given in Table 2. n-alkane distributions show that both samples exhibit the presence of only high molecular weight (C_{24}–C_{36}) n-alkane with odd carbon predominance and odd carbon number as its carbon maximum. Carbon Preference Index calculated for C_{12}–C_{36} (CPI_{12-36}) and C_{24}–C_{36} (CPI_{24-36}) range showed TSP and PM_{10} gave CPI values larger than one. In this study, TSP sample

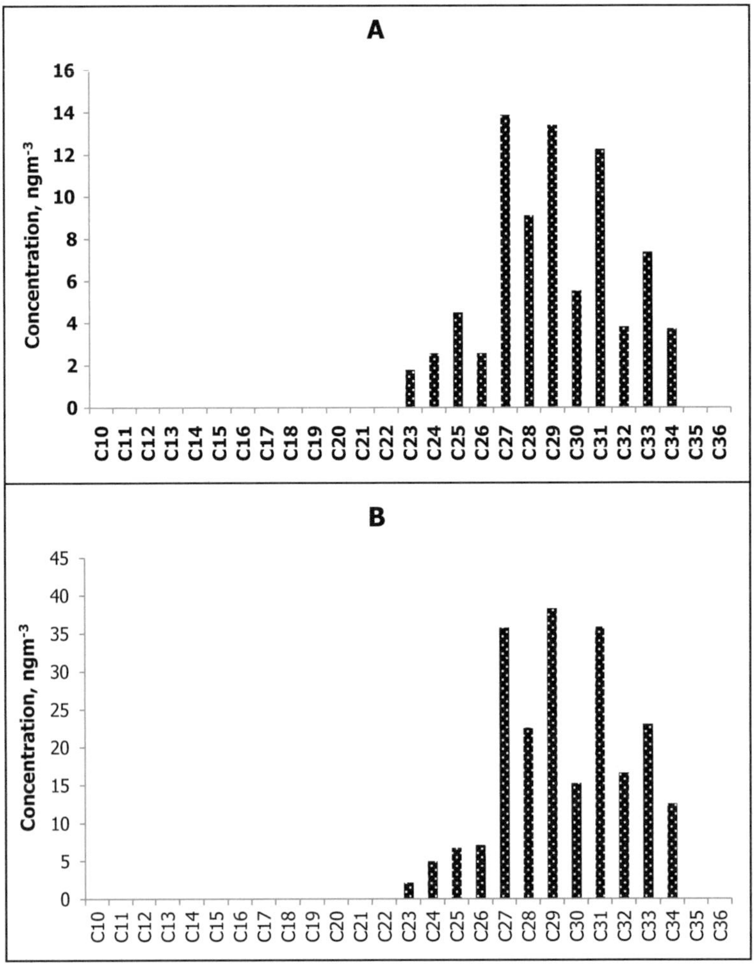

Fig. 2 Distribution of aliphatic hydrocarbon in (**a**) TSP and (**b**) PM_{10}

Table 2 Aliphatic hydrocarbons indices value in air particulate matter studied

Aliphatic hydrocarbon indices	Total suspended particulate (TSP)	Particulate matter size 10 µm (PM_{10})
Carbon maximum, C_{max}	C_{27}	C_{29}
Total Identified Resolved Aliphatic Hydrocarbon, TIRAH (ngm^{-3})	80.8	221.4
Carbon Preference Index, $CPI_{(12–36)}$	1.94	1.79
Carbon Preference Index, $CPI_{(24–36)}$	1.98	1.83

$CPI_{(12-36)} = 0.5^*[(\sum(C_{13} - C_{35})/\sum(C_{12} - C_{36})) + (\sum(C_{13} - C_{35})/\sum(C_{14} - C_{36}))]$,
$CPI_{(24-36)} = 0.5^*[(\sum(C_{23} - C_{35})/\sum(C_{22} - C_{36})) + (\sum(C_{23} - C_{35})/\sum(C_{24} - C_{36}))]$

exhibited C_{max} at C_{27}, $CPI_{12–36}$ of 1.94 and $CPI_{24–36}$ of 1.98 whilst PM_{10} sample exhibited C_{max} of C_{29}, $CPI_{12–36}$ of 1.79 and $CPI_{24–36}$ of 1.83. In addition, PM_{10} gave higher TIRAH concentration (221.4 ngm^{-3}) compared to TSP (80.8 ngm^{-3}). The results obtained clearly suggest that aliphatic n-alkanes obtained in TSP and PM_{10} are mainly derived from terrestrial plant waxes, most likely through leaves abrasion which introduce plant waxes into the surrounding air.

Polycyclic Aromatic Hydrocarbons

Only 18 PAHs compounds were quantified in this study. These compounds were as follows: naphthalene (Naph), acenaphthylene (Acena), acenaphtene (Ace), flourene (Fluo), phenanthrene (Phen), anthracene (Anth), fluoranthene (Flan), pyrene (Py), benz(a)anthracene (BaA), chrysene (Chry), benzo(b)fluoranthene (BbF), benzo(k) fluoranthene (BkF), benzo(e)pyrene (BeP), benzo(a)pyrene (BaP), Perylene (Py), Indeno(1,2,3-cd)pyrene (Indeno), diben(a,h)anthracene (Dibenz) and benzo(g,h,i) perylene (BgP). Sum of all these PAHs compounds were referred to as total PAHs (TPAHs).

In this study, polycyclic aromatic hydrocarbon compounds (PAHs) were only detected in aerosols particulate samples. No PAHs were detected in soil and sediment samples. The distributions of PAHs associated with TSP and PM_{10} aerosols are shown in Fig. 3. In general, the concentration of PAHs compound detected in both aerosols particulate were in part per billion thus can be considered as very low in concentration. Higher TPAHs concentration (561 pg/m^3) was detected in PM_{10} particles size fraction compared to the TSP particle size fraction (239 pg/m^3). The trend observed is consistent with earlier studies which suggested that organic compounds tend to be associated with the finer particulate matter fraction (Fang et al. 2005; Kaushik et al. 2012). The distribution of PAHs in TSP is mainly dominated by the four rings PAHs followed by the five rings PAH. Fluoranthene (Flan) recorded the highest value with concentration of 62.2 pg/m^3 followed by its other four ring member Pyrene (Py), Benz(a)anthracene (BaA) and Chrysene (Chry). Benzo(a)pyrene

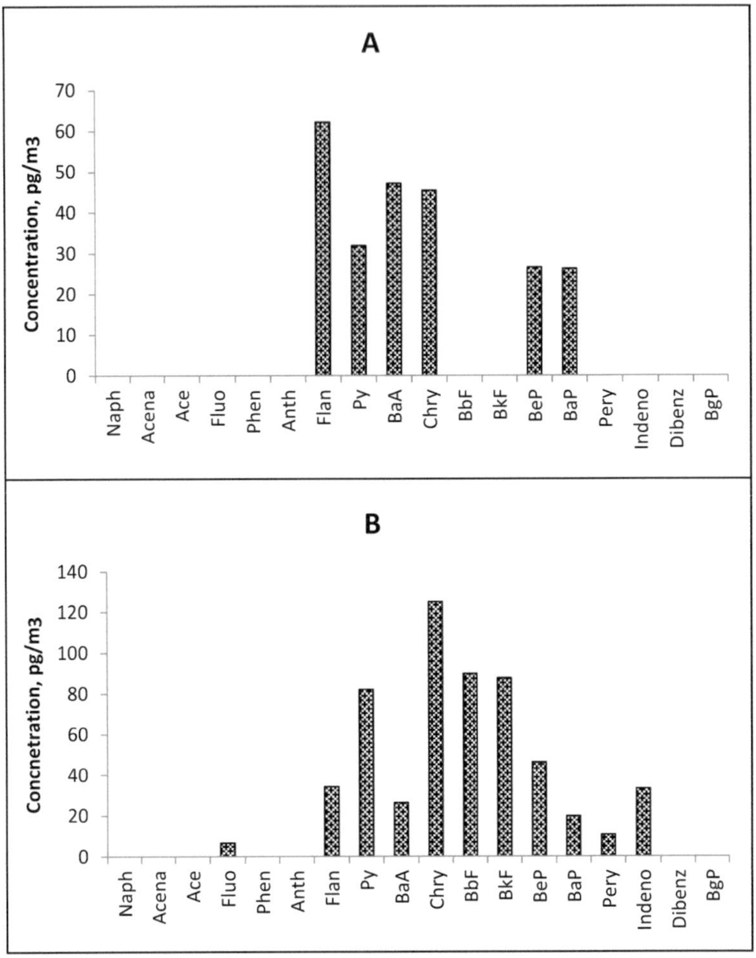

Fig. 3 Polycyclic aromatic hydrocarbons distribution in (**a**) TSP and (**b**) PM_{10}

(BaP) and Benzo(e)pyrene (five rings PAHs) is present in similar amount with noticeable absence of the other rings PAHs of Benzo(b)fluoranthene (BbF) and Benzo(k)fluoranthene (BkF) in the TSP fraction. PM_{10} samples on the other hand exhibited a wider range of PAHs compound compared to TSP with Chrysene yielding the highest concentration of 124.9 pg/m^3. All the five rings PAHs were detected with Benzo(b)fluoranthene (BbF) giving the highest concentration with 89.8 pg/m^3 and Perylene the lowest concentration with 10.6 pg/m^3. Indeno(1,2,3-cd) pyrene (Indeno) was the only 6 ring PAH detected in PM_{10}.

The PAHs ratios are commonly used as diagnostic ratios for determining sources of PAH that contribute to their presence in the particulate matter, clarification of sample by location and estimating the importance of combustion (sometime referred to as pyrolytic) and/or petroleum derived PAH (Yunker et al. 2002a, b). The usual index of combustion and/or anthropogenic input is an increase in the proportion of the less stable and/or kinetically produced parent PAH isomers relative to the thermodynamically stable isomers (e.g. fluoranthene relative to pyrene) or to the molecular mass totals (Yunker et al. 2002b). Such ratio calculations are traditionally restricted to PAH within a given molecular mass to minimize confounding factors such as differences in volatility, water solubility, adsorption, and etc. (Yunker et al. 2002a). Table 3 below shows the selected and commonly used PAH ratios used to distinguish the most probable source of PAHs compounds whilst Table 4 shows the molecular diagnostic ratios calculated for TSP and PM_{10} sample. The entire ratios ∑COMB/∑PAHs, (Flan/(Flan+Py), BaA/(BaA+Chry) and Indeno/(Indeno+BgP) calculated for TSP are indicative of the contribution from combustion in particular biomass burning (grass, wood or coal sources). On the other hand, PM_{10} shows mix sources of petrogenic and combustion sources. The ratio of ∑LMW/∑HMW, Indeno/(Indeno+BgP) and ∑COMB/∑PAHs suggest contribution from pyrogenic

Table 3 Molecular diagnostic ratio of Polycyclic Aromatic Hydrocarbons suggested by previous literatures

PAH ratio	Value	Source	References
∑LMW/HMW	<1	Pyrogenic	Zhang et al. (2008)
	>1	Petrogenic	
∑COMB/PAHs	0.3	Petrogenic	Pandey et al. (1999)
	0.7	Pyrogenic	
Anth/(Anth + Phen)	<0.1	Petrogenic	Pies et al. (2008)
	>0.1	Pyrogenic	
Flan/(Flan+Py)	<0.4	Petrogenic	Yunker et al. (2002b)
	0.4–0.5	Fossil fuel combustion	
	>0.5	Grass, wood, coal combustion	
BaA/(BaA + Chry)	<0.2	Petrogenic	Yunker et al. (2002b)
	>0.35	Combustion	
Indeno/(Indeno + BgP)	<0.2	Petrogenic	Yunker et al. (2002b)
	0.2–0.5	Petroleum combustion	
	>0.5	Grass, wood, coal combustion	

Table 4 Polycyclic aromatic hydrocarbons diagnostic ratio for calculated for TSP and PM_{10} samples

PAHs ratios	TSP	PM_{10}
∑LMW/∑HMW	–	0.01
∑COMB/∑PAHs	0.89	0.89
Flan/(Flan + Py)	0.66	0.29
BaA/(BaA+Chry)	0.51	0.17
Indeno/(Indeno + BgP)	–	1
BeP/BaP	1.01	2.34
Total polycyclic aromatic hydrocarbons (TPAH), pgm^{-3}	239	561

or combustion sources. However, values of Flan/(Flan+Py) and BaA/(BaA+Chry) diagnostic ratios appeared to be slightly lower when compared to the designated diagnostic ratio, thus suggesting possible influence of petrogenic source. However, the high ∑COMB/∑PAHs ratio (0.89) in TSP and PM_{10}, suggest a strong influence of the pyrogenic combustion process. BeP/BaP ratio has been used to assess the extent of chemical degradation upon emission of these PAHs and the present results suggest that PAHs in PM_{10} samples appeared to be more aged compared TSP samples with values of 2.34 over 1.01.

Conclusion

Aliphatic hydrocarbons were found in all three media studied whilst polycyclic aromatic hydrocarbon (PAH) compounds were detected, only in the atmospheric particulate samples, at trace levels. The 24 h- PM_{10} and TSP concentrations is still below the recommended Malaysian Air Quality Guidelines, listed by Department of Environment (2000) for the respective parameter. Analyses of the results suggest that terrestrial plant waxes is the main source of aliphatic hydrocarbons in the study area whereas sources for PAHs in the particulate samples are most likely derived from combustion sources. The detection of PAHs in the aerosol samples also suggest possible anthropogenic input of particulate matter to the study area; however the PAHs levels detected are still well below the values reported for other locations in Malaysia. Sources of particulate matter in the study area could be influenced by local activities and long range air transport, however result from this one-off monitoring study does not allow the differentiation between these two types of sources. Since the trans-boundary input (or long range air transport) of atmospheric particulate matter to the study area could not be mitigated, any mitigating measures should be focus on local sources. For instance, campfires at campsites, open burning of refuse and/or biomass, forest clearing and use of boats or any water transportation within the lake powered by fossil fuel namely gasoline or diesel should be regulated and monitored so as to minimize the impact on surrounding air quality thus ensuring sustainability of the general environmental quality in Tasik Kenyir and the Terengganu National Park.

Acknowledgements The financial and logistic support from UMT through Research Management and Innovation Centre (RMIC) to conduct the survey at Terengganu National Park and Tasik Kenyir is greatly acknowledged. The use of GC-MS from Institute of Oceanography and Environment (INOS) is also acknowledged.

References

Abas, M. R., & Simoneit, B. R. T. (1998). Wax lipids from leaf surfaces of some common plants of Malaysia. *Pertanika Journal of Science & Technology, 6*, 171–182.

Department of Environment. (2000). *A guide to Air Pollutant Index in Malaysia (API)* (4th ed.). Kuala Lumpur: Ministry of Science, Technology and the Environment Malaysia.

Fang, G. C., Wu, Y. S., Chen, J. C., Peter, F. P. C., Chang, C. N., Ho, T. T., & Chen, M. S. (2005). Characteristic study of polycyclic aromatic hydrocarbons for fine and course particulates at Pastureland near Industrial Park sampling site of central Taiwan. *Chemosphere, 60*, 427–433.

Kaushik, C. P., Sangwan, P., & Haritash, A. K. (2012). Association of polycyclic aromatic hydrocarbons (PAHs) with different sizes of atmospheric particulate in Hisar City and its health aspects. *Polycyclic Aromatic Compounds, 32*(5), 626–642. https://doi.org/10.1080/10406638.2012.695318.

Mazurek, M. A., & Simoneit, B. R. T. (1984). Characterization of biogenic and petroleum derived organic matter in aerosols over remote, rural, and urban areas. In L. H. Keith (Ed.), *Identification and analysis of organic pollutants in air* (pp. 353–370). Woburn: Ann Arbor Science

Mohd Tahir, N., Poh, S. C., Suratman, S., Ariffin, M. M., & Shazali, N. A. M. (2009). Determination of trace metals in airborne particulate matter of Kuala Terengganu, Malaysia. *Bulletin of Environmental Contamination and Toxicology, 83*, 199–203.

Mohd Tahir, N., Suratman, S., Foo, T. F., Hamzah, M. S., & Latif, M. T. (2013). Temporal distribution and chemical characterization of atmospheric particulate matter in the eastern coast of Peninsular Malaysia. *Aerosol and Air Quality Research, 13*, 584–595.

Pandey, P. K., Patel, K. S., & Lenicek, J. (1999). Polycyclic aromatic hydrocarbons: Need for assessment of health risks in India? Study of an urban-industrial location in India. *Environmental Monitoring and Assessment, 59*, 287–319.

Pies, C., Hoffmann, B., Petrowsky, J., Yang, Y., Ternes, T. A., & Hofmann, T. (2008). Characterization and source identification of polycyclic aromatic hydrocarbons (PAHs) in river bank soils. *Chemosphere, 72*, 1594–1601.

Ravindra, K., Sokhi, R., & Van Grieken, R. (2008). Atmospheric polycyclic aromatic hydrocarbons: Source attribution, emission factors and regulation. *Atmospheric Environment, 42*, 2895–2921.

Simoneit, B. R. T. (2002). Biomass burning-a review of organic tracers for smoke from incomplete combustion. *Applied Geochemistry, 17*(3), 129–162.

Simoneit, B. R. T., & Mazurek, M. A. (1982). Organis matter of the troposphere-II. Natural background of biogenic lipid matter in aerosols over the rural western United States. *Atmospheric Environment, 16*(9), 2139–2159.

Wang, X. T., Miao, Y., Zhang, Y., Li, Y. C., Wu, M. H., & Yu, G. (2013). Polycyclic aromatic hydrocarbons (PAHs) in urban soils of the megacity Shanghai: Occurrence, source apportionment and potential health risk. *Science of the Total Environment, 447*, 80–89.

Yunker, M. B., Backus, S. M., Pannatier, E. G., Jefferies, D. S., & Macdonald, R. W. (2002a). Sources and significance of alkane and PAH hydrocarbons in Canadian Artic Rivers. *Estuarine, Coastal and Shelf Science, 55*, 1–31.

Yunker, M. B., Macdonald, R. W., Vingarzan, R., Mitchell, R. H., Goyette, D., & Sylvestre, S. (2002b). PAHs in the Fraser River basin: A critical appraisal of PAH ratios as indicators of PAH source and composition. *Organic Geochemistry, 33*, 489–515.

Zhang, W., Zhang, S., Wan, C., Yue, D., Ye, Y., & Wang, X. (2008). Source diagnostics of polycyclic aromatic hydrocarbons in urban road runoff, dust, rain and canopy throughfall. *Environmental Pollution, 153*, 594–601.

Diversity of Bryophytes of Terengganu and Their Ecological Roles in the Environment

Gaik Ee Lee, Ahmad Damanhuri, and Nik Norhazrina

Abstract The diversity of bryophytes of Terengganu and their ecological roles in the environment is reviewed. Thus far, there are 257 taxa of bryophytes are recorded for Terengganu. A total of three species of mosses viz. *Mastopoma perundulatum* (Dixon) Horik. & Ando in Kira & Umesao, *Sematophyllum subhumile* (Müll. Hal.) M.Fleisch., and *Trismegistia complanatula* (Müll. Hal.) Müll. Hal. are reported for the first time for Peninsular Malaysia while 11 other taxa are new additions to the bryoflora of Terengganu.

Keywords Bryophytes · Mosses · Liverworts · Terengganu

Introduction

Bryophytes are a group of non-vascular plants including mosses (Bryophyta), liverworts (Marchantiophyta) and hornworts (Anthocerotophyta), known to be the second largest group among all land plants, surpassed only by the flowering plants (Magnoliophyta), with 450,000 species (Pimm and Joppa 2015). The bryophytes comprised 15,000 (Gradstein et al. 2001) to 25,000 species (Crum 2001) and they are most abundant in tropical rainforests. Unfortunately they have been frequently ignored and also extremely under-collected. They abound in wet, mountainous forest where they often grow luxuriantly in loose mats and patches, or tightly appressed to substrates such as rocks, leaves or bark of trees, and also some are found growing in pendulous festoons or hanging down loosely from tree branches.

Malaysian bryophytes have been collected since the nineteenth century and, to date, there are 2227 species and infraspecific taxa of bryophytes known to occur in

G. E. Lee (✉)
School of Marine and Environmental Sciences, Universiti Malaysia Terengganu,
Kuala Nerus, Terengganu, Malaysia
e-mail: gaik.ee@umt.edu.my

A. Damanhuri · N. Norhazrina
School of Environmental and Natural Resource Sciences, Universiti Kebangsaan Malaysia,
Bangi, Selangor, Malaysia

M. T. Abdullah et al. (eds.), *Greater Kenyir Landscapes*,
https://doi.org/10.1007/978-3-319-92264-5_5

Malaysia (Cheah and Yong 2016; Pócs and Lee 2016; Sukkharak 2014; Furuki et al. 2013; Yong et al. 2013; Katagiri et al. 2012; Chuah-Petiot 2011; Suleiman et al. 2006). The first record of mosses in Terengganu was published by Dixon (1926), where he reported nine taxa which were collected in Kuala Berang, Kuala Telumong (or Kuala Telemong) and Kuala Terengganu, and subsequently Mohamed and Tan (1988), in their checklist of mosses of Peninsular Malaya and Singapore, enumerated 64 species, one subspecies, and one variety of mosses in 42 genera and 20 families in Terengganu. Recently, a new checklist of mosses with updated nomenclatures, synonymy, taxonomic bibliography, and distribution records in Peninsular Malaysia and Singapore has been produced in the framework of the project called The Flora of Peninsular Malaysia. The revised moss checklist recorded 174 taxa of mosses for Terengganu and this represents 31% of the 558 taxa reported for Peninsular Malaysia (Yong et al. 2013). Meanwhile, the first record of liverworts in Terengganu was reported by Yamada (1979) from Kuala Berang i.e. *Radula retroflexa* Taylor and followed by Inoue (1984) who listed one species i.e. *Plagiochila arbuscula* (Brid. ex Lehm. & Lindenb.) Lindenb. Some species of mosses, based on specimens collected in various localities in Terengganu, were also reported in various revisions and monographic studies of mosses (Mohamed and Robinson 1991; Iwatsuki and Mohamed1987; Frahm and Mohamed 1987; Miller and Manuel 1982; Eddy 1977; Johnson 1964). Since then, neither a comprehensive floristic nor taxonomic study on collection of bryophytes from Terengganu has been carried out until the third millennium, particularly the works of Damanhuri et al. (2006, 2008, 2011, 2014). Thus far, the bryophyte flora of Terengganu comprises 255 taxa of mosses, two species of liverworts, and none for hornwort.

The taxa of bryophytes reported in this paper are entirely based on the literatures on bryophytes reported for Terengganu starting from Dixon (1926). Further explanation regarding the specimens collected during the various scientific expeditions and also the deposition of specimens were recorded in Damanhuri et al. (2006, 2008, 2011, 2014). Distribution and references of all the bryophyte taxa reported for Terengganu are summarised in Table 1. Major collection localities are shown in Fig. 1 and listed in the Appendix 1.

Previous work on bryophytes in Terengganu are very scanty and comprehensive study is lacking except that of Damanhuri et al. (2006, 2008, 2011, 2014), in which most of the moss collections were made during scientific expeditions (Gunung Mandi Angin, Bukit Bauk, Gunung Gagau, Gunung Tebu) organised by the Forestry Department of Peninsular Malaysia. In those series of papers, three taxa of mosses are reported as new additions to the moss flora of Peninsular Malaysia viz. *Rhynchostegiella menadensis* (Sande Lac.) E.B.Bartram, *Schistomitrium robustum* Dozy & Molk., and *Syrrhopodon perarmatus* Broth. in Broth & Watts. The species *R. menadensis* was found growing on a tree trunk in Gunung Mandi Angin and hitherto only recorded in China, Indonesia and the Philippines (Liu et al. 2010). *Syrrhopodon perarmatus* occurs in temperate Australasia, Papua New Guinea and Pacific Islands (Fiji, Vanuatu), while *S. robustum* is endemic to Malesia (Eddy 1990).

Table 1 Summary of the number of taxa of mosses and liverworts collected at various localities in Terengganu

Locality	Total taxa (mosses)	Total taxa (liverworts)	References
Kuala Berang, K. Telumong, K. Terengganu	9 taxa	1 species	Dixon (1926) and Yamada (1979)
Terengganu	–	1 species	Inoue (1984)
Terengganu	64 species, 1 subspecies, 1 variety	–	Mohamed and Tan (1988)
Lata Payung Recreational Forest	81 species, 3 subspecies	–	Yushana (2005)[a]
Gunung Mandi Angin	139 species, 1 subspecies, 3 varieties	–	Damanhuri et al. (2006)
Bukit Bauk Urban Forest	67 species, 1 variety	–	Damanhuri et al. (2008)
Gunung Gagau	97 species, 6 subspecies, 17 varieties	–	Damanhuri et al. (2011)
Terengganu	174 taxa		Yong et al. (2013)
Lata Tembakah Recreational Forest	66 species, 3 subspecies, 10 varieties	–	Aishah (2013)[a]
Gunung Tebu	100 species, 7 subspecies, 20 varieties	–	Damanhuri et al. (2014)
Jeram Tanduk Forest Reserve	60 species, 3 subspecies, 15 varieties	–	Syazwana (2016)[a]
Tembat Permanent Forest Reserve	72 species, 3 subspecies, 6 varieties	–	Nurulhuda (2016)[a]

[a]Unpublished data from thesis

In this paper, an additional 81 taxa are added to the latest moss checklist of Terengganu and out of these, another three species were reported which are new to Peninsular Malaysia i.e. *Mastopoma perundulatum* (Dixon) Horik. & Ando in Kira & Umesao, *Sematophyllum subhumile* (Müll. Hal.) M.Fleisch., and *Trismegistia complanatula* (Müll. Hal.) Müll. Hal. and another 11 are new records for Terengganu, based on the unpublished theses by Yushana (2005), Aishah (2013), Nurulhuda (2016) and Syazwana (2016). Moreover, Terengganu is also the second locality for ten rare species of mosses (see Appendix 1 for the list of rare species of mosses), where all of these species were previously reported only from Pahang except one from Kedah (*Clastobryum caudatum*) (Yong et al. 2013). For the liverwort species, *R. retroflexa*, is only known to occur in Terengganu in Peninsular Malaysia and in Mt. Kinabalu in Sabah (Yamada 1979). *Radula retroflexa* is exclusively a rupicolous species. It was collected on rocks and rarely on tree trunks and branches (Yamada 1979) and this species is probably under-collected due to little fieldwork being done in Peninsular Malaysia.

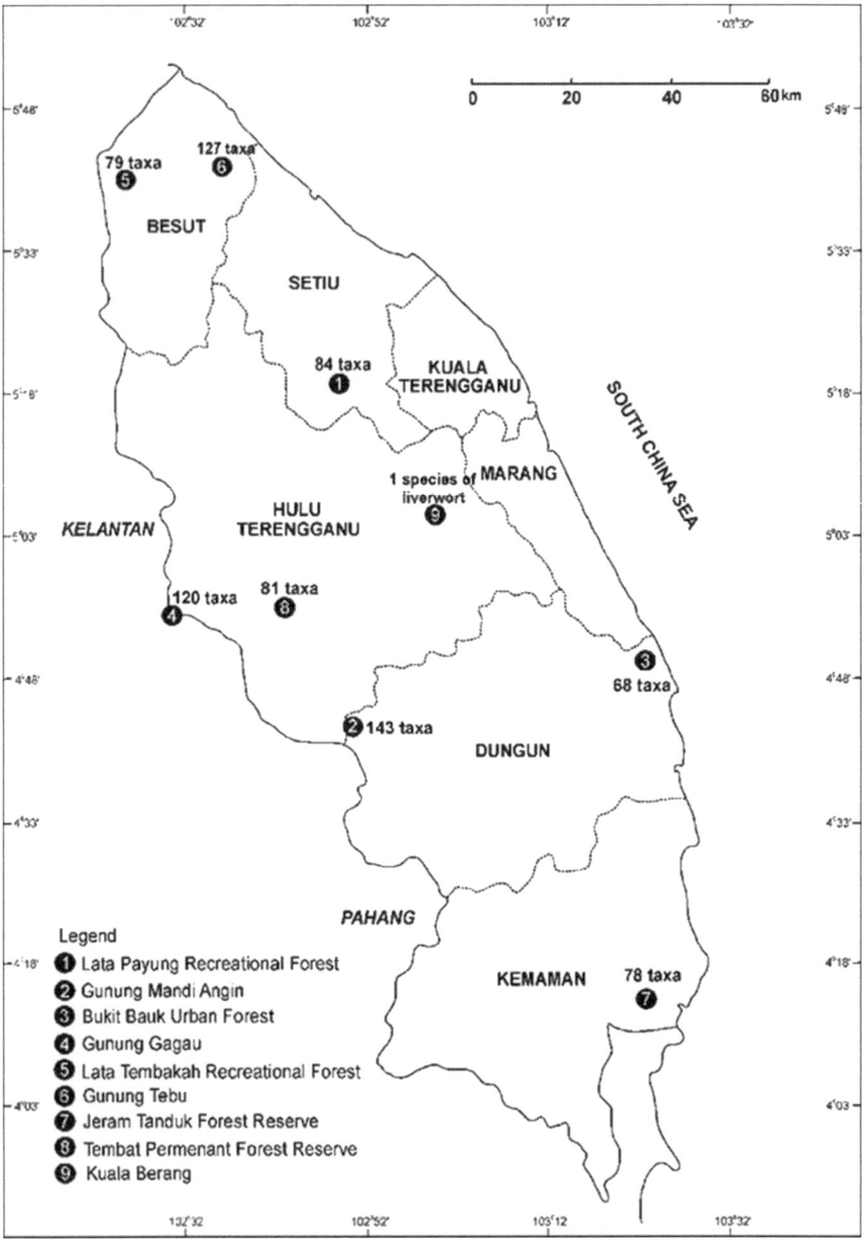

Fig. 1 Map of Terengganu showing the major collection localities and the number of taxa reported for the districts of Terengganu

The Ecological Role of Bryophytes: Water Retention and Environmental Indicators

Many bryophytes species have a higher and more effective water retention capacity as compared to many other plant groups (Proctor 2008). Bryophytes are able to hold a huge quantity of water during heavy rainfall and then gradually release the water from wet to dry area over a much longer period, and can, therefore, enhance and regulate the water flow in a forest or elsewhere. The exceptional water storage ability in bryophytes are due to its unique structure, for example the moss species, *Sphagnum* where the tissues contain empty and specialized water-holding hyaline cells which has incredible ability to absorb 20 or more times their own dry weight in water (Glime 2015; Nichols 1918). This sponge-like characteristic of *Sphagnum* (e.g., three species of *Sphagnum* in Gunung Gagau) definitely play an important and active role in the protection of slopes, banks, and steep hillsides against erosion by monitoring the destructive effects of heavy rainfalls and regulating the rhythm of waterway (Pócs 1980).

Besides, bryophytes are very sensitive to natural fluctuations in relative humidity, climate change and forest fragmentation, and therefore, many bryophytes are very restricted to specific forest microclimates. For example the epiphyllous liverworts, they have been pointed out particularly sensitive to environmental changes and have been assessed as possible indicators of forest ecological conditions (Daniels and Kariyappa 2007; Frego 2007; Pócs 1996).

Conclusion

A total of 257 taxa of bryophytes are recorded for Terengganu. Of these, six species of mosses viz. *Mastopoma perundulatum* (Dixon) Horik. & Ando in Kira & Umesao, *Rhynchostegiella menadensis* (Sande Lac.) E.B.Bartram, *Sematophyllum subhumile* (Müll. Hal.) M.Fleisch., *Schistomitrium robustum* Dozy & Molk., *Syrrhopodon perarmatus* E.B.Bartram, *Trismegistia complanatula* (Müll. Hal.) Müll. Hal. and one species of liverwort i.e. *Radula retroflexa* Taylor are reported to occur only in Terengganu. While ten rare species of mosses previously recorded only in Pahang or Kedah were also found in Terengganu, thus making the latter state the second reported locality in Peninsular Malaysia. In the present paper, a total of three species are reported for the first time for Peninsular Malaysia while 11 other taxa are new additions to the bryoflora of Terengganu. Terengganu harbours quite a number of new records and localities for Malaysian bryophytes as mentioned above and with further exploration and fieldwork it is expected many new discoveries will certainly be made in the future. Therefore, an urgent effort should be undertaken to collect and study the bryophytes, especially the liverworts and hornworts, in Terengganu before the microhabitats of these tiny plants are destroyed. Of special interest is the Kenyir tropical rainforest in Terengganu where bryophytes flourish particularly well in such moist and humid forest and different bryophyte species can be found on tree trunks, bases, branches, twigs, or leaves, or on fallen and rotten logs in various stages of the undisturbed forest.

Acknowledgements We thank the editors for comments and inviting us to contribute our knowledge in this book. Financial support from the grants FRGS/1/2017/STG03/02/2 and GGPM-2017-090 to N. Norhazrina is gratefully acknowledged.

Appendix 1

Species Checklist and new additions to the bryoflora of Peninsular Malaysia and the state of Terengganu are indicated by "**" and "*" respectively. Rare species of mosses are marked with®. Major collection localities are represented in numbers: Lata Payung Recreational Forest (1), Gunung Mandi Angin (2), Bukit Bauk Urban Forest (3), Gunung Gagau (4), Lata Tembakah Recreational Forest (5), Gunung Tebu (6), Jeram Tanduk Forest Reserve (7), Tembat Permanent Forest Reserve (8). Name of taxa are arranged alphabetically.

No.	Name of taxa of mosses	Locality
1	*Acanthorrhynchium papillatum* (Harv.) M.Fleisch.	1, 2, 3, 4, 5, 6, 7, 8
2	*Acroporium adspersum* (Hampe) Broth.	4, 6
3	*Acroporium condensatum* Müll. Hal. ex E.B.Bartram	4, 6
4	*Acroporium convolutum* (Sande Lac.) M.Fleisch. var. *convolutum*	1, 2, 3, 4
5	*Acroporium convolutum* var. *elatum* (Dixon) B.C.Tan	2
6	*Acroporium diminutum* (Brid.) M.Fleisch.	1, 2, 3, 4, 6, 7
7	*Acroporium downii* (Dixon) Broth.	1, 2
8	*Acroporium joannis-winkleri* Broth.	1, 2, 4, 6
9	*Acroporium lamprophyllum* Mitt.	1, 2, 3, 4, 5, 6, 7, 8
10	*Acroporium macroturgidum* Dixon	4
11	*Acroporium pungens* (Hedw.) Broth.	4
12	*Acroporium rigens* (Broth. ex Dixon) Dixon	4, 5, 6
13	*Acroporium rufum* (Reinw. & Hornsch.) M.Fleisch.	2, 4
14	*Acroporium secundum* (Reinw. & Hornsch.) M.Fleisch.	2, 4
15	*Acroporium stramineum* (Reinw. & Horsch.) M.Fleisch. var. *stramineum*	2, 4, 6
16	*Acroporium stramineum* var. *hamulatum* (M.Fleisch.) B.C.Tan	2, 4
17	*Acroporium strepsiphyllum* (Mont.) B.C.Tan	1, 2, 4, 6
18	*Aequatoriella bifaria* (Bosch & Sande Lac.) Touw	2, 3, 8
19	*Aerobryidium aureonitens* (Hook. ex Schwägr.) Broth.	1, 2, 4, 5, 6, 7, 8
20	*Aerobryidium crispifolium* (Broth. & Geh.) M.Fleisch. in Broth.	4, 6
21	*Aerobryopsis leptosigmata* (Müll. Hal. ex Broth. & Geh.) M.Fleisch.	2
22	*Aerobryopsis longissima* (Dozy & Molk.) M.Fleisch	5, 6, 7, 8
23	**Aerobryopsis subleptostigmata* Broth. & Paris	8

No.	Name of taxa of mosses	Locality
24	*Arthrocormus schimperi* (Dozy & Molk.) Dozy & Molk.	1, 2, 3, 4, 5, 6, 7, 8
25	*®*Barbella enervis* (Thwaites & Mitt.) M.Fleisch in Broth.	1
26	*Barbula consanguinea* (Thwaites & Mitt.) A.Jaeger	1, 6
27	*Barbula indica* (Hook.) Spreng.	3
28	*Bryum apiculatum* Schwägr.	3, 7
29	*Bryum clavatum* (Schimp.) Müll. Hal.	2, 7
30	*Bryum coronatum* Schwägr.	1, 3, 5, 6
31	*Caduciella mariei* (Besch.) Enroth	3, 8
32	*Callicostella papillata* (Mont.) Mitt. var. *papillata*	2, 8
33	*Callicostella papillata* var. *prabaktiana* (Müll. Hal.) Streimann	1, 2, 4, 7, 8
34	*Calymperes afzelii* Sw.	1, 2, 3, 5, 6, 8
35	*Calymperes boulayi* Besch.	1, 3, 6
36	*Calymperes crassinerve* (Mitt.) A.Jaeger	2, 5, 6
37	*Calymperes erosum* Müll. Hal.	1, 2, 3, 5, 6, 7, 8
38	*Calymperes fasciculatum* Dozy & Molk.	1, 4, 5, 6, 7
39	*Calymperes graeffeanum* Müll. Hal.	6
40	*Calymperes lonchophyllum* Schwägr. subsp. *lonchophyllum*	1, 2, 3, 4, 5, 6, 7, 8
41	*Calymperes lonchophyllum* subsp. *beccarii* (Hampe) M.Menzel	1, 2, 4, 6, 7, 8
42	*Calymperes moluccense* Schwägr.	1, 2, 3, 4, 5, 6, 7, 8
43	*Calymperes motleyi* Mitt. in Dozy & Molk.	3
44	*Calymperes palisotii* Schwägr.	3, 5, 6
45	*Calymperes porrectum* Mitt.	1, 2, 3, 4, 6, 7, 8
46	**Calymperes schmidtii* Broth. in J.Schmidt	1
47	*Calymperes serratum* A.Braun ex Müll. Hal.	2, 5, 6, 8
48	*Calymperes strictifolium* (Mitt.) G.Roth	4, 8
49	*Calymperes taitense* (Sull.) Mitt.	2, 3, 8
50	*Calymperes tenerum* Müll. Hal.	3
51	*Calyptrochaeta remotifolia* (Müll. Hal.) Z.Iwats, B.C.Tan & Touw	2
52	*Campylopus ericoides* (Griff.) A.Jaeger	6
53	*Campylopus exasperatus* (Nees & Blume) Brid.	6
54	*Campylopus macgregorii* Broth. & Geh.	Gunung Padang
55	*Campylopus serratus* Sande Lac.	4, 5, 6
56	*Chaetomitrium borneense* Mitt.	2
57	*Chaetomitrium leptopoma* (Schwägr.) Bosch & Sande Lac.	2, 7
58	*Chaetomitrium orthorrhynchum* (Dozy & Molk.) Bosch & Sande Lac.	2, 4, 7, 8
59	*Chaetomitrium papillifolium* Bosch & Sande Lac.	Kampung Pasir Raja
60	*Chaetomitrium setosum* Broth. ex Dixon	2, 8
61	*Circulifolium exiguum* (Bosch & Sande Lac.) S.Olsson, Enroth & D.Quandt	2, 3, 7, 8
62	*Circulifolium microdendron* (Mont.) S.Olsson, Enroth & D.Quandt	2, 4, 8
63	*Cladopodanthus heterophyllus* (M.Fleisch.) E.B.Bartram	4
64	*Clastobryophilum bogoricum* (Bosch. & Sande Lac.) M.Fleisch.	4, 5, 6
65	®*Clastobryum caudatum* (Sande Lac.) M.Fleisch.	6, 7

No.	Name of taxa of mosses	Locality
66	*Clastobryum cuculligerum* (Sande Lac.) Tixier	1, 2, 6, 7
67	*Clastobryum epiphyllum* (Renauld & Cardot) B.C.Tan & Touw	2, 4, 6
68	*Clastobryum indicum* (Dozy & Molk.) Dozy & Molk.	3, 4
69	**Clastobryum scalare* (Müll. Hal.) Tixier	7
70	*Cryptopapillaria fuscescens* (Hook.) M.Menzel	2
71	*Ctenidiadelphus plumularia* (Müll. Hal.) M.Fleisch.	3
72	*Cyathophorum spinosum* (Müll. Hal.) M.Fleisch.	2
73	**Dendrohypnum subspininervium* subsp. *arborescens* (Mitt.) N.E.Bell, A.E.Newton & D.Quandt	8
74	*Dendrohypnum subspininervium* (Müll. Hal.) N.E.Bell, A.E.Newton & D.Quandt subsp. *subspininervium*	1, 2, 4, 6
75	*Desmotheca apiculata* (Dozy & Molk.) Lindb.	2
76	*Dicranella coarctata* (Müll. Hal.) Bosch & Sande Lac.	6
77	*Dicranoloma blumii* (Nees) Paris	2
78	*Dicranoloma braunii* (Müll. Hal.) Paris	2
79	*Dicranoloma brevisetum* (Dozy & Molk.) Paris	2
80	*Dimorphocladon borneense* Dixon	2, 4, 7, 8
81	*Diphyscium mucronifolium* Mitt.	1, 2, 4, 5, 6, 7, 8
82	®*Distichophyllidium nymanianum* M.Fleisch.	2
83	*Distichophyllum brevicuspis* M.Fleisch.	2
84	*Distichophyllum cuspidatum* (Dozy & Molk.) Dozy & Molk.	2, 4, 6
85	*Distichophyllum mittenii* Bosch & Sande Lac.	2
86	*Distichophyllum nigricaule* var. *cirratum* (Renauld & Cardot) M. Fleisch.	2, 4, 6
87	*Distichophyllum osterwaldii* M.Fleisch.	2, 4
88	*Distichophyllum schmidtii* Broth.	2, 4, 7
89	*Distichophyllum spathulatum* (Dozy & Molk.) Dozy & Molk.	2
90	*Duthiella wallichii* (Mitt.) Müll. Hal.	Terengganu
91	*Ectropotheciella decrescens* (Sande Lac.) M.Fleisch.	Terengganu
92	*Ectropothecium buitenzorgii* (Bél) Mitt.	1, 4, 5, 6, 7
93	*Ectropothecium dealbatum* (Reinw. & Hornsch.) A.Jaeger	1, 4
94	*Ectropothecium eleganti-pinnatum* (Müll. Hal.) A.Jaeger	2, 4, 7
95	*Ectropothecium ichnotocladum* (Müll. Hal.) A.Jaeger	5, 6
96	*Ectropothecium incubans* (Reinw. & Hornsch.) A.Jaeger	Terengganu
97	*Ectropothecium perminutum* Broth. ex E.B.Bartram	5, 6, 7
98	*Ectropothecium singapurense* Dixon	Terengganu
99	®*Ectropothecium striatulum* Dixon ex E.Bartram	1, 4
100	*Ephemeropsis tjibodensis* K.I.Goebel	4, 6
101	*Exostratum blumii* (Nees ex Hampe) L.T.Ellis	1, 2, 3, 4, 6, 7, 8
102	*Fissidens ceylonensis* Dozy & Molk.	1, 6
103	*Fissidens crassinervis* Sande Lac.	1, 2, 3, 4, 5, 6, 7, 8
104	*Fissidens crenulatus* var. *elmeri* (Broth.) Z.Iwats & Tad.Suzuki	3
105	*Fissidens crispulus* Brid. var. *crispulus*	1, 2, 3, 4, 5, 6, 7, 8
106	*Fissidens crispulus* var. *robinsonii* (Broth.) Z.Iwats. & Z.H.Li	4

No.	Name of taxa of mosses	Locality
107	*Fissidens guangdongensis* Z.Iwats. & Z.H.Li	3, 4, 5, 6, 7
108	*Fissidens hollianus* Dozy & Molk.	1, 2, 3, 4, 6, 7, 8
109	*Fissidens javanicus* Dozy & Molk.	1, 2, 3, 4, 6, 7, 8
110	*Fissidens oblongifolius* Hook.*f* & Wilson	6
111	*Fissidens pellucidus* Hornsch.	2, 3, 6, 7, 8
112	*Fissidens serratus* Müll. Hal.	4
113	*Fissidens subangustus* M.Fleisch.	Terengganu
114	*Fissidens zollingeri* Mont.	1, 8
115	*Floribundaria floribunda* (Dozy & Molk.) M.Fleisch.	2, 4
116	®*Gammiella tonkinensis* (Broth. & Paris) B.C.Tan	6
117	*Garovaglia compressa* Mitt.	4, 7
118	*Garovaglia elegans* (Dozy & Molk.) Hampe ex Bosch & Sande Lac.	2
119	*Garovaglia powellii* Mitt.	2
120	*Gymnostomum recurvirostrum* Hedw.	Terengganu
121	*Himantocladium plumula* (Nees) M.Fleisch.	3, 4, 6, 7, 8
122	*Himantocladium cyclophyllum* (Müll. Hal.) M.Fleisch.	5, 6
123	*Homaliodendron flabellatum* (Sm.) M.Fleisch.	2, 4
124	*Hyophila involuta* (Hook.) A.Jaeger	1, 5, 6
125	*Isocladiella surcularis* (Dixon) B.C.Tan & Mohamed	2, 3, 4, 6, 8
126	*Isopterygium albescens* (Hook.) A.Jaeger	1, 2, 3, 5, 6, 7, 8
127	*Leptotrichella miqueliana* (Mont.) Lindb. ex Broth.	2, 3
128	*Leucobryum aduncum* Dozy & Molk. var. *aduncum*	1, 2, 3, 4, 5, 6, 8
129	*Leucobryum aduncum* var. *scalare* Müll. Hal. ex M.Fleisch.	5, 6
130	*Leucobryum bowringii* Mitt.	1, 2, 3, 4, 5, 6, 7
131	*Leucobryum candidum* (Brid. ex P.Beauv.) Wilson	1, 2, 4, 5, 6, 8
132	*Leucobryum chlorophyllosum* Müll. Hal.	2, 3, 4, 5, 6, 7, 8
133	*Leucobryum javense* (Brid.) Mitt.	2, 4, 6
134	*Leucobryum juniperoideum* (Brid.) Müll. Hal.	2
135	*Leucobryum sanctum* (Nees ex Schwägr.) Hampe	1, 2, 3, 4, 5, 6, 7, 8
136	*Leucobryum sumatranum* Broth. ex M.Fleisch.	2, 4
137	*Leucoloma amoene-virens* Mitt.	1, 2, 5, 6, 8
138	*Leucoloma molle* (Müll. Hal.) Mitt.	2, 4, 6
139	*Leucoloma walkeri* Broth.	2
140	**Leucomium strumosum* (Hornsch.) Mitt.	8
141	*Leucophanes angustifolium* Renauld & Cardot	1, 2, 3, 4, 5, 6, 8
142	*Leucophanes candidum* (Schwägr.) Lindb.	4
143	*Leucophanes glaucum* (Schwägr.) Mitt.	3, 4, 5, 6, 7
144	*Leucophanes octoblepharioides* Brid.	1, 2, 3, 4, 5, 6, 7, 8
145	*Lopidium struthiopteris* (Brid.) M.Fleisch.	2
146	*Macromitrium blumei* Nees ex Schwägr.	2, 4
147	*Macromitrium cuspidatum* Hampe	Terengganu
148	*Macromitrium fuscescens* Schwägr.	2, 4
149	*Macromitrium ochraceum* (Dozy & Molk.) Müll. Hal.	4
150	**Macromitrium orthostichum* Nees ex Schwägr.	8

No.	Name of taxa of mosses	Locality
151	*Macromitrium salakanum* Müll. Hal.	2, 8
152	*Mastopoma brauniana* (Bosch & Sande Lac.) H.Akiyama	2
153	***Mastopoma perundulatum* (Dixon) Horik. & Ando in Kira & Umesao	1
154	*Mastopoma uncinifolium* (Broth.) Broth.	6
155	*Meiothecium microcarpum* (Hook.) Mitt.	5, 6
156	*Meteorium polytrichum* Dozy & Molk.	2
157	*Mitthyridium constrictum* (Sull.) H.Rob.	1, 2, 4, 6, 7, 8
158	*Mitthyridium fasciculatum* subsp. *cardotii* (M.Fleisch.) B.C.Tan & L.T.Ellis	1, 2, 4, 5, 6, 7
159	*Mitthyridium fasciculatum* (Hook. & Grev.) H.Rob. subsp. *fasciculatum*	1, 2, 4, 5, 6, 7
160	*Mitthyridium fasciculatum* subsp. *obtusifolium* (Lindb.) M.Menzel	1, 2
161	*Mitthyridium flavum* (Müll. Hal.) H.Rob.	1, 2, 3, 4, 5, 6, 7, 8
162	*Mitthyridium jungquilianum* (Mitt.) H.Rob.	1, 2, 3, 4, 5, 6, 7, 8
163	*Mitthyridium luteum* (Mitt.) H.Rob.	1, 2, 4
164	*Mitthyridium papuanum* (Broth.) H.Rob.	2, 3, 7, 8
165	*Mitthyridium repens* (Harv.) H.Rob.	2, 3, 4, 5, 6
166	®*Mitthyridium retusum* (Besch.) W.D.Reese	1, 4
167	*Mitthyridium subluteum* (Müll. Hal.) H.K.Nowak	Sekayu Recreational Forest
168	*Mitthyridium undulatum* (Dozy & Molk.) H.Rob.	1, 2, 3, 4, 5, 6, 7, 8
169	*Mitthyridium wallisii* var. *crassum* (Broth.) M.Menzel in M.Menzel & W.Schultze-Motel	1, 2, 4, 5, 6
170	*Mitthyridium wallisii* (Müll. Hal.) H.Rob. var. *wallisii*	1, 2, 4, 5, 6, 7
171	*Mniodendron dendroides* (Brid.) Wijk & Margad.	4, 6
172	*Mniomalia semilimbata* (Mitt.) Müll. Hal.	2, 3, 5, 6
173	*Neckeropsis gracilenta* (Bosch & Sande Lac.) M.Fleisch.	4, 5, 6, 7, 8
174	*Neckeropsis lepineana* (Mont.) M.Fleisch.	Terengganu
175	*Octoblepharum albidum* Hedw.	1, 4, 5, 6
176	*Oedicladium pseudorufescens* (Hampe) B.C.Tan & Mohamed	1, 2, 3, 4, 5, 6, 7
177	*Orthodontium infractum* Dozy & Molk.	6
178	*Papillidiopsis bruchii* (Dozy & Molk.) W.R.Buck & B.C.Tan	4, 6
179	*Papillidiopsis complanata* (Dixon) W.R.Buck & B.C.Tan	1, 2, 4, 5, 6, 7, 8
180	*Papillidiopsis luxurians* (Dozy & Molk.) W.R.Buck & B.C.Tan	5, 6
181	*Papillidiopsis malesiana* W.R.Buck & B.C.Tan	2, 4, 5, 6, 8
182	*Papillidiopsis ramulina* (Thwaites & Mitt.) W.R.Buck & B.C.Tan	2
183	*Pelekium velatum* Mitt.	3, 7, 8
184	*Pinnatella alopecuroides* (Hook.) M.Fleisch.	Terengganu
185	*Pinnatella ambigua* (Bosch & Sande Lac.) M.Fleisch.	Terengganu
186	**Pinnatella kuehliana* (Bosch & Sande Lac.) M. Fleisch.	Batu Biwa Limestone Hill
187	*Pinnatella mucronata* (Bosch & Sande Lac.) M. Fleisch.	2, 3, 5, 6, 7, 8
188	*Pogonatum cirratum* subsp. *fuscatum* (Mitt.) Hyvönen	6

No.	Name of taxa of mosses	Locality
189	**Pogonatum cirratum* subsp. *macrophyllum* (Dozy & Molk.) Hyvönen	6
190	*Pogonatum piliferum* (Dozy & Molk.) Touw	1, 2, 7, 8
191	*Pseudotaxiphyllum pohliaecarpum* (Sull. & Lesq.) Z.Iwats.	6
192	*Pseudotrachypus wallichii* (Brid.) W.R.Buck	4
193	*Pterobryopsis crassicaulis* (Müll. Hal.) M.Fleisch.	2
194	®*Pterobryopsis gedehensis* M.Fleisch.	2, 4
195	*Pyrrhobryum latifolium* (Bosch & Sande Lac.) Mitt.	1, 2, 4, 5, 6, 7, 8
196	*Pyrrhobryum medium* (Besch.) Manuel	4, 6, 7
197	*Pyrrhobryum spiniforme* (Hedw.) Mitt.	2, 4, 5, 6, 8
198	*Radulina borbonica* (Bél.) W.R.Buck	1, 2, 8
199	*Rhaphidostichum bunodicarpum* (C.Müll.) M.Fleisch.	4, 5, 6, 8
200	®*Rhaphidostichum piliferum* (Broth.) Broth.	1, 6
201	*Rhodobryum aubertii* (Schwägr.) Thér.	Terengganu
202	*Rhynchostegiella menadensis* (Sande Lac.) E.B.Bartram	2
203	*Rhynchostegium celebicum* (Sande Lac.) A.Jaeger	Terengganu
204	*Schistomitrium apiculatum* (Dozy & Molk.) Dozy & Molk.	2, 4
205	*Schistomitrium mucronifolium* (Müll. Hal.) M.Fleisch.	Gunung Sembili
206	*Schistomitrium robustum* Dozy & Molk.	4
207	***Sematophyllum subhumile* (Müll. Hal.) M.Fleisch.	5
208	*Sematophyllum subpinnatum* (Brid.) E.Britton	Terengganu
209	*Sphagnum cuspidatum* subsp. *subrecurvum* (Warnst.) A.Eddy var. *subrecurvum*	Terengganu
210	*Sphagnum junghuhnianum* Dozy & Molk.	4
211	*Sphagnum perichaetiale* Hampe	4
212	*Symphysodontella cylindracea* (Mont.) M.Fleisch.	2
213	*Syrrhopodon albo-vaginatus* Schwägr.	1, 2, 3, 4, 5, 6, 7, 8
214	*Syrrhopodon aristifolius* Mitt.	2, 3, 4, 5, 6, 7
215	**Syrrhopodon ciliatus* (Hook.) Schwägr.	1
216	*Syrrhopodon confertus* Sande Lac.	1, 2, 5, 6, 7, 8
217	*Syrrhopodon croceus* Mitt.	1, 2, 3, 4, 5, 6, 7, 8
218	*Syrrhopodon japonicus* (Besch.) Broth.	2
219	*Syrrhopodon loreus* (Sande Lac.) W.D.Reese	1, 2, 4, 5, 6, 7, 8
220	*Syrrhopodon muelleri* (Dozy & Molk.) Sande Lac.	1, 2, 3, 4, 5, 6, 7, 8
221	*Syrrhopodon perarmatus* Broth. in Broth. & Watts	4
222	*Syrrhopodon prolifer* Schwägr. var. *prolifer*	1, 2, 4, 6
223	*Syrrhopodon prolifer* var. *laevis* (Dixon) A.Eddy	4, 7
224	*Syrrhopodon spiculosus* Hook. & Grev. var. *spiculosus*	1, 2, 3, 4, 5, 6, 7
225	®*Syrrhopodon spiculosus* var. *patens* (Dixon) A.Eddy	4
226	®*Syrrhopodon stonae* W.D.Reese	5, 6
227	*Syrrhopodon tjibodensis* M.Fleisch.	2
228	*Syrrhopodon trachyphyllus* Mont.	1, 2, 4, 5, 6, 7, 8
229	*Syrrhopodon tristichus* Nees ex Schwägr.	1, 2, 3, 4, 6, 8
230	**Taxiphyllum taxirameum* (Mitt.) M.Fleisch.	8

No.	Name of taxa of mosses	Locality
231	*Taxithelium instratum* (Brid.) Broth.	2, 3, 5, 6, 7, 8
232	*Taxithelium isocladum* (Bosch & Sande Lac.) Renauld & Cardot	1, 2, 3, 5, 6, 7, 8
233	*Taxithelium kerianum* (Broth.) Broth.	6, 7, 8
234	*Taxithelium lindbergii* (A.Jaeger) Renauld & Cardot	4, 6
235	*Taxithelium nepalense* (Schwägr.) Broth.	1, 2, 3, 6
236	*Taxithelium vernieri* (Duby) Besch.	2
237	*Thuidium cymbifolium* (Dozy & Molk.) Dozy & Molk.	2
238	*Thuidium plumulosum* (Dozy & Molk.) Dozy & Molk.	1, 3, 8
239	*Thuidium pristocalyx* (Müll. Hal.) A.Jaeger	2, 4, 5, 6
240	*Trachyloma indicum* Mitt.	Gunung Padang
241	*Trachypus humilis* Lindb.	2
242	*Trichosteleum boschii* (Dozy & Molk.) A.Jaeger	1, 2, 3, 4, 5, 6, 7, 8
243	*Trichosteleum saproxylophilum* (Müll. Hal.) B.C.Tan, W.B.Schofield & H.P.Ramsay	1, 6
244	*Trichosteleum singapurense* M.Fleisch.	1, 2, 3, 5, 6
245	*Trichosteleum stigmosum* Mitt.	3, 5, 6
246	*Trismegistia calderensis* (Sull.) Broth. var. *calderensis*	2
247	**Trismegistia calderensis* var. *convoluta* (Bosch & Sande Lac.) K.T.Yong, B.C.Tan & B.C.Ho	5
248	**Trismegistia calderensis* var. *subintegrifolia* (Broth.) H.Akiyama	1
249	***Trismegistia complanatula* (Müll. Hal.) Müll. Hal.	5
250	*Trismegistia lancifolia* (Harv. in Hook.) Broth. var. *lancifolia*	1, 2, 4, 5, 6, 7, 8
251	*Trismegistia lancifolia* var. *pseudoplicata* H. Akiyama	6, 7, 8
252	*Vesicularia dubyana* (Müll. Hal.) Broth.	1, 3, 5, 6, 7
253	*Vesicularia miquelii* (Sande Lac.) M.Fleisch.	2, 3, 4, 6, 7, 8
254	*Vesicularia montagnei* (Schimp.) Broth.	3, 5, 6, 7, 8
255	*Vesicularia reticulata* (Dozy & Molk.) Broth.	6, 7, 8
	Name of taxa of liverworts	
256	*Plagiochila arbuscula* (Brid. ex Lehm. & Lindenb.) Lindenb.	Terengganu
257	*Radula retroflexa* Taylor	Kuala Berang

References

Aishah, S. (2013). *Flora lumut jati di Hutan Lipur Lata Tembakah, Besut, Terengganu Darul Iman*. Unpublished Bachelor thesis.

Cheah, Y. H., & Yong, K. T. (2016). New record of *Bazzania* species (Marchantiophyta: Lepidoziaceae) in Peninsular Malaysia with identification key. *Cryptogamie, Bryologie, 37*(2), 199–210.

Chuah-Petiot, M. S. (2011). A checklist of Hepaticae and Anthocerotae of Malaysia. *Polish Botanical Journal, 56*(1), 1–44.

Crum, H. A. (2001). *Structural diversity of bryophytes*. Ann Arbor: The University of Michigan Herbarium.

Damanhuri, A., Yong, K.T., Maideen, H., Hamdan, N., Idris, N., Md. Som, J., Mahusin, C. W., Lee, G.E. (2006). A checklist of mosses of Gunung Mandi Angin, with additional records for Peninsular Malaysia. A. Muda, N. Jaafar, M.R. Sabran, J. Md. Som, M.S. Nizam, A. Latiff, Gunung Mandi Angin, Terengganu: Pengurusan, Persekitaran Fizikal, Kepelbagaian Biologi dan Pelancongan Ekologi (155–172). Kuala Lumpur: Jabatan Perhutanan Semenanjung Malaysia.

Damanhuri, A., Idris, N., Maideen, H., Mohamad, B., Mohamad, O., & Norhazrina, N. (2008). Mosses of Bukit Bauk Urban Forest. In R. Ujang, H. L. Koh, R. Mohd Rahim, N. Jaafar, I. Faridah-Hanum, & A. Latiff (Eds.), *Rimba Bandar Bukit Bauk: Terengganu. Pengurusan Hutan, Persekitaran Fizikal dan Kepelbagaian Biologi* (pp. 139–146). Kuala Lumpur: Jabatan Perhutanan Semenanjung Malaysia.

Damanhuri, A., Norhazrina, N., Suliza, S., Lee, G. E., & Maideen, H. (2011). A preliminary checklist of mosses of Gunung Gagau, Terengganu. In A. Latiff, C. A. Ali, & K. R. Mohamad (Eds.), *Gunung Gagau, Terengganu: Transforming natural assets into an ecotourism product* (pp. 93–111). Kuala Lumpur: Akademi Sains Malaysia.

Damanhuri, A., Maideen, H., Nur-Aqidah, N., Aishah, S., & Norhazrina, N. (2014). Flora Lumut Jati di Hutan Simpan Gunung Tebu dan Kawasan Sekitarnya, Terengganu. In A. R. A. Rahim, M. N. A. Hassan, A. Nordin, M. B. N. Hasliza, & A. Latiff (Eds.), *Hutan Simpan Gunung Tebu, Terengganu: Pengurusan Hutan, Persekitaran Fizikal dan Kepelbagaian Biologi* (pp. 123–136). Kuala Lumpur: Jabatan Perhutanan Semenanjung Malaysia.

Daniels, A. E. D., & Kariyappa, K. C. (2007). Bryophyte diversity along gradient of human disturbance in the southern Western Ghats. *Current Science, 93*, 976–982.

Dixon, H. N. (1926). A list of mosses of the Malay Peninsula. *The Gardens' Bulletin, Straits Settlements, 4*, 1–46.

Eddy, A. (1977). Sphagnales of tropical Asia. *Bulletin of the British Museum (Natural History), Botany, 5*, 359–445.

Eddy, A. (1990). *A handbook of Malesian mosses. 2. Leucobryaceae to Buxbaumiaceae*. London: Natural History Museum Publications.

Frahm, J. P., & Mohamed, M. A. H. (1987). A survey of *Campylopus* and *Bryohumbertia* (Dicranaceae) in Malaysia. *Memoirs of the New York Botanical Garden, 45*, 470–491.

Frego, K. A. (2007). Bryophytes as potential indicators of forest integrity. *Forest Ecology and Management, 242*, 65–75.

Furuki, T., Yong, K. T., & Mohamed, M. A. H. (2013). *Riccardia deguchii* Furuki & K.T.Yong (Marchantiophyta, Aneuraceae), sp. nov. from Malaysia. *Hikobia, 16*, 285–288.

Glime, J.M. (2015). Bryophyte ecology. Water relations: Physiological adaptations. Retrieved from http://www.bryoecol.mtu.edu.

Gradstein, S. R., Churchill, S. P., & Salazar-Allen, N. (2001). Guide to the bryophytes of tropical America. *Memoirs of the New York Botanical Garden, 86*, 1–577.

Inoue, H. (1984). *The genus Plagiochila (Dumort.) Dumort. in Southeast Asia*. Tokyo: Academia Scientific Book.

Iwatsuki, Z., & Mohamed, M. A. H. (1987). The genus *Fissidens* in Peninsular Malaysia and Singapore (A preliminary study). *Journal of the Hattori Botanical Laboratory, 62*, 339–360.

Johnson, A. (1964). An account of the Malaysian Leucobryaceae (lumut putih). *The Garden's Bulletin Straits Settlements, 20*, 315–360.

Katagiri, T., Suleiman, M., & Deguchi, H. (2012). Taxonomic studies of the Trichocoleaceae in Southeast Asia II. A new species of *Eotrichocolea* from Malaysia. *The Bryologist, 115*(4), 518–522.

Liu, Q., Wang, Y. F., & Zuo, Q. (2010). Notes on Brachytheciaceae (Bryophyta) of China. *Journal of Wuhan Botanical Research, 28*(1), 10–15.

Miller, N. G., & Manuel, M. G. (1982). *Trachyloma* (Bryophytina, Pterobryaceae): A taxonomic monograph. *Journal of the Hattori Botanical Laboratory, 51*, 273–321.

Mohamed, M. A. H., & Robinson, H. (1991). A taxonomic revision of the moss families Hookeriaceae and Hypopterygiaceae in Malaya. *Smithsonian Contribution to Botany, 80*, 1–44.

Mohamed, H., & Tan, B. C. (1988). A checklist of mosses of Peninsular Malaya and Singapore. *The Bryologist, 91*(1), 24–44.

Nichols, G. E. (1918). War work for bryophytes. *The Bryologist, 21*, 53–56.

Nurulhuda, M. (2016). *Flora lumut jati (Bryophyta) di Hutan Rizab Kekal Tembat, Hulu Terengganu, Terengganu*. Unpublished Bachelor thesis.

Pimm, S. L., & Joppa, L. N. (2015). How many plant species are there, where are they, and at what rate are they going extinct? *Annals of the Missouri Botanical Garden, 100*, 170–176.

Pócs, T. (1980). The epiphytic biomass and its effect on the water balance of two rain forest types in the Uluguru Mountains (Tanzania, East Africa). *Acta Botanica Academiae Scientiarum Hungaricae Budapest, 26*, 143–167.

Pócs, T. (1996). Epiphyllous liverwort diversity at worldwide level and its threat and conservation. *Anales del Instituto de Biologia de la Universidad Nacional Autónama de México, Serie Botanica, 67*(1), 109–127.

Pócs, T., & Lee, G. E. (2016). Data to the Malaysian liverworts flora II. *Cryptogamie, Bryologie, 37*(1), 39–52.

Proctor, M. C. F. (2008). Physiological ecology. In B. Goffinet & A. J. Shaw (Eds.), *Bryophyte biology* (pp. 237–268). Cambridge: Cambridge University Press.

Sukkharak, P. (2014). Studies on the genus *Mastigolejeunea* (Marchantiophyta: Lejeuneaceae): *Mastigolejeunea gradsteinii* Sukkharak sp. nov. *Journal of Bryology, 36*(1), 56–60.

Suleiman, M., Akiyama, H., & Tan, B. C. (2006). A revised catalogue of mosses reported from Borneo. *Journal of the Hattori Botanical Laboratory, 99*, 107–184.

Syazwana, N. (2016). *Flora lumut jati di Jeram Tanduk, Hutan Simpan Sungai Nipah, Kemaman, Terengganu*. Unpublished Bachelor thesis.

Yamada, K. (1979). A revision of Asian taxa of *Radula*, Hepaticae. *Journal of the Hattori Botanical Laboratory, 45*, 201–322.

Yong, K. T., Tan, B. C., Ho, B. C., Ho, Q. Y., & Mohamed, H. (2013). *A revised moss checklist of Peninsular Malaysia and Singapore*. Kepong: Forest Research Institute of Malaysia.

Yushana, Y. (2005). *Flora lumut jati di Hutan Lipur Lata Payung, Terengganu, Malaysia*. Unpublished Bachelor thesis.

Macrofungi of Tasik Kenyir

Aqilah Mohammad, Lee Shyen Yee, and Ayu Kalsum Kasran

Abstract Surveys on macrofungi have been carried out in several areas of Tasik Kenyir at different occasions between October 2013 and March 2016. A total of 31 species belonged to 26 genera, and 14 families were documented from Air Terjun Saok, Tanjung Mentong, Taman Tropika Kenyir and Taman Orkid in Tasik Kenyir. Among all sites, Taman Orkid Tasik Kenyir shown a higher number of macrofungal species collected. This study marked the first report of the macrofungi surveyed in the largest human-made island in the Southeast Asia.

Keywords Macrofungi · Tasik Kenyir

Introduction

Tasik Kenyir is the biggest human-made lake in Southeast Asia which area was once hilltops and highlands surrounded by a vast area of tropical rainforest. The advantage of being surrounded by tropical forest has made Tasik Kenyir a perfect natural sanctuary for thousands of species of flora and fauna including fungi, the tiny degraders in the forest ecosystems.

Macrofungi are best known for several of their vital roles in most ecosystems. Their services include cycling and uptake of nutrients, buffering trees from toxic minerals and establish mycorrhiza association with trees by providing them nourishments (Tibuhwa 2011). Also, macrofungi also serve as an important decomposer in many forest ecosystems. As a decomposer, they breakdown all organic matters of all kind, including wood. It has been estimated that about 90% of the nutrient cycling processes in the terrestrial ecosystem is conducted by fungal activities (Spooner and Roberts 2005). Macrofungi are good indicators of the forest condition. Their diversity in an area which includes species richness and abundance is closely related to the nutrient's turnover of the area and also the rate of recycling of

A. Mohammad (✉) · L. S. Yee · A. K. Kasran
School of Marine and Environmental Science, Universiti Malaysia Terengganu, Kuala Nerus, Terengganu, Malaysia
e-mail: aqilahmohammad@umt.edu.my

M. T. Abdullah et al. (eds.), *Greater Kenyir Landscapes*,
https://doi.org/10.1007/978-3-319-92264-5_6

macrofungi (Osemwegie and Okhuoya 2009). Apart from this, they are an essential element due to their nutrient and chemical contents which used as food and medicine (Sanmee et al. 2003 as cited in Manzi et al. 1999; Ashwani et al. 2013).

Macrofungi can grow on different substrates and usually categorised in different functional groups. They include saprophytic, mycorrhizal and parasitic. Saprophytic macrofungi are the largest group of fungi, which grow on the dead organic matter including fallen leaves, dead wood, dead animals and insects, etc. Meanwhile, mycorrhizal is where the fungi and roots of plants are associated, which can be defined as the symbiosis relationship between plant and fungus (Spooner and Roberts 2005). Lastly, parasitic macrofungi grow on living plants and absorb the nutrient from the plants.

Early documentations on macrofungal diversity were carried out in the 1800s by a foreign researcher in Peninsular Malaysia (Chipp 1921 in Lee et al. 2012) and some identified species kept increasing since then. To the latest, Lee et al. (2012) have made an effort to compile all list of related publications from entire regions that made up Malaysia today and were able to report nearly 4000 species of fungi for Malaysia including 1141 ascomycetes, 1820 species of Basidiomycetes and 728 species of mitosporic fungi. Given that, very few fungal records were documented in the east coast of Peninsular Malaysia particularly in Terengganu (Pang et al. 2010; Zainuddin et al. 2010a, b; Corner 1966, 1987, 1991b). Furthermore, no scientific documentation on macrofungal species and its diversity in Tasik Kenyir, Terengganu has been published. Thus, this study is a first attempt to document macrofungal species at selected areas in the largest human-made lake in Southeast Asia.

Study Area

Fungal forays were made at several areas in Tasik Kenyir including Air Terjun Saok, Tanjung Mentong, Taman Tropika Kenyir and Taman Orkid Tasik Kenyir (which constituted by three small islands; Pulau Belit, Pulau Hulu Selimbar and Pulau Hilir Selimbar). Collections were made at different periods, between October 2013 and March 2016. Most study areas were characterised by sloping hills where there were vegetation covers, and forest floors were covered by litters. Fruit body occurrence was observed within areas surrounding main trails (include both nature and human-made trails) in each area.

Data Collection and Species Identification

Only fresh fruit bodies were observed and randomly collected along forest trails/walking pathways in each study site. Photos of fresh fruit bodies within their habitat were taken before being removed from its substrate, and every sample was tied up with jeweller's tag that contains information such as the name of collector, date of collection, location and its substrate. Meanwhile, smaller size fruit bodies were kept in a collecting box with a few compartments while the larger size fruit bodies were kept in paper bags. All samples were brought back to base camp for macroscopic observations.

Morphological characteristics such as shape, size and colour of each part of the fruit body such as pileus (cap), lamellae (structure underneath pileus) stalk (stem) were observed and recorded. For pileus, the shape, apex, surface and margin were observed. For lamellae, the attachment and its arrangement were also observed. Besides that, the attachment of stipe to cap and substrate, shape and surface of stipe also were observed. Every observation was recorded in field notebook for macroscopic identification. Lastly, each sample was photographed in the field for herbarium purpose. Colours of the specimens were determined by referring to Royal Botanic Garden Edinburgh's Colour Identification Chart (Royal Botanic Garden, Edinburgh 1969). Spore prints were made from large size fresh samples. The specimens were then dried at 40 °C in a portable mushroom dryer overnight. Dried samples were packed and transported back to Mycology Laboratory at Universiti Malaysia Terengganu for further microscopic examination such as determination of spore size, spore shape and occurrence of basidia and cystidia for identification purposes. Several local reference books have been used in identifying species including Thi et al. (2011), Zainuddin et al. (2010a, b), and Chandrasrikul et al. (2008).

Macrofungal Species at Tasik Kenyir

A total of 31 species belonged to 26 genera, and 14 families were collected and documented from Air Terjun Saok, Tanjung Mentong, Taman Tropika Kenyir and Taman Orkid in Tasik Kenyir. Macrofungi collected from each site are listed in Table 1, with some species that were only observed and not collected were also

Table 1 The number of specimens of macrofungi collected in Tasik Kenyir

Study site	Species
Air Terjun Saok	*Auricularia* sp., Boletus sp., *Polyporus* sp.
Tanjung Mentong	*Ganoderma* sp., *Hygrocybe* sp. *Inocybe* sp., *Microporus* sp., *Coriolopsis* sp., *Pycnoporus* sp., *Trametes* sp., *Scleroderma* sp.
Taman Tropika Kenyir	*Fomitopsis* sp., *Ganoderma* sp., *Microporus* sp., *Daedalea* sp., *Pycnoporus* sp., *Polyporus* sp., *Trametes* sp., *Schizophyllum commune*
Taman Orkid Tasik Kenyir	*Agaricus* sp., *Amanita* sp. 1, *Amanita sp.* 2, *Cyathus* sp., *Lepiota* sp., *Lycoperdon* sp., *Auricularia* sp., *Boletus* sp., *Fomitella* sp., *Fomitopsis* sp., *Amauroderma* spp., *Ganoderma* spp., *Hydnum* sp., *Phallus indusiatus*, *Phallus* sp., *Microporus* sp., *Pycnoporus* sp., *Trametes* sp., *Russula* sp., *Scleroderma* sp., *Pisolithus* sp., *Xylaria* sp.

reported in this paper. Results demonstrated the different abundance of macrofungal species was recorded from each site. The difference was expected as it might be due to the different sampling frequency conducted at different areas which cause the difference in types of the macrofungal species found. Besides that, the difference in sampling time between trips also could alter the probability of different macrofungi species found as this may be linked to different seasonal macrofungal growth as the condition might vary from time to time depending on the weather during the sampling period. Among all sites, Taman Orkid Tasik Kenyir (TOTK) displayed a higher number of specimen collected in Tasik Kenyir (n = 22). Taman Orkid is one of many islands in Tasik Kenyir that actively operating and planted various of orchid species including local and international breeds. Based on our observation, the operator of Taman Orkid watered their orchids almost everyday and used organic fertilisers to enhance orchid's growth. The presence of moisture and nutrients from fertiliser and the surrounding areas that consist of many forest litters may influence the rapid growth of macrofungi within the area.

Family Polyporaceae displayed the highest number of the genus in this study (n = 6) (Table 2). Family Polyporaceae or commonly known as the polypores, is the most important group of the wood-inhabiting fungi which usually grow on woods; living trees or dead wood (Fig. 1). They usually have the large and tough texture of their fruit bodies (McKnight and Mcknight 1998) that distinguish this group from other macrofungal species. The fact that Tasik Kenyir is surrounded by dipterocarp trees, the occurrence of polypores were expected.

The other common family was Agaricaceae where there were five genera were recorded in this study; *Agaricus*, *Amanita*, *Cyathus, Lepiota* and *Lycoperdon* which was only found on one site, Taman Orkid Tasik Kenyir (Fig. 2). The family contained various shapes and colours of fruitbodies, which are widespread, and mostly appear in woodland as a saprobe (Cannon and Kirk 2007). Based on-field observation, the surroundings of Taman Orkid Tasik Kenyir was shady and partially covered with trees, mostly with grass and shrubs. Therefore, these conditions might have favoured the growth of different fungal species from the family Agaricaceae.

This study also has discovered two species of *Phallus* namely *P. indusiatus* Vent. (Fig. 3a) and the other *Phallus* that has yet to be confirmed (which morphologically resemble *P. multicolour*) (Fig. 3b). These findings were unique to Tasik Kenyir as *P. indusiatus* was only recorded by Olridge et al. (1985) in Taman Negara, Rigit Blau, Pahang (Lee et al. 2012).

Macrofungi grow abundantly on the favourable wet environment especially during monsoon season with high rainfalls and the restriction to visit Tasik Kenyir during monsoon season might decrease the number of macrofungi species recorded in this study. A few studies also supported that the best time to collect macrofungi samples is in the wet season as the rainfall and humidity increases which favour the growth of macrofungi (Zainuddin et al. 2010a, b; Andrew et al. 2013). Besides that, there might also be some overlooked tiny macrofungi during the sampling trip given that macrofungi can grow in various sizes and in any place that provides an ideal condition for them to grow. Some macrofungi might grow in a less visible place such as on decayed tree branches hid under the forest floor fills with leaf litters.

Table 2 List of macrofungi collected in Tasik Kenyir based on family

Family	Genus/species
Agaricaceae	*Agaricus* sp.
	Amanita sp. 1
	Amanita sp. 2
	Cyathus sp
	Lepiota sp.
	Lycoperdon sp.
Auriculariaceae	*Auricularia* sp.
Boletaceae	*Boletus* sp.
Fomitopsidaceae	*Fomitella* sp.
	Fomitopsis sp.
Ganodermataceae	*Amauroderma* sp. 1
	Amauroderma sp. 2
	Ganoderma sp. 1
	Ganoderma sp. 2
Hydnaceae	*Hydnum* sp.
Hygrophoraceae	*Hygrocybe* sp.
Inocybaceae	*Inocybe* sp.
Phallaceae	*Phallus indusiatus*
	Phallus sp. 1
Polyporaceae	*Microporus* sp.
	Coriolopsis sp.
	Daedalea sp.
	Pycnoporus sp.
	Polyporus sp.
	Trametes sp.
Russulaceae	*Russula* sp.
Schizophyllaceae	*Schizophyllum commune*
Sclerodermataceae	*Scleroderma* sp.
	Pisolithus sp.
Xylariaceae	*Xylaria* sp.1
	Xylaria sp.2

Conclusion

A further study is needed to discover more macrofungi species in Tasik Kenyir as there is still limited diversity study being conducted in this human-made lake. Future studies on mycology in Tasik Kenyir also may increase the understanding and knowledge on macrofungi.

Fig. 1 Some polypores found in Tasik Kenyir (**a**) *Pycnoporus* sp., (**b**) *Macroporus* sp. and (**c**) *Trametes* sp.

Fig. 2 Common fruitbodies found in Tasik Kenyir that belong to the family of Agaricaceae (**a**) *Agaricus* sp., (**b**) *Amanita* sp., (**c**) *Cyathus* sp., (**d**) *Lepiota* sp. and (**e**) *Lycoperdon* sp.

Fig. 3 Two species of *Phallus* (stinkhorns) found in Tasik Kenyir (**a**) *Phallus indusiatus* and (**b**) *Phallus* sp.

Acknowledgements We would like to thank School of Marine and Environmental Sciences, Universiti Malaysia Terengganu (UMT) for project funding as well providing us with research facilities. Special thank also goes to UMT Kenyir Research Institute for their invitation to join Tanjung Mentong's Scientific Expedition in 2014. We would also like to acknowledge the operators of Taman Orkid Tasik Kenyir and Taman Tropika Kenyir for their permission to enter their sites.

References

Andrew, E. E., Kinge, T. R., Tabi, E. M., Thiobal, N., & Mih, A. M. (2013). Diversity and distribution of macrofungi (mushroom) in the Mount Cameroon Region. *Journal of Ecology and the Natural Environment, 5*(10), 318–334.

Ashwani, T., Rejesh, K., & Shailesh, P. (2013). Diversity and frequency of macrofungi associated with wet evergreen tropical forest in Assam, India. *Biodiversitas, 14*(2), 73–78.

Cannon, P. F., & Kirk, P. M. (2007). *Fungal families of the world*. Oxfordshire: CABI Publishing.

Chandrasrikul, A., Suwanarit, P., Sangwanarit, U., Morinaga, T., Nishizawa, Y., & Murakami, Y. (2008). *Diversity of mushrooms and macrofungi in Thailand*. Bangkok: Kasetsart University.

Chipp, T. F. (1921). A list of the fungi of the Malay Peninsula. *Gardens Bulletin of the Straits Settlements, 2*, 311–418.

Corner, E. J. H. (1966). *A monograph of Cantharelloid fungi*. Oxford: Oxford University Press.

Corner, E. J. H. (1987). Ad Polyporaceas IV. The genera *Daedalea, Flabellophora, Flavodon, Gloeophyllum, Heteroporus, Irpex, Lenzites, Microporellus, Nigrofomes, Nigroporus, Oxyporus, Paratrichaptum, Rigidoporus, Scenidium, Trichaptum, Vanderbylia, and Steccherinum. Beih. Nova Hedwig, 86*, 1–265.
Corner, E. J. H. (1991). *Trogia (Basidiomycetes). Gardens' Bulletin, Singapore, 2*, 1–97.
Lee S. S., Alias S. A., Jones E. G. B., Zainuddin N., & Chan H. T. (2012). Checklist of fungi of Malaysia, Issue/No. 132. Forest Research Institute Malaysia (FRIM), Malaysia. pp. 556.
McKnight, K. H., & Mcknight, V. B. (1998). *A field guide to mushrooms of North America*. New York: Houghton Mifflin Company.
Oldridge, S. G., Pegler, D. N., Reid, D. A., & Spooner, B. M. (1986). A collection of fungi from Pahang and Negeri Sembilan (Malaysia). *Kew Bulletin, 41*(4), 855–872.
Osemwegie, O. O., & Okhuoya, J. A. (2009). Diversity of macrofungi in oil palm agroforests of Edo State, Nigeria. *Journal of Biological Sciences, 9*(6), 584–593.
Pang, K. L., Sharuddin, S. S., Alias, S. A., Nor, N. A., & Awaluddin, H. H. (2010). Diversity and abundance of lignicolous marine fungi from the east and west coasts of Peninsular Malaysia and Sabah (Borneo Island). *Botanica Marina, 53*(6), 515–523.
Royal Botanic Garden Edinburgh. (1969). *Flora of British fungi; colour identification chart*. Edinburgh: Her Majesty's Stationary Office.
Spooner, B., & Roberts, P. (2005). *Fungi*. London: Harper Collins Publishers.
Thi, B. K., Lee, S. S., Zainuddin, N., & Chan, H. T. (2011). *A guidebook to the macrofungi of Fraser's Hill*. Kepong: Forest Research Institute Malaysia.
Tibuhwa, D. D. (2011). Diversity of macrofungi at the University of Dar es Salaam Mlimani main campus in Tanzania. *International Journal of Biodiversity and Conservation, 3*(11), 540–550.
Zainuddin, N., Lee, S. S., Chan, H. T., & Thi, B. K. (2010a). *A guidebook to the macrofungi of Tasik Bera*. Kepong: Forest Research Institute Malaysia.
Zainuddin, N., Lee, S. S., Chan, H. T., Thi, B. K., & Alias, S. A. (2010b). Macrofungi of Pulau Redang, Terengganu and Pulau Aur, Johor in the South China Sea. *Journal of Science and Technology in the Tropics, 6*, S120–S125.

Investigation on Bioactive Potential of Selected Wild Ginger, Genus *Etlingera* from Tasik Kenyir, Terengganu

Thilahgavani Nagappan, Yannick Tatin, and Jana-Leong Skornickova

Abstract The genus Zingiberaceae includes four subfamilies and six tribes: Siphonochiloideae (Siphonochileae tribe), Tamijioideae (Tamijieae tribe), Alpinioideae (Alpinieae and Riedelieae tribes), Zingiberoideae (Zingibereae and Globbeae tribes) making it among the diverse angiosperm group to date. About 200 species belongs to 19 genera are recorded in Peninsular Malaysia with most of them are endemic and posses medicinal properties. Hence, four *Etlingera* species (*E. megalocheilos, E. metriocheilos, E. punicea, E. pauciflora*) found in sites of Tasik Kenyir, Terengganu were studied through *in-vitro* assays. The crude extract of *E. punicea* exhibited the best antimicrobial activity against *S. aureus, Microccocus* sp., *S. uberis* and *K. pneumoniae* (inhibition zone of 15.2 mm, 25.2 mm, 18.5 mm and 22.4 mm respectively at 100 mg mL^{-1} of crude extract) while crude extract of *E. punicea* exhibited the best cytotoxicity (LC_{50} = 18 μg mL^{-1}) followed by *E. metriocheilos* (LC_{50} = 29.2 μg mL^{-1}), *E. megalocheilos* (LC_{50} = 33.4 μg mL^{-1}) and *E. pauciflora* (LC_{50} > 100 μg mL^{-1}) against MCF-7 cells. This findings reveals that Tasik Kenyir could potentially be "home" for many unexplored and biologically active phytochemicals that will benefit mankind in years to come.

Keywords *Etlingera* · Tasik Kenyir · Antimicrobial · Cytotoxicity

T. Nagappan (✉)
School of Marine and Environmental Sciences, Universiti Malaysia Terengganu, Kuala Nerus, Terengganu, Malaysia

Institute of Marine Biotechnology, Universiti Malaysia Terengganu, Kuala Nerus, Terengganu, Malaysia
e-mail: thila.vani@umt.edu.my

Y. Tatin
Tropimundo Erasmus-Mundus, Vrije Universiteit Brussels (VUB), Brussels, Belgium

J.-L. Skornickova
The Herbarium, Singapore Botanical Gardens, Singapore, Singapore

M. T. Abdullah et al. (eds.), *Greater Kenyir Landscapes*,
https://doi.org/10.1007/978-3-319-92264-5_7

Introduction

Known to be pantropical plants, members of family Zingiberaceae is widely distributed in South East Asia region. This perennial monocot plants are mostly found in wet and shady areas of primary rainforest as well as few species found in secondary forest. Some species of this family has high tolerance against sunlight and grows well in high elevation (Larsen 2007). Commonly known as "gingers", this tropical family have multiple uses in Asia as spices, food, perfumes, medicines, dyes and ornaments (Skornickova and Newman 2015; Nontasit et al. 2014). The rhizome is generally aromatic, fleshy with tuber-bearing roots. Highly sought after commodity from this genus is *Zingiber officinale* Roscoe (ginger), *Curcuma longa* (L.) Roxb. (tumeric) and *Elettaria cardamomum* (L.) Maton (cardamom). These three species have an important economic value and extensive researches were conducted pertaining to medicinal properties in curing many ailments and diseases such as diarrhoea, chronic bronchitis, rheumatism, neuralgia, constipation, digestive problems, cardiovascular diseases, cough and bacterial infections (Ong and Nordiana 1999; Basak et al. 2010; Victorio 2011; Skornickova and Newman 2015).

Etlingera is a genus of ginger from the *Alpineae* tribe that produces colourful inflorescence, flowers and fruits. Gingers from this genus are considered tall forest plants with some of its kind reaching great heights of as tall as 6 m (Chan et al. 2007). It is also a perennial ginger native to Indo-Pacific region and consisting more than 100 species growing at different altitudes, from sea level to 2500 m. Revision by Poulsen (2006) brings the number of *Etlingera* to 40 with majority of these ginger species are endemic to Malaysia, Sumatera and Indonesia. Plants from these genus have diverse traditional and commercial uses. Compared to the other members of gingers, *Etlingera* are mostly edible and *Etlingera elatior* (Jack) R.M. Sm, is perhaps the only known *Etlingera* that is widely consumed, particularly in Thailand and Malaysia. *Etlingera elatior* or also known as 'torch ginger' or 'bunga kantan' is often used as part of the ingredient in local delicacies in Malaysia such as in *laksa asam, nasi kerabu,* and *nasi ulam* (Subramaniam et al. 2010). The hearts of young shoots, inflorescences, fruits and flower buds are consumed by the various indigenous communities as condiment, eaten as salad, cooked, used as traditional medicine for illness such as stomach ache, earache, wound cleanser and used by post-partum women for bathing to remove body odor (Heim et al. 2002).

Several studies showed antibacterial activities of many *Etlingera* species against various bacterial strains (Ud-Daula et al. 2016; Chan et al. 2007; Norajit et al. 2007; Thomson et al. 2002). Recent studies also revealed potential anticancer activities on different human and non-human cancer cell lines (Chew et al. 2012; Sukari et al. 2010; Sun et al. 2009). However, the bioactive potentials of species in this genus in Malaysia are not extensively explored compared to other Asian countries. Thus, we aim to investigate the bioactive potentials of few *Etlingera* species found in Tasik Kenyir, Terengganu as it is known to house many unique species of Zingiberaceae.

The largest man-made Tasik, Tasik Kenyir is located between the coordinates 4°43′N to 5°15′N and 102°30′E to 102°55′E in Terengganu, on the East coast of

Peninsular Malaysia. The *Etlingera* specimens were collected in the vicinity of Universiti Malaysia Terengganu (UMT) Tasik Kenyir Field Station, approximately 5 km from the Pengkalan Gawi jetty at the coordinates 5°08′33″ N and 102°45′37″ E. Quadrats of 10 × 10 m quadrats were chosen randomly closer to the Tasik area as wet soil is favoured by gingers for growth. Mature species were then collected and transported to Laboratory for Ecological studies, UMT for further processing. All bags were labelled with a specific number to indicate each specimen. The voucher specimens of the identified species were deposited in the herbarium at Central Laboratory Complex, UMT. Four rhizomes of four positively identified *Etlingera* species' was subjected to extraction to obtain crude extract (Vairappan et al. 2012). Antibacterial and cytotoxicity potency of these extracts were evaluated via in-vitro assays using disc diffusion technique to observed inhibition of bacterial zone (Balouiri et al. 2016) and via MTT assay to observed the cytotoxicity of extract against breast cancer cells, MCF-7 (Yang et al. 2010).

Zingiberaceae of Tasik Kenyir

A total of 24 species of Zingiberaceae were found during the exploration of study area and nine was positively identified with the assistance of Dr. Jana-Leong Skornikova, botanist and Ginger expert from Singapore Botanical Garden. The identified species were *Alpinia javanica* (Roxb.) Blume, *Amomum aff. uliginosum* J. Koenig, *Etlingera pauciflora* (Ridl.) R.M. Smith, *E. punicea* (Roxb.) R.M Smith, *E. megalocheilos* (Griff.) A.D. Poulsen, *E. metriocheilos* (Griff.) R.M. Smith, *E. maingayi* (Baker) R.M. Smith, *Honrstedtia conica* Ridl., and *Zingiber spectabile* Griff.. Identification of specimens were done using identification keys and description provided by Holttum (1950), Theilade (1996, 1999, Khaw (2001), Poulsen (2006), Skornikova and Newman (2015) (Figs. 1, 2, and 3).

Bioactive Potentials of Selected *Etlingera*

As for bioactive potentials, only four strains of human pathogenic bacteria, *Staphylococcus aureus, Micrococcus* sp., *Streptococcus uberis* and *Klebsiella pneumoniae* exhibit sensitivity against 25 mg uL^{-1} of all crude extract tested. At 100 mg uL^{-1} of crude extract, *Micrococcus* sp. was inhibited to a higher degree and crude extract of *E. megalocheilos* exhibit antimicrobial activities against all four strains given at this same concentration. The results of inhibition are displayed in Table 1.

The range of inhibition zone exhibited by all crude extracts against *S. aureus* falls between 7.0 and 15.2 mm while against *S. uberis* were from the range of 5.6 to 18.5 mm for all three concentrations tested. It was observed that extract of *E. megalocheilos* were susceptible to all the strains tested at lowest concentration of 25.0 mg uL^{-1} meanwhile the highest concentration of this extract inhibits *Micrococcus* sp.

Fig. 1 (**a**) The fruits of *Alpinia javanica*; (**b**) Leaf arrangement of *Amomum aff. uliginosum*; (**c**) Inflorescence of *Etlingera punicea*

Fig. 2 (**a**) The petals from inflorescence of *Etlingera pauciflora*; (**b**) Leaf arrangement of *Etlingera metriocheilos*; (**c**) Young leaf arrangement of *Etlingera megalocheilos*

Fig. 3 (**a**) The old inflorescence of *Etlingera maingayi*; (**b**) Young leaf arangement of *Zingiber spectabile*; (**c**) Old inflorescence of *Hornstedia conica*

Table 1 Different degrees of antibacterial activities against four strains displayed by crude extract of *E. pauciflora*, *E. punicea*, *E. megalocheilos* and *E. metriocheilos*

	Diameter of inhibition (mm)											
Concentration tested	*Staph. aureus*			*Micrococcus* sp.			*Strep. uberis*			*Kleb. pneumoniae*		
(mg uL-1)	25	50	100	25	50	100	25	50	100	25	50	100
E. punicea	–	11.5	15.2	20	22	25.2	–	13	18.5	–	18.8	22.4
E. pauciflora	–	7	10.1	10	16.3	20.9	–	–	–	–	16	18.3
E. megalocheilos	9	10.7	12	18.1	22	25	9	11.3	14.2	13	18.5	20
E. metriocheilos	–	7.8	9.3	15	16	20.7	5.6	10.1	16.9	–	16.8	21

NB: Diameter of inhibition of gentamicin against *Stap. aureus*: 27 mm; *Micrococcus* sp.: 30 mm; *Strep. uberis*: 16 mm; *Kleb. pneumoniae*: 30 mm. (–) indicated no activity; Sterile saline water was set as negative control

creating 25 mm in diameter. On contrary, extract of *E. pauciflora* was found to be inactive against *S. uberis*. The results also revealed that *Micrococcus* sp. and *K. pneumoniae* are much more sensitive to these four *Etlingera* species than *S. aureus* and *S. uberis*. Indeed, the values of the inhibition zone at 100 mg uL^{-1} of crude extract are relatively close to the values of the inhibition zone of gentamicin. The

Table 2 The LC_{50} values exhibited by crude extracts of four *Etlingera* species against MCF-7 cells

Total crude extract	LC_{50} ($\mu g\ mL^{-1}$)
E. punicea	18.0 ± 0.07
E. metriocheilos	29.2 ± 1.03
E. megalocheilos	33.4 ± 0.19
E. pauciflora	>100

NB: Test was carried out in triplicate. Murine fibroblast cell was set as positive control

antibacterial activity of these ginger could be closely related to the presence of high amounts of oxygenated monoterpenes in found in the rhizome. Previous studies reported that oxygenated monoterpenes displays a wide range of antibacterial activity though weaker than penicillin as a standard antibiotic. The possible presence of borneol and camphor could possibly be the cause of the potent activity since this volatile has been reported to possess antibacterial activity over a range of microbes (Ud-daula et al. 2016; Faridahanim et al. 2007; Martins et al. 2001). Dominance of α-pinene and β-pinene in volatile composition of many *Etlingera* species also shows effectiveness against Gram positive bacteria and fungi (Afolayan and Ashafa 2009).

The cytotoxic evaluation revealed that *E. punicea* had the best LC_{50} value against MCF-7 cells at a concentration of 18.0 $\mu g\ mL^{-1}$ followed by *E. metriocheilos* at 29.2 $\mu g\ mL^{-1}$ while *E. megalocheilos* at 33.4 $\mu g\ mL^{-1}$. The extract of *E. pauciflora* were found to be least active with LC_{50} value more than 100 $\mu g\ mL^{-1}$. Vairappan et al. (2012) had investigated the cytotoxicity of *E. megalocheilos'* essential oil against the similar cell line obtained the LC_{50} value of 30 $\mu g\ mL^{-1}$. Hence, the results obtained *via* MTT assay for four species of *Etlingera* against MCF-7 cells are very encouraging. Indeed, it showed that three of the four plants tested have cytotoxic effect on these cells line (Table 2).

It is believed that presence of gingerol analogues in the extracts of tested *Etlingera* species could influence the cytotoxic effect against the MCF-7 cells. A variety of gingerol analogues, gin-gerols, zingiberene and shogaols comprising hydroxyl functional moieties (-OH) found to be responsible for pungent fresh taste of rhizome displayed high antioxidant activities and superoxide radical capacity (Semwal et al. 2015). Position of these hydroxyl moieties in chemical skeleton also reported to influence its bioactivity against skin tumorgenesis in mice, human liver carcinoma cells, and human neuroblastoma cells (Shukla et al. 2007). According to Dugasani et al. (2010), gingerol analogoues with –OH group at 6-, 8-, 10- position displayed antioxidant activities with IC_{50} values ranging from 8.05 to 26.3 μM for the DPPH radical scavenging, 0.85–4.05 μM for the superoxide radical and 0.72–4.62 μM for the hydroxyl radical reduction. Yang et al. (2010) postulated the variations in composition of volatile mono- and sesquiterpenes and their synergism most likely will influence the cytotoxic activities.

Conclusion

In recent years, nutraceutical compounds derived from botanical sources had gained wide acceptance as preferred alternatives to various synthetic drugs available on the market. These is clearly evident as few investigated species from Tasik Kenyir, Terengganu exhibits a promising results to be developed into new nutraceutical entity. The family Zingiberaceae could be potential candidate in exploration for new antimicrobial and anti-cancer agents. Further studies should be conducted to focus on the identification and isolation of the active molecules that inhibit or kill the cancer cells as *E. punicea* displayed a promising cytotoxicity activity against MCF-7 cells. Possible pathway of structure-related activity derived from chemicals of genus *Etlingera* should also be evaluated in discovering new natural products to address raising cancer issues.

Acknowledgements This work was financially supported by research funds scheme (RAGS) by Ministry of Higher Education, Malaysia. Authors are would like to thank Mr. Ahmad Fakhurazzi Mokhtar and highly indebted to Prof. Emer. Dr. Faizah Mohd Sharoum from Kenyir Research Institute, UMT for the logistic support and access to field facilities as well as to Mr. Tuan Haji Muhammad Razali Salam from School of Marine Science and Environment, UMT for the assistance in voucher specimen preparation.

References

Afolayan, A., & Ashafa, A. (2009). Chemical composition and antimicrobial activity of the essential oil from *Chrysocoma ciliata* L. leaves. *Journal of Medicinal Plant Research, 3*, 390–394.

Balouiri, M., Sadiki, M., & Ibnsouda, S. K. (2016). Methods for in vitro evaluating antimicrobial activity: A review. *Journal of Pharmaceutical Analysis, 6*, 71–79.

Basak, S., Sarma, G. C., & Rangan, L. (2010). Ethnomedical uses of Zingiberaceous plants of Northeast India. *Journal of Ethnopharmacology, 132*(1), 286–296.

Chan, E. W., Lim, Y. Y., & Omar, M. (2007). Antioxidant and antibacterial activity of leaves of *Etlingera* species (Zingiberaceae) in Peninsular Malaysia. *Food Chemistry, 104*(4), 1586–1593.

Chew, A. J., Abidin, N. Z., & Wahab, N. B. A. (2012). Anti-proliferation and anti-migration activities of ten selected Zingiberaceae species against MDA-MB-231 cells. *Journal of Medicinal Plant Research, 6*(21), 3711–3723.

Dugasani, S., Pichika, M. R., Nadarajah, V. D., Balijepalli, M. K., Tandra, S., & Korlakunta, J. N. (2010). Comparative antioxidant and anti-inflammatory effects of [6]- gingerol, [8]-gingerol, [10]-gingerol and [6]-shogaol. *Journal of Ethnopharmacology, 127*, 515–520.

Faridahanim, M. J., Che, P. O., Nor Hadiani, I., & Khalijah, A. (2007). Analysis of essential oils of leaves, stems, flowers, and rhizomes of *Etlingera elatior* (Jack) R. M. Smith. *The Malaysian Journal of Analytical Sciences, 11*(1), 269–273.

Heim, K. E., Tagliaferro, A. R., & Bobilya, D. (2002). Flavonoid antioxidants: Chemistry, metabolism and structure-activity relationships. *The Journal of Nutritional Biochemistry, 1*, 572–584.

Hottlum, R. E. (1950). The Zingiberaceae of the Malay Peninsula. *Garden's Bulletin Singapore, 13*, 1–249.

Khaw, S. H. (2001). The genus *Etlingera* (Zingiberaceae) in Peninsular Malaysia including a new species. *Garden's Bulletin Singapore, 53*(2), 191–239.

Larsen, K. (2007). The exploration of gingers in SE Asia-some milestones and perspectives. *Garden's Bulletin Singapore, 59*(2), 1–12.

Martins, A. P., Salgueiro, L., Gon Asalves, M. J., Cunha, A. P., Villa, R., Caigunel, S., Mazzoni, V., Tomi, F., & Cassanova, J. (2001). Essential oil composition and antimicrobial activity of three Zingiberaceae from S. Tome e Principe. *Planta Medica, 67*, 580–584.

Nontasit, N., Kanlayanapaphon, C., Mekanawakul, M., & Nualmangsar, O. (2014). Taxonomic studies and traditional uses of Zingiberaceae in Khao Luang National Park, Nakhon Si Thammarat Province, Thailand. *Walailak Journal of Science and Technology, 12*(8), 643–658.

Norajit, K., Laohakunjit, N., & Kerdchoechuen, O. (2007). Antibacterial effect of five Zingiberaceae essential oils. *Molecules, 12*(8), 2047–2060.

Ong, H. C., & Nordiana, M. (1999). Malay ethno-medico botany in Machang, Kelantan, Malaysia. *Fitoterapia, 70*(5), 502–513.

Poulsen, A. D. (2006). *Etlingera of Borneo*. Kota Kinabalu: Natural History Publication.

Semwal, R. B., Semwal, D. K., Combrinck, S., & Viljoen, A. M. (2015). Gingerols and shogaols: Important nutraceutical principles from ginger. *Phytochemistry, 117*, 554–568.

Shukla, Y., Prasad, S., Tripathi, C., Singh, M., George, J., & Kalra, N. (2007). In vitro and in vivo modulation of testosterone mediated alterations in apoptosis related proteins by [6]-gingerol. *Molecular Nutrition Food Research, 51*, 1492–1502.

Skornickova, J., & Newman, M. (2015). *Gingers of Cambodia, Laos & Vietnam*. Singapore: National Parks Board, Singapore Botanical Gardens.

Subramaniam, J. T. L., Sreenivasan, S., Vello, S., & Zakaria, Z. (2010). Pharmacological activity, phytochemical analysis and toxicity of methanol extract of *Etlingera elatior* (torch ginger) flowers. *Asian Pacific Journal of Tropical Medicine, 3*(10), 769–774.

Sukari, M. A., Tang, S. W., Neoh, B. K., & Rahmani, M. (2010). Antileukemic activity and chemical constituents of some Zingiberaceae species. *Asian Journal of Chemistry, 22*(10), 7891.

Sun, X. Y., Zheng, Y. P., Lin, D. H., Zhang, H., Zhao, F., & Yuan, C. S. (2009). Potential anti-cancer activities of Furanodiene, a Sesquiterpene from *Curcuma wenyujin*. *The American Journal of Chinese Medicine, 37*(03), 589–596.

Theilade, I. (1996). Revision of the genus Zingiber in Peninsular Malaysia. *Garden's Bulletin Singapore, 48*(1–2), 207–236.

Theilade, I. (1999). A synopsis of the genus Zingiber (Zingiberaceae) in Thailand. *Nordic Journal of Botany, 19*(4), 389–410.

Thomson, M., Al-Qattan, K. K., Al-Sawan, S. M., Alnaqeeb, M. A., Khan, I., & Ali, M. (2002). The use of ginger (*Zingiber officinale* Rosc.) as a potential anti-inflammatory and antithrombotic agent. *Prostalandins, Leukotrienes & Essential Fatty Acids, 67*(6), 475–478.

Ud-Daula, A. F. M., Demirci, F., Abu Salim, K., Demirci, B., Lim, L. B. L., Kemal Husnu, C., & Ahmad, N. (2016). Chemical composition, antioxidant and antimicrobial activities of essential oils from leaves, aerial stems, basal stems, and rhizomes of *Etlingera fimbriobracteata* (K. Schum.) R. M. Sm. *Industrial Crops and Products, 84*, 189–198.

Vairappan, C. S., Nagappan, T., & Palaniveloo, K. (2012). Essential oil composition: Cytotoxic and antibacterial activities of five *Etlingera* species from Borneo. *Natural Products Communications, 7*, 239–242.

Victorio, C. P. (2011). Therapeutic value of the genus *Alpinia*, Zingiberaceae. *Revista Brasileira de Farmacognosia, 21*(1), 194–201.

Yang, Y., Yue, Y., Runwei, Y., & Goulin, Z. (2010). Cytotoxic, apoptotic and antioxidant activity of the essential oil of *Amomum tsao-ko*. *Bioresource Technology, 101*, 4205–4211.

Genetic Diversity of Seed Storage Protein in Selected Melastomataceae and Fagaceae from Tasik Kenyir

Wan Bayani Wan Omar, Hazlina Ahamad Zakeri, Jamilah Mohd Salim, Ummi Fahetah Mohd Fisall, and Muhammad Azhar Amran

Abstract This paper presents the results of a study on genetic diversity from seed storage proteins of six species of plants from two families using sodium dodecyl sulphate polyacrylamide gel electrophoresis (SDS-PAGE). The two families were; Melastomataceae (Senduduk); *Melastoma malabathricum, Clidemia hirta* and *Lijndenia laurina* and Fagaceae (Berangan, chesnut or timber); *Lithocarpus wallichianus*, *Castanopsis lucida* and *Castanopsis schefferiana.* The total numbers of polypeptide bands of Melastomataceae which resolved in 14% gel SDS were 33 bands, where three bands (9%) were monomorphic, while 30 bands (91%) were polymorphic with the size ranging from 12.89 to 95.24 kDa. In Fagaceae a total of 55 polypeptides bands were produced. Out of these one band (2%) was monomorphic among all three species and 54 (98%) were polymorphic. Cluster analysis for seed storage protein clearly distinguished between six species. Seed storage protein profiling of Melastomatecea and Fagaceae from this study highlighted a high degree of genetic diversity within the families. The high polymorphism shown by each family revealed that all the species are genetically variable. Therefore, it is recommended that the species should be conserved in order to establish in their natural habitat.

Keywords Fagacea · Melastomatacea · Genetic diversity · Seed storage protein · Tasik Kenyir

---

W. B. Wan Omar (✉) · U. F. M. Fisall · M. A. Amran
School of Marine and Environmental Sciences, Universiti Malaysia Terengganu,
Kuala Nerus, Terengganu, Malaysia
e-mail: bayani@umt.edu.my

H. A. Zakeri
School of Fundamental Science, Universiti Malaysia Terengganu,
Kuala Nerus, Terengganu, Malaysia

J. M. Salim
School of Marine and Environmental Sciences, Universiti Malaysia Terengganu,
Kuala Nerus, Terengganu, Malaysia

Kenyir Research Institute, Universiti Malaysia Terengganu,
Kuala Nerus, Terengganu, Malaysia

M. T. Abdullah et al. (eds.), *Greater Kenyir Landscapes*,
https://doi.org/10.1007/978-3-319-92264-5_8

Introduction

Genetic diversity is the heritable variation within and between population of organisms and the conservation of plant genetic diversity is essential for present and future human wellbeing (Rao and Hodgkin 2002). The loss of genetic diversity of plants can effect in the loss of valuable and desirable traits and reduce options to use unexplored resources for food production, industry and medicine (Falk et al. 2001). A better understanding of genetic diversity of plant and its distribution is essential for its conservation and use. At the same time will improve our understanding of the taxonomy and origin and evolution of plant species of interest, therefore, some population need a sufficient genetic diversity to survive to a new condition for adaptation (Rao and Hodgkin 2002; Falk et al. 2001).

Morphological trait can be used for assessing genetic diversity in plants but it is often influenced by the environmental factors (Siddique and Naz 2009). Biochemical markers such as proteins can served as genetic markers as they are direct products of active genes and are quite polymorphic, heritable and highly independent of environmental fluctuations (Chittora and Purohit 2012; Gepts 1990). In order to study the genetic diversity of plant, sodium dodecyl sulphate polyacrylamide gel electrophoresis (SDS-PAGE) is one of the technique can be used to analyse the variation of polypeptide band of seeds storage protein (Ungureanu et al. 2007). In this study, two plant families have been chosen in order to study their genetic diversity; (a) Melastomataceae and (b) Fagaceae. Melastomataceae plants originate in the tropic and subtropic regions, with a total of more than 4000 species in the world and 22 species found in Southeast Asian region including Malaysia (Joffry et al. 2011). It is ubiquitous, species-rich, and dominant in the forest area (Silveira et al. 2013) either as herbs or shrubs or tress and the leaf lamina is dorsiventral or centric (Watson and Dallwitz 1992). Three species of Melastomataceae that have been chosen in this research were *Melastoma malabathricum* (*Mm*, Senduduk Ungu), *Clidemia hirta* (*Ch*, Senduduk Bulu) and *Lijndenia laurina* (*Ll*, Nipis Kulit). *M. malabathricum* has ethno-pharmacological value where certain parts of the plant can be used in medicinal treatment such as the leaves, shoots, barks, seeds and roots (Joffry et al. 2011). *M. malabathricum* (Fig. 1a) has gained herbal status and have been used to treat diarrhoea, dysentery, haemorrhoids, cuts and wounds, toothache, and stomach-ache (Joffry et al. 2011) while *C. hirta* (Fig. 1b) also has been used as medicinal treatment for traditional folk medicine to treat some bacterial infection (Dianita et al. 2011), treat for skin infection (Franca et al. 1996), and treat for venom fever (Latiff and Mat-Salleh 2002). Whereas, there is no known study on the medicinal status of *L. laurina* (Fig. 1c).

Another plant family have been studied was Fagaceae or chestnut family which have four genera and 64 species were found in the Malay Peninsula (Ng 1991) either as trees or shrubs and the leaf lamina is dorsiventral to bifacial (Watson and Dallwitz 1992). Many species of Fagaceae have important economic uses such as oak, chestnut, and beech are commonly used as timber for floors, furniture, cabinets, and wine barrels and it is widespread in the Northern hemisphere, with a centre

Fig. 1 *Melastoma malabathricum, Mm* (**a**), *Clidemia hirta, Ch* (**b**) and *Lijndenia laurina, Ll* (**c**)

of diversity found in tropical Southeast Asia (Nixon 1993; Manos et al. 2001). To understand more about the genetic diversity of Fagaceae, there were three species of Fagaceae have been studied in this research such as *Castanopsis lucida (Cl)* (Fig. 2a), *Castanopsis schefferiana (Cs)* (Fig. 2b) and *Lithocarpus wallichianus (Lw)* (Fig. 2c) were used in this study.

Information regarding to the application of molecular technique in genetic study of Melastomatacea and Fagacea is very little. Therefore, this study was conducted to assess the genetic diversity of this two plants by using sodium dodecyl sulphate polyacrylamide gel electrophoresis (SDS-PAGE) with two different protein buffers; (a) Phosphate buffer and (b) Tris-HCL buffer (Chittora and Purohit 2012). The polypeptide bands were analyzed by using Dendrogram Unweighted Pair Group Method with Arithmetic (D-UPGMA) software.

Polymorphism of Polypeptide Bands

In this study, the protein samples were extracted by using two different buffers which were phosphate buffer and Tris-HCl extraction buffer. The range of the protein concentration from 0.1 to 2.0 mg/ml. Based on Jaccard similarity matrix for binary data, the relationships among species have been determined. Figure 3 shows

Fig. 2 *Castanopsis lucida, Cl* (**a**), *Castanopsis schefferiana, Cs* (**b**) and *Lithocarpus wallichianus, Lw* (**c**). (From Wan Bayani et al. 2013)

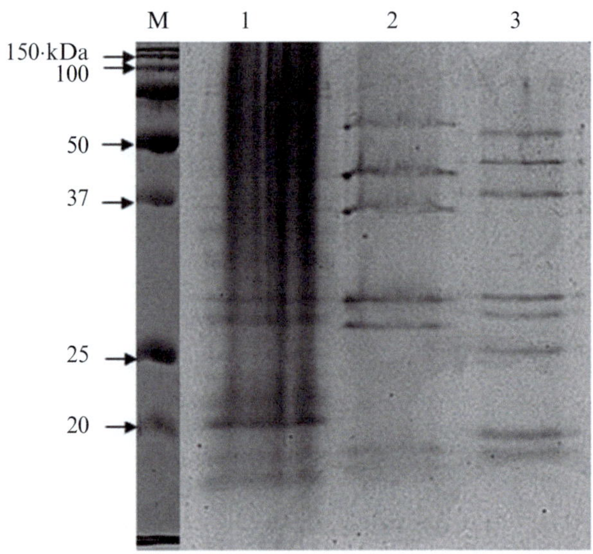

Fig. 3 Protein profile of 14% gel SDS-PAGE. Lane 1: *Mm*, Lane 2: *Ch*, Lane 3: *Ll*. M: protein marker

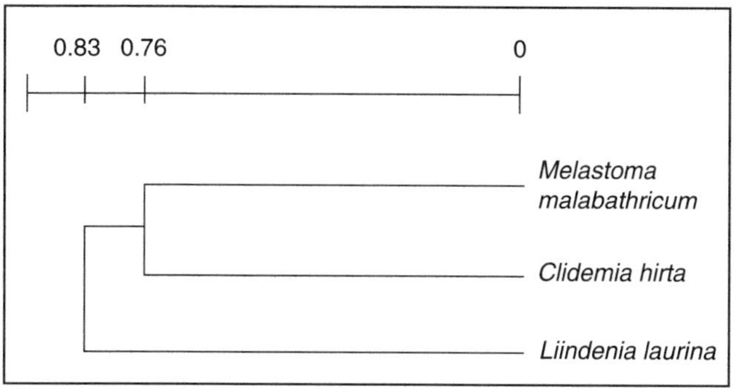

Fig. 4 Dendrogram obtained from polypeptide bands analysis using D-UPGMA showing genetic distance among three species of Melastomataceae

the protein profile of Melastomatacea. The total protein band of seed protein extracts of three species of Melastomataceae from both buffers was 33 protein bands (range from 12.9 to 95.2 kDa), 30 were polymorphic (91%) and only three bands were monomorphic (9%) or common (32.12 kDa, 17.09 kDa and 15.34 kDa) among three species of Melastomataceae. However, three bands have been shared between *Mm* and *Ch* (51.86 kDa, 45.49 kDa and 28.17 kDa), two shared bands between *Mm* and *Ll* (63.33 kDa and 29.48 kDa), whereas between *Ch* and *Ll* only shared one band (64.42 kDa). The cluster analysis was done by using Unweighted Pair-Group Method of Arithmetic (UPGMA) to produce dendrogram (Fig. 4). When the lower similarity, it is means that higher genetic distance among species. *Mm* and *Ch* were clustered in the same group which the genetic distance was 0.76 whereas the genetic distance between *Mm* and *Ch* with *Ll* was 0.83 (the highest genetic distance). The similarity between *Mm* and *Ch* was more than between *Mm* and *Ll* or the most closely related species were between *Mm* and *Ch,* may be from their morphology, *Mm* and *Ch* look similar to each other.

In Fagacea a total of 55 polypeptide bands were recorded from three species by using both extraction buffers (range from 151 to 380 kDa). Figure 5 shows the protein profile of Fagacea. Fifty-four bands were polymorphic (98%) and only one band (190 kDa) was monomorphic (2%) or common among three species. Five bands have been shared between *Cl* and *Cs* (330 kDa, 303 kDa, 180 kDa, 175 kDa and 166 kDa), four shared bands between *Cl* and *Lw* (196 kDa, 194 kDa, 193 kDa and 186 kDa), whereas between *Cs* and *Lw* shared two band (172 kDa and 182 kDa). The genetic distance between *Cl* and *Cs* was 0.77. Meanwhile the genetic distance between *Cl* and *Cs* to *Lw* was 0.88 (Fig. 6). The similarity between *Cl* and *Cs* was more than between *Cl* and *Lw* or the most closely related species were between *Cl* and *Cs* may be from the appearance of their seed, *Cl* and *Cs* look similar to each other.

As the comparison the genetic diversity by using SDS-PAGE in some plants indicated the following results; 5–11% polymorphism in three genotypes of *Abrus precatorius* (Chittora and Purohit 2012), 5–84% polymorphism in ten genotypes of

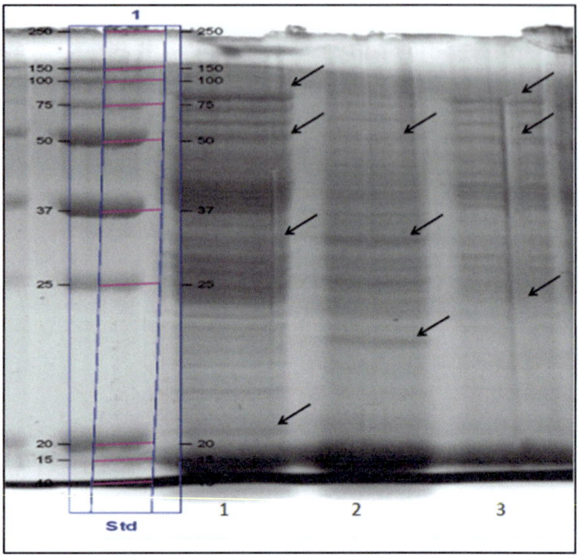

Fig. 5 Protein profile of 12% gel SDS-PAGE. Lane 1: *Cl*, Lane 2: *Cs*, Lane 3:*Lw*. Std: protein marker

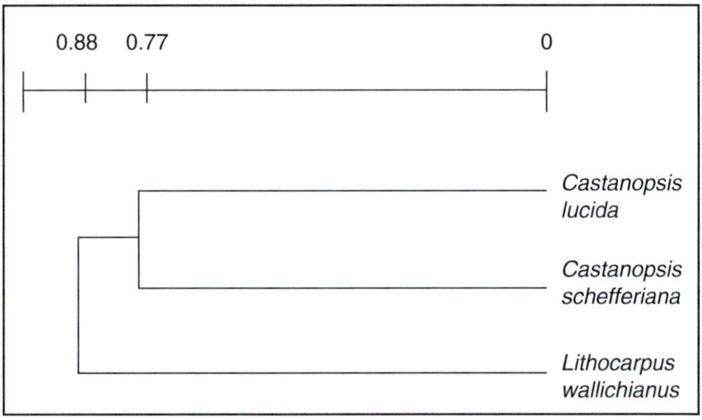

Fig. 6 Dendrogram of three species of Fagaceae. (From Wan Bayani et al. 2013)

wheat (Siddique and Naz 2009), 0–60% polymorphism in twenty genotypes of walnut (Khan et al. 2010), 0–80% polymorphism in eleven genotypes of *Oryza sativa* (Inamullah et al. 2010) and 0–100% polymorphism in nineteen genotypes of *Capsicum* (Akbar et al. 2010).

Conclusion

In conclusion, the genetic diversity of six species of two plants family were successfully determined from seed storage proteins by using SDS-PAGE. The seeds have an important genetic information of the plant and it must be conserved to avoid genetic loss by environmental factors or some diseases. This information will be useful as the basic genetic information for the species. The Melastomatacea is very useful as ethnomedicine since many races using this as their medicine and for Fagaceae, this family has important economic uses especially in timber industry.

Acknowledgements The author wish to acknowledge University Malaysia Terengganu (UMT) for providing the opportunity to pursue this study. I would also like thanks MARDI at Tasik Kenyir for providing the samples and also wish thank you to Mr. Muhamad Razali Bin Salam for identify the samples.

References

Akbar, A., Ahmed, H., Ghafoor, S., & Khan, A. I. (2010). Phylogeny and genetic diversity studies in capsicum using seed storage proteins. *Current Research Journal Biological Science, 72*, 250–252.

Chittora, M., & Purohit, S. D. (2012). Optimization and analysis of seed protein diversity in *Abrus precatorius* genotypes with different seed coat colour. *International Journal of Life Sciences Biotechnology and Pharma Research, 1*, 268–277.

Dianita, R., Ramasamy, K., & Ab Rahman, N. (2011). Antibacterial activity of different extracts of *Clidemia hirta* (L.) D. Don leaves. *Planta Medica, Journal of Medical Plant and Natural Product Research*, 77, https://doi.org/10.1055/s-0031-1282769.

Falk, D. A., Knapp, E., & Guerrant, E. O. 2001. An introduction to restoration genetics. Society for Ecological Restoration. Retrieved from http://www.gov/pubs/rsetgene/restgenepdf. [12 February 2014].

Franca, F., Lago, E. L., & Marsden, P. D. (1996). Plants used in the treatment of leishmanial ulcers due to *Leishmania* (Vannia) *brazilliensis* in an endemic area of Bahia, Brazil. *Revista da Sociedade Brasileira de Medicina Tropical, 9*(3), 229–232.

Gepts, P. (1990). Genetic diversity of seed storage proteins in plants. In A. H. D. Brown, M. T. Clegg, A. L. Kahler, & B. S. Weir (Eds.), *Plant population genetic, breeding and genet resource* (pp. 64–82). Sunderland: Sinauer Associates.

Inamullah, K. I. A., Ahmed, H., Ghafoor, S., Gul, S., Muhammad, I., & Iiyas, M. (2010). Seed storage protein profile of rice varieties commonly grown in Pakistan. *Asian Journal of Agricultural Science, 4*, 120–123.

Joffry, S. M., Yob, N., Rofiee, M., Affandi, M., Suhaili, Z., Othman, F., Akim, A. M., Desa, M. N. M, & Zakaria, Z. (2011). *Melastoma malabathricum* (L.) Smith ethnomedicinal uses, chemical constituents, and pharmacological properties: A review. *Evidence-based Complementary and Alternative Medicine, 2012*, https://doi.org/10.1155/2012/258434.

Khan, M. W., Khan, I. A., Ahmed, H., Ali, H., Ghafoor, S., Afzal, M., Khan, F. A., Shah, M., & Afridi, S. G. (2010). Estimation genetic diversity in walnut. *Pakistan Journal of Botany, 42*, 1791–1796.

Latiff, A., & Mat-Salleh, K. (2002). *Medicinal plants of Malaysia*. Bangi: Universiti Kebangsaan Malaysia.

Manos, P. S., Zhou, Z. K., & Cannon, C. H. (2001). Systematics of Fagaceae: Phylogenetic tests of reproductive trait evolution. *International Journal of Plant Sciences, 162*(6), 1361–1379.
Ng, F. S. P. (1991). Variety of forest fruits, seeds & seedling. *Malayan Forest Record, 1*(34), 85–87.
Nixon, K. C. (1993). Infrageneric classification of Quercus (Fagaceae) and typification of sectional names. *Annals of Forest Science, 50*, 25–34.
Rao, V. R., & Hodgkin, T. (2002). Genetic diversity and conservation and utilization of plant genetic resources. *Plant Cell, Tissue and Organ Culture, 68*(1), 1–19.
Siddique, M. F., & Naz, N. (2009). Protein landmarks for diversity assessment in wheat genotypes. *African Journal of Biotechnology, 8*(9), 1855–1859.
Silveira, F. A. O., Fernandes, G. W., & Lemos-Filho, J. P. (2013). Seed and seedling ecophysiology of neotropical Melastomataceae: Implications for conservation and restoration of savannas and rainforests. *Annals of the Missouri Botanical Garden, 99*(1), 82–99.
Ungureanu, E., Artenie, V., & Ungureanu, A. (2007). SDS-PAGE analysis of protein from the seed of honeylocust (*Gleditsia Triacanthos*) during germination. Analele Stiintifice ale Universitatii "Alexandru Ioan Cuza" din Iasi Sec. II a. *Genetica si Biologie Moleculara, 8*, 31–35.
Wan Bayani, W. O., Hazlina A. Z., & Azhar, A. M. (2013). Optimization of SDS-PAGE conditions and analysis of seed protein diversity in Fagaceae (Chestnut) based on seed storage protein pattern from Taman Tropika Kenyir. The 13th Symposium of the Malaysian Society of Applied Biology. Cherating, Pahang, 8–10 June 2013. Malaysian Society of Applied Biology, Kuala Lumpur.
Watson, L., & Dallwitz, M. J. (1992). The families of the flowering plants: Descriptions, illustrations, identification and information. Retrieved from http://www.delta-intkey.com/angio.html.

Parasitic Plants at the Coastal of Setiu, Terengganu: Distribution and Its Association with Host Trees

Shahrudin Rohani, Ai Lim Teh, and Muhamad Razali Salam

Abstract Parasitic plants have long been perceived as unwanted species due to the habit of destructing their hosts. However, parasite may play a significant roles in determining the species composition in one ecosystem. Therefore, this study was conducted to identify the parasitic plants and their hosts in the coastal forest in Setiu, Terengganu. We recorded seven species of parasites, with *Cassytha filiformis* was found dominating both sites. The inland site showed a higher species number with seven species, while beach site recorded only three species. Parasitic plants was found infested a total of 21 plant species, and *Syzygium zeylanicum* was the host tree that highly parasitized.

Keywords Cover percentage · Host specificity · Infestation

Introduction

Parasitic plants can be defined as plants that live on other plant where the plants, parasites, will harm or killed the host. They can be categorised into four main categories, which are obligate parasites, facultative parasites, holoparasites and hemiparasites (Nickrent and Musselman 2004). These parasites might be stem parasites, root parasites or both. Parasitism is an excellent and highly successful life strategy (Westwood et al. 2010). It is estimated that approximately 1% of the angiosperm evolved the ability to gain their nutritional needs through parasitizing on other plants (Bouwmeester et al. 2007).

The distribution of the parasitic plants is determined by several factors such as pollination, seed dispersal and also host characters (Okubamichael et al. 2016). They normally invade host plants that are quite similar to them both in size and physiological needs. The pattern of host preference is related to the local abundance of hosts. The number of potential host plants may varies among parasite species,

S. Rohani (✉) · A. L. Teh · M. R. Salam
School of Marine and Environmental Sciences, Universiti Malaysia Terengganu, Kuala Nerus, Terengganu, Malaysia
e-mail: rohanishahrudin@umt.edu.my

M. T. Abdullah et al. (eds.), *Greater Kenyir Landscapes*,
https://doi.org/10.1007/978-3-319-92264-5_9

Fig. 1 Map of Peninsular Malaysia showing the study sites (beach plot and inland plot) in Mengabang Panjang, Setiu. (Source: adaptation from Google Earth)

depending on the factors such as host availability, quality and resistance against parasitism and parasite preference (Parker and Riches 1993). Those species with large host range are known as generalist, while species with narrow host range are known as specialist.

This study was initiated to document the parasitic plants that occur in an area at the east coast of Peninsular Malaysia. It is interesting to know about the richness of parasitic plant in the coastal area since this type of ecosystem is still poorly studied. The aims of this study are to identify the abundance and distribution of parasitic plants, and also to identify the plants that associate with the parasites. Transects were established near to the beach and also in inland. Parasites and host plants were identified within the transects (Fig. 1). Visual estimation was used to determine the percentage cover of the parasitic plants. Percentage cover of parasitic plants was estimated in eight cover classes using modified Braun-Blanquet cover classes according to Von Holle and Motzkin (2007).

Parasitic Plants Distribution

A total of seven parasitic plant species were found at both studied sites, which comprised of six genera and three families (Table 1). Out of seven species, five of them were mistletoe while the other two were parasitic vine species. For comparison, five species, four genera and two families were found in the study of the roadside parasitic plants in Penang, Malaysia (Rahmad et al. 2014). However, the roadside study

Table 1 List of parasitic plants in the study area

Parasite species	Family	Common name
Cassytha filiformis	Lauraceae	Dodder vines
Dendrotrophe varians	Santalaceae	–
Dendrophthoe pentandra	Loranthaceae	Malayan mistletoe
Scurrula ferruginea	Loranthaceae	Rusty-leaf mistletoe
Viscum articulatum	Santalaceae	Leafless mistletoe
Macrosolen retusus	Loranthaceae	Pink-flowered mistletoe
Macrosolen cochinchinensis	Loranthaceae	Common Chinese mistletoe

Table 2 The occurrence of parasitic plants in both area (beach and inland) with their percentage cover

Parasitic species	Beach site	Inland site
Cassytha filiformis	4.9%	33.7%
Dendrotrophe varians	34.8%	41.2%
Dendrophthoe pentandra	0.4%	9.6%
Scurrula ferruginea	0	9.6%
Viscum articulatum	0	2.4%
Macrosolen retusus	0	1.2%
Macrosolen cochinchinensis	0	1.2%

was only focused on Loranthaceae family, thus might be the reason for the lower species number as compared to the coastal study.

When comparing between beach site and inland site, three species were found occurred at both sites (*Cassytha filiformis*, *Dendrotrophe varians* and *D. pentandra*) while four species were found only at the inland site (*Scurrula ferruginea*, *Viscum articulatum*, *Macrosolen retusus* and *M. cochinchinensis*) (Table 2). This shows that inland site was having a higher species of parasitic plant than the beach site. Nickrent (2002) stated that majority of the parasitic plant species occur in undisturbed ecosystem. Thus, the low species number of parasitic plants at the beach site might be due to the fragmented area of vegetation due to human activities like road constructions as well as wild fire that reduces the abundance and diversity of host species.

Species that dominated the beach area was *C. filiformis*, meanwhile *D. varians* was found dominating the inland site. *Cassytha filiformis* was found widely distributed compared to other parasitic species. The wide distribution of *C. filiformis* was not surprising since this species has a pantropical distribution, inhabiting coastal vegetation, including sandy dunes, sandy beach and coastal woodlands (Nelson 2008). The dispersion mode of this species, which is by different agents including birds, water current and strong wind (Nelson 2008; Balasubramanian 1990), might be the reason why it was found more abundant at the coastal area like Setiu.

The difference between the percentage cover of *C. filiformis* and *D. varians* was very huge where the percentage cover of *C. fliformis* almost doubled that of the *D. varians*. *Dendrophthoe pentandra* was considered rare at the beach site as their percentage of cover was less than 1%. The parasitic plant that dominated the inland site was *D. varians*, followed by *C. filiformis*. However, there was only small difference between the percentage cover of *D. varians* and *C. filiformis*. *Viscum articulatum*, *M. retusus* and *M. cochinchinensis* were considered rare at the inland site as both of their percentage of cover was less than 3%. Only one individual of both *M. retusus* and *M. cochinchinensis* were found at the inland site.

Parasitic Plants-Host Associations

Twenty-one plant species were found hosting the parasites in this area (Table 3). *Syzygium zeylanicum* was found most frequently being parasitized by five parasitic species (*C. filiformis D. varians*, *D. petandra,* and *M. retusus*). The high frequency of infestation on *S. zeylanicum* might due to a high abundance of this species in the study area.

All parasites found in this study were having a large host range except *V. articulatum*, *M. retusus and M. cochinchinensis*, where these species were found infested only a single species. However, it is a premature decision to say either these species are host-specific or not due to a small sampling area even though *M. retusus* was reported to be a specialist (Barlow 1997). Therefore, it is strongly suggested for future study to increase the sampling area for a better understanding of the ecological aspects for parasitic plants.

This study has successfully documented the list of parasitic plant species that can be found in the Setiu coastal area. The richness of the parasitic plants in this area might contribute to the dynamics of the ecosystem in few ways. Firstly, parasitism will affect the growth and reproduction of host plant, which lead to a high competition between parasite and host. In consequence, the damages on host plant can affect the abundance and diversity of herbivores, pollinators and seed dispersers. Secondly, some parasites do play a significant role in the ecosystem as an 'ecosystem engineer'. This owe to their ability to change the physical state in their abiotic environment, such as alteration in soil water status, nutrient cycling and canopy temperature. Several studies have shown how parasitic plants are able to suppress the dominant plant, which later encourage the increased in plant diversity and also facilitated the growth of rare species (Decleer et al. 2013; Pennings and Callaway 1996). Despite of being perceived as destructor, the occurrence of parasitic plant may be an indicator. However, to determine what is the impact of parasitic plant in this area, further study need to be done to understand the factors regulating their abundance and distribution, and also on the effect after the infestation.

Table 3 List of the host plants that associate with parasitic plants in the study area

Host plant species	Frequency of host plants infested by parasites						
	Cassytha filiformis	*Dendrotrophe varians*	*Dendrophthoe pentandra*	*Scurrula ferruginea*	*Viscum articulatum*	*Macrosolen retusus*	*Macrosolen cochinchinensis*
Acronychia pedunculata	1	0	0	0	0	0	0
Ardisia sp.	1	0	0	0	0	0	0
Baeckea frutescens	1	0	0	0	0	0	0
Calophyllum rupicola	0	1	0	0	0	0	0
Carallia brachiata	1	0	0	0	0	0	0
Champereia manillana	0	0	2	3	2	0	0
Leptocarpus disjunctus	2	0	0	0	0	0	0
Melaleuca cajuputi	1	0	1	0	0	0	0
Memecylon edule	1	0	0	0	0	0	0
Myrica esculenta	0	1	0	0	0	0	0
Olea brachiata	2	0	0	0	0	0	0
Parastemon urophyllus	2	1	0	0	0	0	0
Psychotria obovata	0	0	2	0	0	0	0
Rapanea porterian	0	1	1	0	0	0	0
Rhodomyrtus tomentosa	0	0	1	0	0	0	0
Styphelia malayana	0	0	1	0	0	0	0
Syzygium grande	0	1	0	0	0	0	0

(continued)

Table 3 (continued)

Host plant species	Frequency of host plants infested by parasites						
	Cassytha filiformis	*Dendrotrophe varians*	*Dendrophthoe pentandra*	*Scurrula ferruginea*	*Viscum articulatum*	*Macrosolen retusus*	*Macrosolen cochinchinensis*
Syzygium palembanicum	3	3	0	1	0	0	1
Syzygium sp.	1	0	0	0	0	0	0
Syzygium zeylanicum	**15**	**7**	**1**	**0**	**0**	**1**	**0**
Vaccinium littoreum	0	0	1	1	0	0	0

The most frequent infested species was indicated by bold letters

Acknowledgements The authors would like to thank School of Marine and Environmental Sciences, Universiti Malaysia Terengganu, for the financial and facilities support for this research.

Appendix

(**a**) *Cassytha filiformis* hanging on its host (left). Fruits of *C. filiformis* is a drupe that enclosed by the enlarged, inflated, succulent calyx tube (right). (**b**) *Dendrotrophe varians* with inflorescence (left) and fruits (right). (**c**) *Dendrophthoe pentandra* shows different colours of leaves, which might influenced by factor such as light intensity. (**d**) The pendulous branches of *Scurrula ferruginea* (left), and closed-up picture (left) of a branch showing the structure of haustorium (in the red circle). (**e**) *Viscum articulatum* on its host, *Champereia manillana.* (**f**) *Macrosolen cochinchinensis* with fruits. (**g**) *Macrosolen retusus* with inflorescence (left) and fruits (right)

a

b

c

d

e
f
g

References

Balasubramanian, P. (1990). Seed dispersal of *Cassytha filiformis* at Point Calimere, Tamil Nadu. *Journal of the Bombay Natural History Society, 87*(3), 472.

Barlow, B. A. (1997). Loranthaceae. *Flora Malesiana-Series 1, Spermatophyta, 13*(1), 209–401.

Bouwmeester, H. J., Roux, C., Lopez-Raez, J. A., & Becard, G. (2007). Rhizosphere communication of plants, parasitic plants and AM fungi. *Trends in Plant Science, 12*(5), 224–230.

Decleer, K., Bonte, D., & Van Diggelen, R. (2013). The hemiparasite *Pedicularis palustris*: 'Ecosystem engineer' for fen-meadow restoration. *Journal for Nature Conservation, 21*(2), 65–71.

Nelson, S. C. (2008). *Cassytha filiformis. Plant Disease, 42*, 1–10.

Nickrent, D. L. (2002). Parasitic plants of the world. In J. A. López Sáez, P. Catalán, & L. Sáez (Eds.), *Parasitic plants of the Iberian Peninsula and Balearic Islands* (pp. 7–27). Madrid: Mundi-Prensa.

Nickrent, D. L., & Musselman, L. J. (2004). Introduction to parasitic flowering plants. *The Plant Health Instructor, 13*, 300–315.

Okubamichael, D. Y., Griffiths, M. E., & Ward, D. (2016). Host specificity in parasitic plants-perspectives from mistletoes. *AoB Plants*. https://doi.org/10.1093/aobpla/plw069.

Parker, C., & Riches, C. R. (1993). *Parasitic weeds of the world: Biology and control*. Wallinford: CAB International.

Pennings, S. C., & Callaway, R. M. (1996). Impact of a parasitic plant on the structure and dynamics of saltmarsh vegetation. *Ecology, 77*, 1410–1419.

Rahmad, Z. B., Addo-Fordjour, P., Asyraf, M., & Nik Rosely, N. F. (2014). Mistletoe abundance, distribution and their associations with trees along roadside in Penang, Malaysia. *Tropical Ecology, 55*(2), 255–262.

Von Holle, B., & Motzkin, G. (2007). Historical land use and environmental determinants of nonnative plant distribution in coastal southern New England. *Biological Conservation, 136*, 33–43.

Westwood, J. H., Yoder, J. I., & Timko, M. P. (2010). The evolution of parasitism in plants. *Trends in Plant Science, 15*(4), 227–235.

Observations on Flowers of the Genus *Bruguiera* from Setiu Mangrove Areas, Terengganu

Nur Syamimi Wahab, Nur Athirah Ismail, Ahmad Fadhli Adzemi, and Mohamed Nor Zalipah

Abstract The purpose of this research was to study the flower morphology of mangrove trees from the genus *Bruguiera* (Rhizophoraceae) at Setiu mangrove areas, Terengganu. The objective of this study is to determine the pollination strategy of each species in this genus from their floral traits. A total of five *Bruguiera* species was recorded in this study, including the new locality record of *Bruguiera rhynchopetala*. A total of 33 flowers from 11 trees were collected for morphometric study. The results showed that *B. gymnorhiza* and *B. rhynchopetala* were the largest flowers as compared to *B. sexangula. Bruguiera rhynchopetala* however, showed overlap traits between the parents, and thus suggested the same pollination strategy. Meanwhile, *B. cylindrica* showed the smallest flower from all the measurements recorded and *B. hainesii* showed intermediate size. The small herkogamy in *B. cylindrica* indicates self-pollination, while the large herkogamy in the other four species might indicates cross-pollination by pollinating agents, possibly birds.

Keywords Flower morphometric · Herkogamy · Mangrove · Pollination

Introduction

Globally, mangrove vegetation constitute approximately 80 species from 20 families (Duke 1992). These species are mainly from two main orders, the Myrtales and Rhizophorales, which comprise about 25% of all mangrove families, and almost 50% of all species. The better known taxa of mangroves however are the members of order Rhizopholares, family Rhizophoraceae, consisting of *Rhizophora, Bruguiera, Ceriops* and *Kandelia* (Hogart 2007). *Bruguiera* is the largest genus in the family Rhizophoraceae. In Malaysia, genus *Bruguiera* consists of five species

N. S. Wahab · N. A. Ismail · A. F. Adzemi · M. Nor Zalipah (✉)
School of Marine and Environmental Science, Universiti Malaysia Terengganu, Kuala Nerus, Terengganu, Malaysia
e-mail: zalipah@umt.edu.my

M. T. Abdullah et al. (eds.), *Greater Kenyir Landscapes*,
https://doi.org/10.1007/978-3-319-92264-5_10

which are *Bruguiera cylindrica, B. gymnorhiza, B. hainesii, B. parviflora* and *B. sexangula*. Based on its flower size, these species is categorised into two groups which are the large-flowered group that includes *B. gymnorhiza, B. hainesii* and *B. sexangula* while the small-flowered group consists of *B. cylindrica* and *B. parviflora*.

Flower characteristics such as morphology, colour, nectar and odour, will provide the best strategy for the plant to attract specific pollinators (Nagarajan et al. 2010). For example, the large-flowered group are usually with recurved pedicels and solitary flower, which obviously suitable for the nectarivorous birds to access the flowers and also to prevent the rainfall from diluting the nectar (Tomlinson 1986). In addition, the large flowers also produce nectar in large quantity that is secreted in a deep calyx cup. For the small-flowered group, the flowers are not recurved and small in size, flat calyx cup and with a small quantity of nectar. Therefore based on the characteristics, the large-flowered group is thought to be pollinated by birds while the small-flowered group is considered as insect-pollinated.

Other than the flower size, the relationship between floral traits and its pollinators in genus *Bruguiera* is still poorly understood. *Bruguiera hainesii* for example, reported to be pollinated by birds from their regular visitation to the flowers and also large amount of nectar produced by the flower (Noske 1993). The characteristics of the flower however showed combination of characters such as relatively large flower size and in multiple-flowered inflorescences (Jamilah et al. 2013; Sheue et al. 2005). Recent study by Ono et al. (2016) showed that *B. hainesii* is a hybrid between the small-flowered *B. cylindrica* (maternal parent) and the large-flowered *B. gymnorhiza* (paternal parent). As hybridisation could occur from the sharing of pollinators, this showed that bird visitations to the small-flowered *B. cylindrica* or insect visitation to the big-flowered *B. gymnorhiza* resulted in fertilisation of the ovules. Therefore, other flower characteristics is needed to describe the pollination strategy in this genus. Thus, we examined the flower attributes in genus *Brugueira*, to infer the pollination strategy of each *Brugueira* species in Setiu.

Setiu is located in the northeast of Peninsular Malaysia in the state of Terengganu (5° 40′ N, 102° 43′ E). The mangrove areas of Setiu consist mainly of nipa palm, *Nypa* (Aracaceae) and some stands of true mangrove, *Rhizophora* (Rhizophoraceae) (Nakisah and Fauziah 2003). Previous study in Setiu mangrove areas recorded only three of the five *Bruguiera* species reported to occur in Malaysia (Mohd Lokman and Sulong 2012), namely *Bruguiera cylindrica, B. gymnorrhiza* and *B. sexangula*. The critically endangered *B. hainesii* however, was recently discovered in Setiu mangrove area by Jamilah et al. (2013), making the total to four *Bruguiera* species recorded at Setiu. The four *Bruguiera* species can be distinguished based on the colour of their calyx tube. The calyx tube is yellow-green in *B. cylindrical*, red to pink in *B. gymnorrhiza*, varied from pink-red to greenish *B. hainesii*, and yellow in *B. sexangula* (Tomlinson 1986).

Flower Morphology

During collections of flowers in May 2015, we encountered a huge tree with redish calyx flowers which we firstly identified as *B. gymnorhiza*. Further observations showed that the flowers are *B. rhynchopetala*, characterise from the presence of petal bristles (Fig. 1). *Bruguiera rhynchopetala* is a hybrid previously known to occur in China and Australia (Duke and Ge 2011). In Malaysia, the first record of this species was from Pulau Langkawi, Kedah (Wan Juliana et al. 2014), therefore the finding of this study marked the second locality record of this species in Malaysia (Mohd Razali Salam, pers. comm.). The flowers observed in this study were with light red calyx tube, although reported previously to have rosy blush to all green calyx tube by Tomlinson (1986).

The flower attributes (mean ± SD) for each *Bruguiera* species were as summarised in Table 1. Among species, *B. rhynchopetala* recorded the largest size of flower with petal length of 11.74 ± 0.29 mm, calyx tooth length of 17.63 ± 0.31 mm, calyx length of 34.28 ± 0.95 mm, style length of 20.06 ± 0.79 mm, stamen length of 11.65 ± 0.55 mm and flower disk depth of 6.55 ± 0.52 mm. *Bruguiera gymnorhiza* on the other hand, showed the largest in corolla width (5.97 ± 0.37 mm), herkogamy (9.09 ± 0.84 mm), stigma width (0.50 ± 0.04 mm), and highest count in number of calyx (13.44 ± 0.29 mm), number of petals (12.55 ± 0.60 mm) and number of stamen (21.44 ± 1.13 mm). Meanwhile, *B. cylindrica* showed the smallest size in all data recorded.

From the Principal Component Analysis (PCA), results showed that the first two components explained 85.8% of the total variance combined. The first component

Fig. 1 A flower and a close bud of *Bruguiera rhynchopetala* found at Setiu mangrove area, Terengganu. (Photograph by Mohamed Nor Zalipah)

Table 1 Mean (± SD) of flower attributes of five *Bruguiera* species recorded in Setiu, Terengganu

Characteristics/species	*B. cylindrica* (N = 9)	*B. gymnorhiza* (N = 9)	*B. hainesii* (N = 6)	*B. rhynchopetala*(N = 3)	*B. sexangula*(N = 6)
Petals length (mm)	2.44 ± 0.18^{a}	11.59 ± 2.20^{c}	5.32 ± 0.39^{b}	11.74 ± 0.50^{c}	$11.\ 23 \pm 0.35^{c}$
Calyx tooth length (mm)	4.58 ± 0.21^{a}	16.15 ± 2.78^{c}	8.88 ± 0.78^{b}	17.63 ± 0.53^{c}	16.36 ± 0.57^{c}
Calyx length (mm)	9.82 ± 0.27^{a}	34.10 ± 3.37^{c}	20.39 ± 2.08^{b}	34.28 ± 1.65^{c}	32.26 ± 0.84^{c}
Corolla width (mm)	2.18 ± 0.37^{a}	5.97 ± 1.10^{d}	3.96 ± 0.64^{b}	$5.55 \pm 0.83^{c,d}$	$4.53 \pm 0.16^{b,c}$
Style length (mm)	3.99 ± 0.26^{a}	19.33 ± 2.87^{d}	9.70 ± 0.67^{b}	20.06 ± 1.37^{d}	16.34 ± 0.55^{c}
Stamen length (mm)	2.79 ± 0.14^{a}	10.24 ± 2.20^{c}	4.70 ± 0.57^{b}	11.65 ± 0.95^{c}	9.76 ± 0.62^{c}
Flower disk depth (mm)	3.24 ± 0.35^{a}	6.02 ± 1.12^{b}	3.64 ± 1.04^{a}	6.55 ± 0.89^{b}	4.36 ± 0.37^{a}
Stigma width (mm)	0.22 ± 0.07^{a}	0.50 ± 0.13^{b}	0.21 ± 0.08^{a}	0.38 ± 0.04^{b}	0.26 ± 0.02^{a}
Herkogamy (mm)	1.20 ± 0.31^{a}	9.09 ± 2.53^{c}	5.00 ± 0.64^{b}	8.41 ± 2.12^{c}	$6.58 \pm 0.98^{b,c}$
Number of petals	7.56 ± 0.53^{a}	12.55 ± 1.81^{c}	10.00 ± 0.63^{b}	12.00 ± 0.00^{c}	10.33 ± 0.33^{b}
Number of calyx	7.67 ± 0.50^{a}	13.44 ± 0.88^{d}	10.00 ± 0.26^{b}	11.67 ± 0.58^{c}	10.50 ± 0.22^{b}
Number of stamen	11.11 ± 2.71^{a}	21.44 ± 3.40^{b}	14.83 ± 2.32^{a}	$15.67 \pm 11.02^{a,b}$	11.83 ± 2.51^{a}

Different letters indicate significant differences from one another (multiple comparison following significant result from Kruskal-Wallis test. Significant results were recorded at $P < 0.001$ for all characteristic)

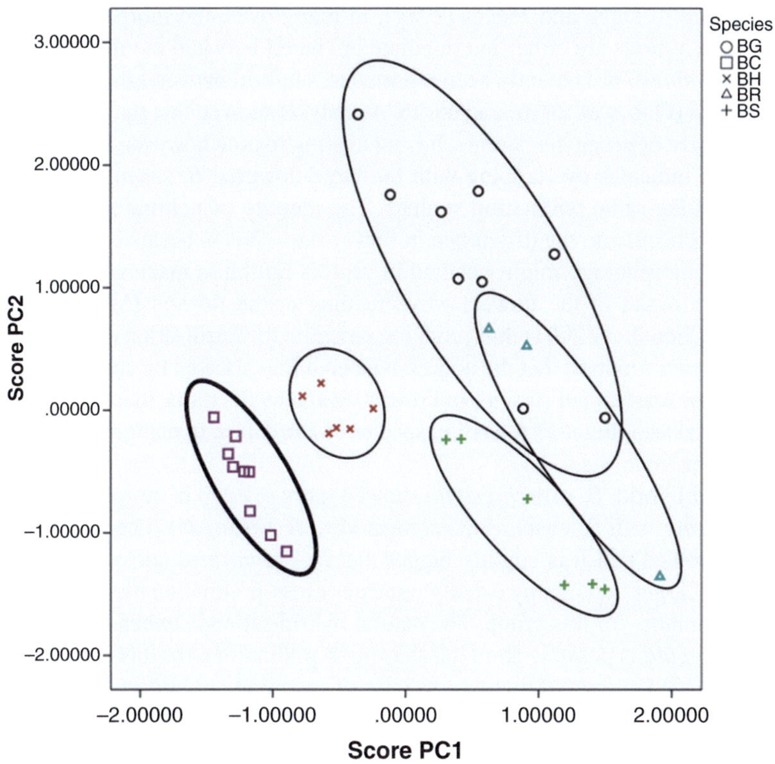

Fig. 2 Ordinations of *Bruguiera* flowers on the first two principal component axes of principal component analysis (PCA)

revealed the height of the flower which indicate by the petal length, calyx tooth length, calyx length and style length of the flowers. The second component explained the gynoecium (stigma width) and androecium (number of anthers) in the flowers. The flower morphology of *Bruguiera* species plotted in morphospace defined by the first two components of PCA showed that *B. cylindrica, B. hainesii* and *B. sexangula* were totally different with no overlap between species (Fig. 2). However *B. gymnorhiza* and *B. rhynchopetala* showed overlapping in the characteristics measured and recorded.

The smallest flower among *Bruguiera* species in Setiu, *B. cylindrica* can be found in abundance at the mangrove area (Nakisah and Fauziah 2003). The trees can be found beside the river and in swampy areas of the mangrove. With its small flowers, this species was found to be pollinated by thrips (Nagarajan et al. 2008) while Tomlinson et al. (1979) reported butterflies as their pollinators. The adaption for insect pollination in *B. cylindrica* is similar with the other small-flowered *B. parviflora*, except for the shorter floral cup and with distinct calyx cup. However, the small herkogamy (anther-stigma separation) of this species indicates efficient pollen transfer, thus self-pollination in this species is possible (Ushimaru and Nakata 2002), although reported to be insect pollinated.

According to Duke and Jackes (1987), in mangroves, the morphological attributes of the hybrids are either intermediate or shared between the maternal plants. *Bruguiera hainesii* has recently been reported as a hybrid between *B. cylindrica* and *B. sexangula* (Ono et al. 2016) and our PCA analysis showed that the flower is intermediate in size between the parents. It is interesting to note however, herkogamy for this species indicates overlapping with the large-flowered *B. sexangula,* therefore might adopt the same pollination strategy. The identity of pollinator species however, might differ from the difference in flower size. This is because, the mismatch with flower morphology might resulted in visitors failing to make contact with the reproductive organ of the flowers while feeding at the flowers (Muchhala 2003; Pandit and Choudhury 2001) thus reducing potential for fertilisation to occur. Noske (1993) however reported that the aggressive behaviour showed by sunbirds is likely the reason for selection of *Bruguiera* flowers visited by the birds, therefore restricted their potential as pollinators to certain species of *Bruguiera* in mangroves elsewhere of Peninsular Malaysia.

The other hybrid, *B. rhynchopetala* showed more overlap in morphology including herkogamy with *B. gymnorhiza* compared to *B. sexangula.* The measurements recorded showed that it is slightly bigger that *B. gymnorhiza* and with lighter red calyx tube. Larger herkogamy usually indicates cross-pollination therefore required animal pollinators for this group. The natural hybridisation between *B. gymnorhiza* and *B. sexangula* is possible from the sharing of pollinators, for this case, the birds. Wee et al. (2014) reported that two species of common sunbird in Singapore, the olive-backed sunbird (*Nectarinia jugularis*) and the copper-throated sunbird (*Nectarinia calcostetha*) as pollinators for *B. sexangula. Bruguiera sexangula* usually occurs in upstream reaches of freshwater dominated estuaries, while *B. gymnorhiza* is more marine and tolerates to high levels of salinity (Duke and Ge 2011). Although habitat differentiation between the two species may reflect their different level of salt tolerance, considerable overlap in geographic distribution and habitats provides opportunities for hybridisation. Other factor contributing to natural hybridisation includes the long and partially overlapping flowering periods (Duke and Jackes 1987).

Conclusion

In Setiu, based on flower morphometric analysis, the large-flowered group consists of *Bruguiera gymnorhiza. B. sexangula*, and their hybrid, *B. rhynchopetala*, while the small-flowered group was only represented by *B. cylindrica. Bruguiera hainesii* on the other hand, is intermediate in size between the two groups. Although previous study reported possible bird pollination for this species, their flower morphology suggests that this species might be pollinated by smaller size pollinators as compared to the other three large-flowered *Bruguiera.* Larger herkogamy in the large-flowered group including *Bruguiera hainesii,* indicates cross-pollination by birds, while small herkogamy suggested self-pollination in *B. cylindrica.* To confirm pollination by pollinators however, more study on floral functional aspects (such as floral persistence, anthesis time, pollen production, nectar production etc.) of this genus is needed. A better understanding of the pollination biology of this important mangrove taxa is fundamental to conserving these important coastal ecosystems.

Acknowledgements We would like to thank School of Marine and Environmental Sciences for equipment and logistic support to conduct this research. We would also like thank Tuan Haji Muhamad Razali B. Salam, senior lab assistant of the School of Marine and Environmental Sciences for identification of *Bruguiera* species, particularly the newly recorded *Bruguiera rhynchopetala*.

Appendix

Eigenvectors of the first two components extracted in Principal Component Analysis (PCA) for each variable.

Variable	Component 1	Component 2
Petals length	0.948	−0.242
Calyx tooth length	0.941	−0.296
Calyx length	0.968	−0.210
Corolla width	0.868	0.139
Style length	0.979	−0.162
Stamen length	0.921	−0.201
Flower disk depth	0.851	−0.069
Stigma width	0.748	0.366
Herkogamy	0.884	−0.096
Number of petals	0.874	0.270
Number of calyx	0.930	0.198
Number of stamen	0.561	0.683

The flower of *Bruguiera hainesii*, *B. gymnorhiza* and *B. sexangula* (left to right) showed different colour of calyx (Photograph by Nur Athirah Ismail)

The difference in flower size of *Bruguiera cylindrica, B. hainesii*, *B. gymnorhiza* and *B. rhynchopetala* (left to right) (Photograph by Nur Syamimi Wahab)

Sunbird observed assessing the flowers of *Bruguiera rhynchopetala* at Setiu mangrove area (Photographs by Ahmad Fadhli Adzemi)

References

Duke, N. C. (1992). Mangrove floristics and biogeography. In A. I. Robertson & D. M. Alongi (Eds.), *Tropical mangrove ecosystems* (pp. 63–100). Washington, DC: American Geophysical Union.

Duke, N. C., & Ge, X. J. (2011). *Bruguiera* (Rhizophoraceae) in the Indo West Pacific: A morphometric assessment of hybridization within single flowered taxa. *National Herbarium Nederland, 56*, 36–48.

Duke, N. C., & Jackes, B. R. (1987). A systematic revision of the mangrove genus *Sonneratia* (Sonneratiaceae) in Australasia. *Blumea, 32*, 277–302.

Hogart, P. T. (2007). *The biology of mangrove and sea grasses* (2nd ed.). Oxford: Oxford University Press.

Jamilah, M. S., Razali, M. S., & Putri Nurnaim, J. (2013). Discovery of the critically endangered *Bruguiera hainesii* (Berus Mata Buaya) in Pulau Layat, Setiu Wetlands, Terengganu. *Conservation Malaysia, 18*, 1–3.

Mohd Lokman, H., & Sulong, I. (2012). *Mangroves of Terengganu* (2nd ed.). Kuala Terengganu: Penerbit Universiti Malaysia Terengganu.

Muchhala, N. (2003). Exploring boundary between pollination syndromes: Bats and hummingbirds as pollinators of *Burmeistera cyclostigmata* and *B. tenuiflora* (Campanulaceae). *Oecologia, 134*, 373–380.

Nagarajan, B., Pandiarajan, C., Krishnamoorthy, M., & Sophia, P. (2008). Reproductive fitness and success in mangroves: Implication on conservation. In M. Sengupta & R. Dalwani (Eds.). Taal the 12th World Lake Conference, Coimbatore, India, November 2007. International Environment Lake Committee, Shiga, Japan.

Nagarajan, B., Krishnamoorthy, M., Padmini, S., & Daniel, A. (2010). Nectar robbery in bird pollinated mangrove. *Current Science, 98*, 603–604.

Nakisah, M. A., & Fauziah, A. H. (2003). *Setiu Wetlands, tranquility amidst plenty*. Kuala Terengganu: Kolej Universiti Sains dan Teknologi Malaysia.

Noske, R. A. (1993). *Bruguiera hainesii*: Another bird-pollinated mangrove? *Biotropica, 25*, 481–483.

Ono, J., Yong, J. W., Takayama, K., Saleh, M. N. B., Wee, A. K., Asakawa, T., Yllano, O. B., G-Salmo, S., III, Suleiman, M., Tung, N. X., Soe, K. K., Meenakshisundaram, S. H., Watano, Y., Webb, E. L., & Kajita, T. (2016). *Bruguiera hainesii*, a critically endangered mangrove species, is a hybrid between *B. cylindrica* and *B. gymnorhiza* (Rhizophoraceae). *Conservation Genetics, 17*, 1137–1144.

Pandit, S., & Choudhury, B. C. (2001). Factors affecting pollinator visitation and reproductive success in *Sonneratia caseolaris* and *Aegiceras corniculatum* in a mangrove forest in India. *Journal of Tropical Ecology, 17*, 431–447.

Sheue, C. R., Yong, J. W. H., & Yang, Y. P. (2005). The *Bruguiera* (Rhizophoraceae) species in the mangroves of Singapore, especially on the new record and the rediscovery. *Taiwania, 50*, 251–260.

Tomlinson, P. B. (1986). *The botany of mangrove*. Cambridge: Cambridge University Press.

Tomlinson, P. B., Primack, R. B., & Bunt, J. S. (1979). Preliminary observations on floral biology in mangrove Rhizophoraceae. *Biotropica, 11*, 256–277.

Ushimaru, A., & Nakata, K. (2002). The evolution of flower allometry in selfing species. *Evolutionary Ecology Research, 4*, 1217–1227.

Wan Juliana, W. A., Razali, M. S., Ehwan, N., Faizi, Z., Azmil, R., Mohd-Arrabe, A. B., Abdul Ghani, H., Norhayati, A., & Latiff, A. (2014). Langkawi Geopark is home to rare mangrove flora. *Malaysian Naturalist, 67*, 50–53.

Wee, A. K. S., Low, S. Y., & Webb, E. L. (2014). Pollen limitation affects reproductive outcome in the bird-pollinated mangrove *Bruguiera gymnorrhiza* (Lam.) in a highly urbanized environment. *Aquatic Botany, 120*, 240–243.

Morphological, Nest Architecture and Colony Characteristics of Stingless Bees (Hymenoptera; Apidae; Meliponini) from Tasik Kenyir, Terengganu

Wahizatul Afzan Azmi, Roziah Ghazi, and Iylia Syazwanie Nasharuddin

Abstract The objectives of this study were to investigate the morphological characteristics, nest architecture and colony characteristics of two common stingless bees, *Heterotrigona itama* and *Geniotrigona thoracica* from Taman Tropika Kenyir, Tasik Kenyir, Hulu Terengganu. Variations were found in terms of the colours, length and width of the body parts. All the measurements of the morphological characters in *G. thoracica* were significantly greater than *H. itama* ($p < 0.05$). The average body size of *H. itama* was 4.7 ± 1.55 mm whilst *G. thoracica* was 7.44 ± 2.05 mm ($N = 50$). *Heterotrigona itama* are black in colour with grey wings while on the contrary, *G. thoracica* are brown in colour with dark brown wings and white tips at the apex of the wings. Both species have corbiculated hind legs fringed with sensilla, ten flagellum antennae segments, six abdomen segments and three ocelli. Three nest cavities of both species were also measured and observed to investigate the nest architecture between the two species. The nest measurements showed that *G. thoracica* nest was significantly larger compared to *H. itama* ($p < 0.05$). Both of the nests also have different unique characteristics from each other in terms of the colour, shape and smell. Both of the colonies shared similar behaviours especially on their foraging, dumping and nesting activities. Nevertheless, the defense mechanism between the two species was totally different from each other where *H. itama* colonies were observed to be more aggressive compared to *G. thoracica* colonies.

Keywords Stingless bee · Morphological characteristics · Nest architecture · Colony behaviour · Tasik Kenyir

W. A. Azmi (✉) · I. S. Nasharuddin
School of Marine and Environmental Sciences, Universiti Malaysia Terengganu, Kuala Nerus, Malaysia
e-mail: wahizatul@umt.edu.my

R. Ghazi
Institut Agropolis Unisza, Universiti Sultan Zainal Abidin Kampus Tembila, Kuala Terengganu, Malaysia

M. T. Abdullah et al. (eds.), *Greater Kenyir Landscapes*,
https://doi.org/10.1007/978-3-319-92264-5_11

Introduction

Stingless bees (Hymenoptera: Apidae: Meliponini) are closely related to the common honey bees and are widely distributed in tropical and subtropical regions (Salim et al. 2012). There are about 700 species have been recorded worldwide and stingless bees are the smallest of the honey producing bees (Heard 1999). They are highly social insects that living in permanent colonies, nesting in old walls, logs, crevices and such other concealed places. The sting of stingless bee is greatly reduced without an effective tip. Hence, the defense behaviour is by chasing the intruders and enemies by biting, becoming entangled in the intruder's hairs and getting into the nose, ears and the eyes (Danaraddi 2007).

Malaysia is home to diverse species of stingless bees which consists of ~33 species (Mohd Norowi et al. 2008). Stingless bees are native to Malaysia, thus they are resistant to the diseases and parasites of European honey bees, *Apis mellifera* (Delfinado et al. 1989). Besides, stingless bees are considered domesticated species because they are being reared for their products such as honey, propolis, bee bread (which is the pollen and other ingredients stored in the brood cells) and bee wax.

Stingless bees are known to be important pollinators in tropical rainforest (Eltz et al. 2003) and also good candidates for providing pollination services in agricultural ecosystem such as starfruits, mango, durian, watermelon, guava and coconut (Slaa et al. 2006). The value of insect pollination service of these crops was estimated more than USD 19 million (Heard 1999). At least four species of stingless bees, *H. itama, G. thoracica, Tetragonilla atripes* and *Tetrigona peninsularis* can be potentially domesticated for pollination service in Malaysia agricultural ecosystem. Their small size allows them to have access to many kinds of flowers whose openings are too narrow to permit penetration by other bees and they are common visitors to flowering plants in the tropics (Heard 1999). Other than pollination, stingless bees also play important roles in health care through the production of medicinal hive products such as propolis, beebread, and also honey which is believed to have higher nutrients content compared to *Apis mellifera* (Kwapong et al. 2010).

To date, there is limited collection of voucher specimens for taxonomic purposes on the stingless bees from Tasik Kenyir, Terengganu. Descriptions of most of the known species of stingless bee from Peninsular Malaysia and Borneo, including collections from earlier work carried out in mid 1950s can be found in Schwarz (1939) and Eltz et al. (2003). Therefore, it is important to stress that the taxonomic work on Tasik Kenyir stingless bees is in exhaustive and particular attention must be given, now that the taxa are considered for its economic and health properties. Even though considerable studies have been conducted on the use of stingless bees for commercial pollination purposes, however to date, very little attention has been made to study the colony and morphological characteristics, and the nest architecture of Malaysian stingless bees. Outcome of this study is hoped would give an insight on their biology, ecology and behaviour which are absolutely essential for the sustainability of meliponiniculture industry in Malaysia.

Morphological Characteristics

The most common stingless bee species in Tasik Kenyir, *H. itama* and *G. thoracica* were collected from Taman Tropika Kenyir (TTK), Pulau Tekak Besar, Tasik Kenyir, Terengganu (5°02′ N, 102°44′E). Samplings of the stingless bees were carried out from August until October 2013 (N = 50). Twenty five individuals of each species (*H. itama* and *G.thoracica*) were selected for the morphological characteristics study. The external morphological characters between the species such as the length and width of the head, thorax, abdomen, fore wing, hind wing, and leg parts were measured using the digital vernier callipers (mm). Images and measurement for each morphological character were captured using Dino-Eye microscope camera. Morphological characteristics such as the colour of the wing and body were also observed.

In this study, the leg part of the stingless bees is divided into coxa, trochanter, femur, tibia, and tarsus. The front leg of both species *G. thoracica* and *H. itama* have broader tibia which is subclavate compared to the tibia on the hind leg (Fig. 1). The front and hind legs of *G. thoracica* are almost similar with *H. itama* except that the legs of *G. thoracica* (7.04 ± 0.09 mm) are longer than *H. itama* (4.25 ± 0.05 mm). Both of the species leg parts are covered with sensillum, a sense organ in insects, typically consisting of a receptor organ in the integument connected to sensory neurons. Stingless bees are depending on the sensilla for choosing profitable food source, resins, water sources, and also for nestmate recognition (de Brito Sanchez 2011). Another unique feature on the hind leg part is the pollen basket (corbiculum). The pollen basket is formed by the outer and inner rows of long, curved hairs. The pollen basket is used to collect and also to carry pollens. Pollen grains, which are collected between the dense hairs on the body of the worker, are removed and compressed into pellet form; the pellet is placed into the pollen basket which is located

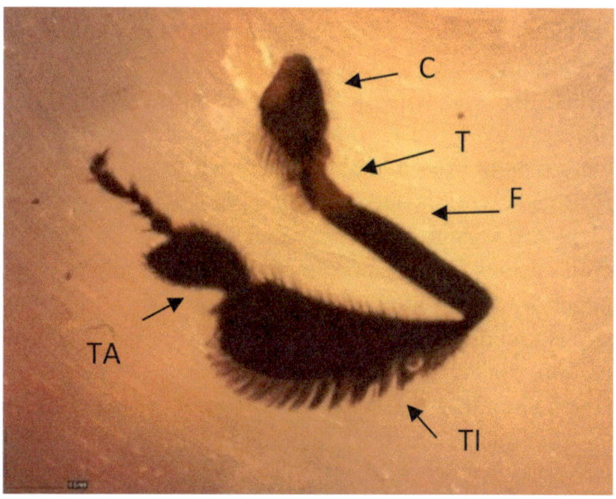

Fig. 1 The front leg of *G. thoracica* which are divided into coxa, trochanter, femur, tibia and tarsus at 71.5× magnifications. *C* Coxa, *T* Trochanter, *F* Femur, *TI* Tibia, *TA* Tarsus

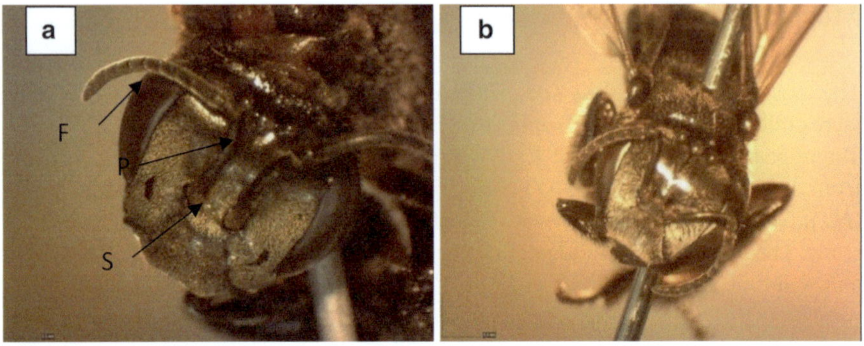

Fig. 2 The antennae part of *G. thoracica* (**a**) which are divided into scape, pedicel, and flagellum and *H. itama* (**b**) geniculate antenna with ten flagellum segments at 71.5× magnifications. *S* Scape, *P* Pedicel, *F* Flagellum

on the outer face of each hind leg and once the baskets are fully loaded, the worker returns to the hive to store the pollen for later use (Arita et al. 1989).

The antennae are paired segmented appendages located on the head usually between or below the compound eyes. The antennae arise from the antennae sclerite in front of the anterior margin of the compound eye at the frontal sutures. The three basic segments of the stingless bee antenna are the scape (base), the pedicel (stem), and finally the flagellum (Fig. 2a). The antennae of *G. thoracica* and *H. itama* shared similar shape and number of flagellum segments (ten segments). The antennae of both species are geniculate in which they are hinged or bent like an elbow (Fig. 2b). Apart from that, similarly to the leg part, gustatory sensilla are also abundant on the antennae which act as a sensory organ (de Brito Sanchez 2011).

Geniotrigona thoracica and *H. itama* have two pairs of wings which are the fore wings and hind wings that are attached on the mesothorax and metathorax respectively (Fig. 3a, b). The forewings are relatively larger and broader compared to the hind wing. The fore wing is leathery and it is used to protect the hind wing whilst the hind wing is thinner and more papery. The wings of *G. thoracica* are different from the *H. itama* in terms of colour and size. *Geniotrigona thoracica* wing is light brown in colour with white tips in the end of it while on the contrary, *H. itama*'swing is dark grey in colour. The wing size of *G. thoracica* is also larger (7.34 ± 0.25 mm) than *H. itama* (4.24 ± 1.03 mm). However, the hind wings of both species have a unique characteristic for wings adaptation and modification which is called hamuli. Hamuli are the tiny hooks that hold the front and the hind wings together and there are eight tiny hooks altogether attached on the hind wings of both species (Fig. 3c). Besides that, the wings also have tiny sensory hairs attached on the surface.

Abdomen is the third functional region or tagma of the bee's body; the abdomen is located just behind the thorax (Fig. 4a, b). Thorax is the second (middle) tagma of

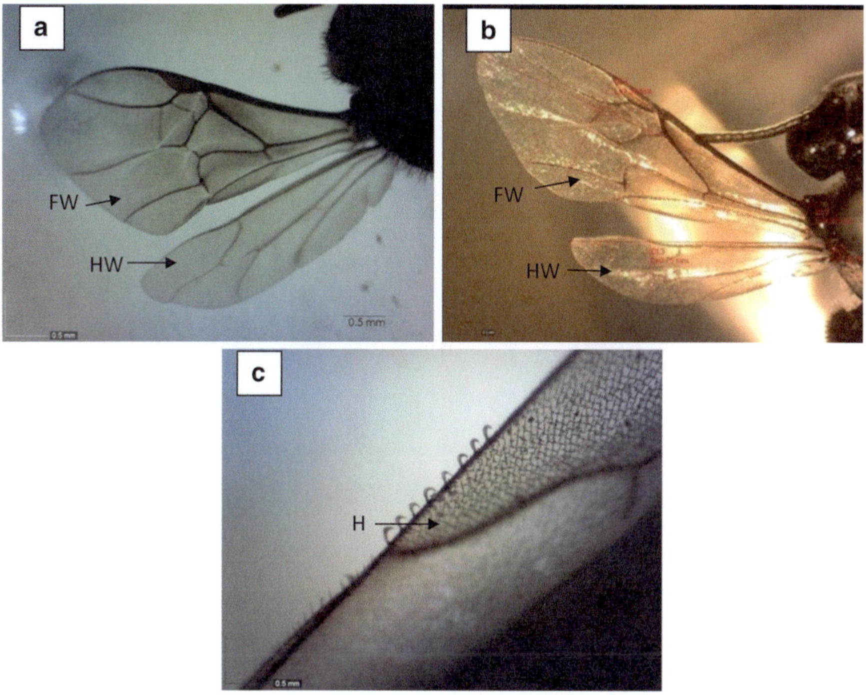

Fig. 3 The forewing and hind wing of *H. itama* (**a**) and *G. thoracica* (**b**) at 71.5× magnifications. The hamuli (**c**) on the hind wing which attach the forewing and the hind wing together at 82.9× magnifications. *FW* Forewing, *HW* Hind wing, *H* Hamuli

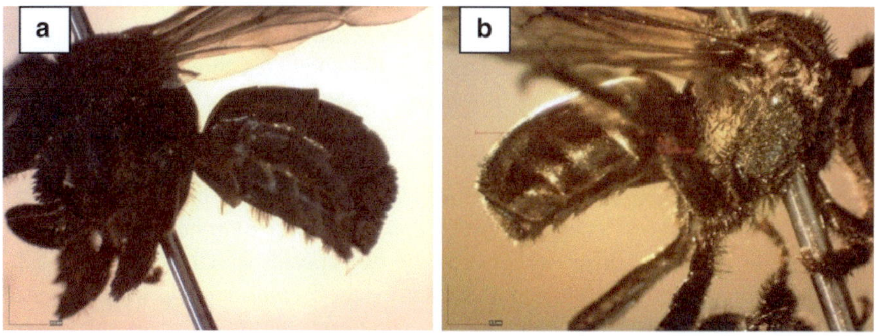

Fig. 4 The abdomen of *G. thoracica* (**a**) and *H. itama* (**b**) at 71.5× magnifications

an insect's body. This region is almost exclusively adapted for locomotion where it contains three pairs of walking legs and two pairs of wings. Structurally, the thorax is composed of three body segments which are the prothorax, mesothorax, and metathorax.

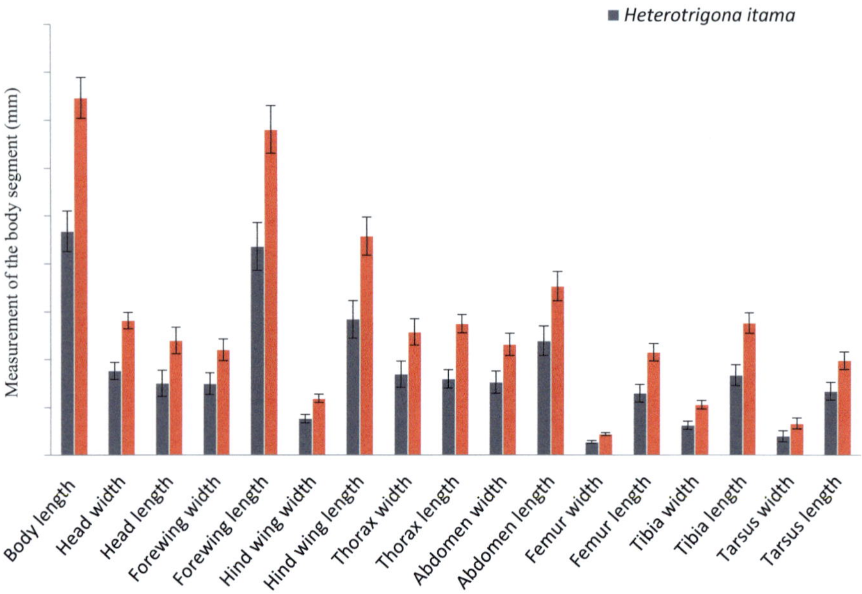

Fig. 5 Mean and standard deviation of *H. itama* and *G. thoracica* body segments' measurements (N = 50)

The junction between the thorax and abdomen is petiolate giving the appearance of a "wasp-waist" (Meyer 2005). Each segment of the abdomen consists of a dorsal sclerite, the tergum and a ventral sclerite, the sternum, joined to one another laterally by a pleural membrane. The thorax and abdomen of *G. thoracica* is brownish in colour compared to *H. itama* which has black body. However, the thorax and abdomen surface of both species are covered with sensilla. Uniquely, the sensilla are longer, more plumose and rough on the lower abdomen area compared to the upper part of the abdomen where the sensilla are much shorter and softer.

Figure 5 shows the measurements for 25 individuals of *H. itama* and *G. thoracica* that were collected from Taman Tropika Kenyir. The length and width of the body, legs, thorax, abdomen, head and wings of *G. thoracica* are relatively larger than *H. itama.* The t-test analysis revealed that all of these morphological measurements are significantly different between the two species ($p < 0.05$). This is supported by Smith (2012), where he reported that *G. thoracica* is larger in size compared to *H. itama* and probably the largest species in Asia.

Nest Architecture and Colony Characteristics

Three nests' cavities of each species were opened and observed to compare the nest architecture between the two species. The length and width of the nests were taken using the electronic vernier calipers. Other than that, shape, colour, scent of nests and location of the brood cells, entrance tube, honey and pollen pots were also measured and described. The behaviours of the stingless bees were observed without influencing the bees for about 72 h at the same time from 0800 to 1800 h. The behaviours observed include of (i) foraging activity (the act of searching for food and provisions); (ii) defense mechanism (observed by opening and disturbing their nest by inserting small stick in the entrance); (iii) territory and guarding; and (iv) nesting and dumping activities.

Generally, the length and width of entrance tube, honey pots, pollen pots and brood cells of of *G. thoracica* nest is larger than *H. itama* (Table 1). The nests of stingless bees are usually consist of an external tube, internal tunnel, resin dump, waste dumps, and food pots for storing pollen and honey. Brood cells, honey and pollen pots are arranged in separate clusters. Brood cells and food pots are made of cerumen which is a mixture of wax and resin. The cells of combs will be in contact with one another, sometimes connected by small pillars or connectives of soft cerumen (Danaraddi 2007).

The characteristics of the nests between the two species are also quite different from each other (Table 2). The colour and surface texture of the entrance tube of *G. thoracica* is dark brown in colour and has a very coarse and sticky surface compared

Table 1 The mean and standard deviation of the nest characteristics from both species *H. itama* and *G. thoracica* recorded from Tasik Kenyir

No	Nest parameters	Mean ± standard error mean (mm)	
		Heterotrigona itama	*Geniotrigona thoracica*
1	Entrance tube length*	114.817 ± 50.378	41.591 ± 2.364
2	Entrance tube width*	15.807 ± 3.566	28.060 ± 1.797
3	Honey/pollen pots length	20.364 ± 2.511	25.920 ± 4.576
4	Honey/pollen pots width*	14.023 ± 0.740	21.517 ± 2.541
5	Brood cells length	4.663 ± 0.133	5.753 ± 0.901
6	Brood cells width	3.303 ± 0.133	4.010 ± 0.424

* = $p < 0.05$

Table 2 The characteristics of the nest entrance tube, honey/pollen pots, and also the brood cells of *H. itama* and *G. thoracica*

No	Nest parameters	*Heterotrigona itama*	*Geniotrigona thoracica*
1	Entrance tube shape	Cylindrical/tube-like	Cylindrical/tube-like
2	Entrance tube colour	Light brown	Dark brown
3	Entrance tube surface	Very soft and brittle	Very coarse and sticky
4	Honey/pollen pots shape	Oval/capsule-like	Oval/capsule-like
5	Honey/pollen pots colour	Dark brown	Dark brown
6	Brood cells shape	Round/bead- like shape	Round/bead-like shape
7	Brood cells colour	Yellow	Yellow
8	Other characteristics	Have no sweet scent	Have strong sweet scent

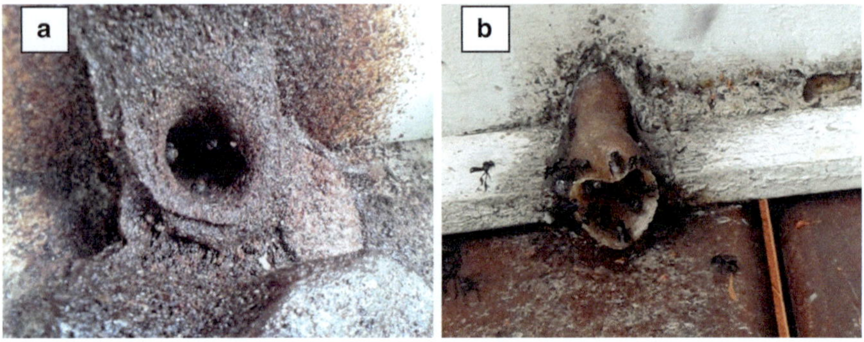

Fig. 6 The nest entrance tube of *G. thoracica* (**a**) which is larger and coarser than *H. itama* (**b**)

to *H. itama* which is more smooth, brittle, and lighter in colour (Fig. 6a, b). The nest of *G. thoracica* also has a very strong sweet smell (vanilla-like smell) which is obtained from the propolis and resins. Nevertheless, although the nest of *H. itama* also has propolis attached to it, the nest is somehow do not smell sweet at all. Despite the differences, they also shared some similarities especially in terms of the shape and colour of the brood cells and honey or pollen pots as well as the entrance tube shape (Fig. 7a, b).

The colony characteristics and behaviour of each species such as defense mechanisms, nesting and dumping activities, and their foraging behaviour were observed. In this study, the defense mechanism between the two species is very different from each other. The guard bees of *G. thoracica* were very timid and passive. They were withdrawing when a small stick was inserted in the entrance. They also did not crawl in hairs or clothes and they did not bite at all even when the nest was opened. They stayed calm for the whole day from early morning to late evening. On the contrary, *H. itama* were very active and the guard bees were usually guarding in the opening. They deposited a lot of sticky resin droplets around the inner and outer sides of the opening when a small stick was inserted. Numerous landed on hairs and skin and they performed disturbing biting when the nest was opened. As for the nesting and dumping activities, the workers from both species carried waste pellets (small leaves/foreign materials) outside of the nest. As a result, a pile of waste dump was found in front of the nest entrance. Some workers of *G. thoracica* were found cooling their nest by flapping their wings vigorously which was to improve their nest ventilation.

Nests 'breathe' in the sense that tidal gas exchange occurs frequently, although the entrance tube is the only connection to outside (Roubik 2006). Circulation is accomplished by workers that fan their wings while facing outward toward the entrance. This behavior is called 'nest-cooling' where the stingless bees cool down their nest by wing fanning. There was no 'nest-cooling' activity found within

Fig. 7 The nest brood cells and honey pots of *G. thoracica* (**a**) and also the brood cells (**b**) and honey pots (**c**) of *H. itama*. *HP* Honey pots, *BC* Brood cell

H. itama colonies probably because their nest are still small. Both species were actively nesting in the morning and decrease gradually in time in which they were more relaxed in the evening. Both species were very active during morning and early afternoon (0800–1200) in which they were actively foraging. They were seen carrying pollens and propolis inside and outside of the nest. They also could be seen perching on flowers during that time range (Fig. 8). However, both species were a lot calmer in the late afternoon and evening (1300–1800) in which the number of workers carrying pollens and propolis were declining. In the late evening, very few bees were spotted foraging and most of them were resting inside their nest.

The stingless bees communicate locations of forage sources by secreting chemical scents (pheromones) and through the use of the sun's direction (Kwapong et al. 2010). The bees were not foraging during the late evening where the sun is down. Workers begin foraging activities as early as dawn and end by dusk depending upon weather conditions and availability of flower.

Fig. 8 *G. thoracica* (**a**) and *H.itama* (**b–d**) were spotted perching on flowers carrying along pollens on their legs in the morning and early afternoon

Conclusion

Heterotrigona itama and *G. thoracica* can be distinguished by the size and also the colour of their body. In general, *H. itama* are black in colour with grey wings and also known as the black jet species whilst *G. thoracica* has brown body and dark brown wings with white tips in the end. *Heterotrigona itama* is relatively larger in size compared to *G. thoracica*. Both species has corbiculated hind legs fringed with sensilla. They also have ten flagellum segments of antennae, six abdomen segments and three ocelli. The nests of *G. thoracica* are larger in size compared to *H. itama*. The nests of both species have different unique characteristics from each other in terms of colour, shape, nest surface, and also smell. Nevertheless, they also share two similarities in terms of the brood cells shape (bead-like shape) and colour (yellow) and also the honey and pollen pots which are capsule-like and dark brown in colour. As for the colony characteristics, both of the colonies were similar in terms on their foraging behaviour, dumping and nesting activities. However, it seemed that the defense mechanisms between the two species were very different from each other in which *H. itama* colonies were observed to be more aggressive when compared to *G. thoracica* colonies. For the best utilization of the stingless bees, research on their diversity, biology, ecology and behaviour is absolutely essential. Therefore, through this study, we hope to establish a baseline of the indigenous stingless bees in promoting Tasik Kenyir as a potential area of sustainable apiculture industry particularly in Terengganu state.

Acknowledgements We would like to acknowledge School of Marine and Environmental Sciences, Universiti Malaysia Terengganu (UMT) for the financial support of this study, Malaysian Agricultural Research and Development Institute (MARDI) for giving the permission to conduct this study in Taman Tropika Kenyir, Tasik Kenyir, Terengganu and to the laboratory assistants of Histology Laboratory and General Biology Laboratory (MBU) for the technical support and field works.

References

Arita, H. L., Sakai, W. S., & Moniz, J. J. (1989). Pollen types collected by honey bees at three localities on the Island Hawaii. Part I: Flora. *Source Identification, 29*, 125–132.

Danaraddi, C. S. (2007). Studies on stingless bee, *Trigona iridipennis* Smith with special reference to foraging behaviour and melissopalynology at Dharwad, Karnataka. (M.Sc. dissertation). University of Agricultural Sciences, Dharwad, India.

de Brito Sanchez, M. G. (2011). Taste perception in honey bees. *Chemical Senses, 36*(8), 675–692.

Delfinado, B. M., Baker, E. W., & Phoon, A. C. G. (1989). Mites (Acari) associated with bees (Apidae) in Asia, with description of a new species. *Annual Bee Journal, 129*(9), 609–613.

Eltz, T., Brühl, C. A., Imiyabir, Z., & Linsenmair, K. E. (2003). Nesting and nest trees of stingless bees (Apidae: Meliponini) in lowland dipterocarp forests in Sabah, Malaysia, with implications for forest management. *Forest Ecology and Management, 172*, 301–313.

Heard, T. A. (1999). The role of stingless bees in crop pollination. *Annual Review of Entomology, 44*, 183–206.

Kwapong, P., Aidoo, K., Combey, R., & Karikari, A. (2010). *Stingless bees: Importance, management, and utilisation. A training manual for stingless beekeeping*. Ghana: Unimax Macmillan Ltd.

Meyer, J. R. (2005). External anatomy: The abdomen, general entomology. Retrieve from https://projects.ncsu.edu/cals/course/ent425/library/tutorials/external_anatomy/abdomen.html.

Mohd Norowi, H., Sajap, A. S., Rosliza, J., Mohd Fahimie, J., & Suri, R. (2008). *Conservation and sustainable utilization of stingless bees for pollination services in agricultural ecosystems in Malaysia* (p. 5). Kuala Lumpur: Department of Agriculture.

Roubik, D. W. (2006). Stingless bee nesting biology. *Apidologie, 37*(2), 124–143.

Salim, H. M. W., Ahmad, D. D., Harrison, R. D., Fletcher, C., Kassim, A. R., & Potts, M. D. (2012). Stingless bee (Hymenoptera: Apidae: Meliponini) diversity in dipterocarp forest reserves in peninsular Malaysia. *The Raffles Bulletin of Zoology, 60*(1), 213–219.

Schwarz, H. F. (1939). The Indo-Malayan species of *Trigona*. *Bulletin of the American Museum of Natural History, 76*, 83–141.

Slaa, E. J., Chaves, L. A. S., Malagodi-Braga, K. S., & Hofstede, F. E. (2006). Stingless bees in applied pollination: Practice and perspectives. *Apidologie, 37*(2), 293–315.

Smith, D. R. (2012). *Key to workers of Indo-Malayan stingless bees* [Powerpoint slides]. Retrieve from https://ses.library.edu.usa/bitstream/2123/111123/4/Poster122.pdf.

A Checklist of the Moths of Tasik Kenyir

Ju Lian Chong, Muhammad Hafiz Sulaiman, and Nuraishah Abdul Manaf

Abstract In Peninsular Malaysia, there were not much research conducted on the diversity and composition of moths. Thus, a study was conducted to investigate the abundance and diversity of the moths species at Kenyir Lake, situated in the Taman Negara Terengganu. Sampling was conducted continuously for three night starting from 10 March 2011 to 12 March 2011 with a light trap that was set up before dusk until midnight. A total of 179 individuals from 34 species, and 33 genus were identified from 13 families. Majority of the moth were from the family Lymantriidae (76%), followed by Arctiidae (4%), Geometridae (4%), Limacodidae (3%), Sphingidae (3%), Saturniidae (2%), Bombycidae (1%), Lasiocampidae (1%) and Nolidae (1%). Three families were recorded as the least abundant family and were represented by singletons during the study namely Drepanidae (0.6%), Notodontidae (0.6%) and Uraniidae (0.6%). In addition, results shown that the genus Nygmia (72.5%) was the dominant genus found at Kenyir. The Shannon and Fisher Alpha Diversity Index calculated were $H' = 1.53$ and α-diversity 12.44 respectively, with species richness (D) and species evenness (J) calculated were $D = 2.54$ and $J = 0.43$. For this study, the preliminary checklist of the moth species at Kenyir was established and an initial result of the moth diversity and composition obtained. As such, this study will be a baseline to initiate more studies conducted on moth diversity in Terengganu.

J. L. Chong (✉)
School of Marine and Environmental Sciences, Universiti Malaysia Terengganu, Kuala Nerus, Terengganu, Malaysia

Institute of Tropical Biodiversity and Sustainable Development, Universiti Malaysia Terengganu, Kuala Nerus, Terengganu, Malaysia
e-mail: julian@umt.edu.my

M. H. Sulaiman
Institute of Tropical Biodiversity and Sustainable Development, Universiti Malaysia Terengganu, Kuala Nerus, Terengganu, Malaysia

N. A. Manaf
School of Marine and Environmental Sciences, Universiti Malaysia Terengganu, Kuala Nerus, Terengganu, Malaysia

M. T. Abdullah et al. (eds.), *Greater Kenyir Landscapes*,
https://doi.org/10.1007/978-3-319-92264-5_12

Keywords Moth · Diversity · Composition · Kenyir · Taman Negara Terengganu · Malaysia

Introduction

Insects are important organisms because of their diversity, ecological role, and influence on agriculture, human health, and natural resources (Foottit and Adler 2009). Among the insects, butterflies and moth tend to elicit a range of emotion from humans, from wonder at their incredible beauty, their ability to metamorphose or their unbelievable feats of physical achievement, to dismay at their voracious appetites for our crops and clothing (Balmer 2007).

The Lepidoptera namely, butterflies and moths, are a diverse and abundant insect group in many different ecosystems, playing roles as herbivores, pollinators, prey (Barlow and Woiwod 1989) and pest (Ithnin et al. 2008). Lepidoptera are important in biodiversity studies because they are the major group of plant-feeding insects.

Although moths comprise the vast majority of the order of Lepidoptera (Chong and Choo 2011; Ramel 2008; New 2004; Ricketts et al. 2001; Birkinshaw and Thomas 1999), with some 111,000 species of recorded moths, they generally are not considered as interesting as their brethren, the butterflies. However, moth studies are important as moth inventories can indicate the ecological health of plant communities, and long-term inventories can show changes in these communities over time (Foottit and Adler 2009). As moths exhibit host-plant specialization as a mean to ensure their existence and survival, thus they can act as biological indicators due to their sensitivity to the environment changes (Abang and Karim 2005; New 2004).

In Malaysia, only a few studies on moths were conducted with most in Borneo such as Abang and Karim (2002, 2005), Fiedler and Schulze (2004), and Chey (2000), while in Peninsular Malaysia, there is not much records on the diversity and composition of moths especially on the east coast region particularly in Terengganu. Taman Negara Terengganu is part of the three-state Taman Negara National Parks, and was established under the Taman Negara Enactment (Terengganu) No.6, 1939 *[En.6 of 1358]* (Pakhriazad et al. 2009). Additionally, Tasik Kenyir, the largest man-made lake in Southeast Asia is located in the national park.

The objectives of this study were to investigate the diversity and composition of moth in Tasik Kenyir, Taman Negara Terengganu and to establish the checklist of the moth species that found at the study site. This study was conducted as part of the Program Preliminary Research Baseline Data Preparation at Taman Negara Terengganu, Taman Negara Terengganu, with sampling conducted at a site 500 m from the trail at Tanjung Mentong, Taman Negara Terengganu for three consecutive night sampling, from 10 March 2011 to 12 March 2011. During this study, a single light trap with a 160 w mercury vapour bulb as the light source was set up from 1900 to 2400 h for each night. For sample collection, three individuals were assigned to collect the moths by placing the moths in a container containing ethyl acetate to euthanize the samples.

The samples were then transferred into a paper envelope to temporarily protect the soft body and moth wings from being damaged and also to prevent the mix-up of scales from different moths before brought back for preservation in the laboratory. The specimens were identified to species level by referring to Holloway (1986, 1997, 1999), Robinson et al. (1994), Barlow (1982), and 'The Moths of Borneo' website (2017 http://www.mothsofborneo.com/). Diversity indices including Shannon Index (Shannon and Weaver 1949), species richness and evenness and Fisher's alpha indices were used in this study.

Moth Diversity and Composition

A total of 179 individuals belonging to 34 species, 33 genera and 13 families were sampled. In the first sampling night, only 44 moth individuals were collected. However, the number of moths collected increased during the second and third sampling night. The second night of sampling collected 53 individuals, followed by the third sampling night, which had the greatest number of moth collected among the sampling nights with 82 individuals. Figure 1 shows the number of moths collected during the study while Table 1 shows the species recorded.

Figure 2 showed the moth composition percentage according to family. Majority of the moth collected in this study were from the family Lymantriidae (75.98%), followed by family Arctiidae (3.91%), Geometridae (3.91%), Limacodidae (3.35%), Sphingidae (3.35%), Noctuidae (2.23%) and Saturniidae (2.23%). While Bombycidae, Lasiocampidae and Nolidae were recorded as the least abundant family found at the study site (1.12% respectively). The Drepanidae, Notodontidae and

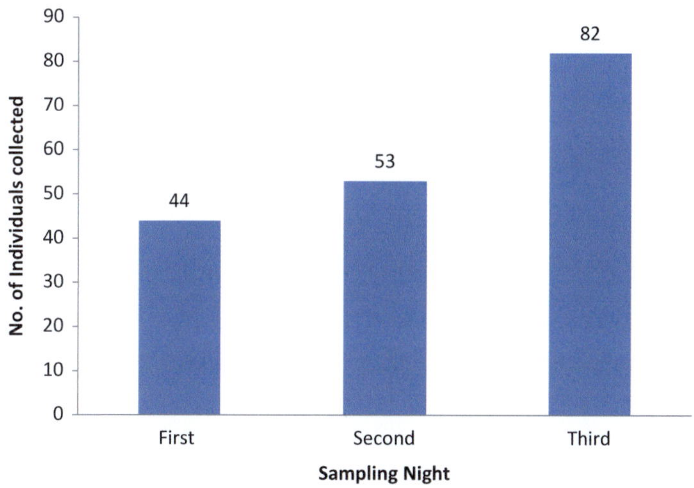

Fig. 1 Number of moths collected according to sampling nights

Table 1 Moth species collected

Family	Species	Number of individuals
Arctiidae	*Barsine rubricostata*	1
	Adites cornutata	3
	Lyclene biseriata	1
	Cyana perornata	2
Bombycidae	*Mustilia dierli*	2
Drepanidae	*Drapetodes matulata*	1
Geometridae	*Ornithospila cincta*	1
	Ornithospila submonstrans	3
	Hypochrosis pyrrhophaeata	3
Lasiocampidae	*Hallicarnia albipectus*	1
	Kunugia drakei	1
Limacodidae	*Narosa concinna*	4
	Scopelodes albipalpalis	1
	Griseothosea cruda	1
Lymantriidae	*Somena aurantiacoides*	1
	Nygmia fumosa	129
	Carriola ecnomoda	4
	Parapellucens aphrasta	1
	Arctornis pseudungula	1
Noctuidae	*Anuga rotunda*	1
	Thyas honesta	1
	Tamba delicata	1
	Eudocima homaena	1
Nolidae	*Siglophora bella*	2
Notodontidae	*Euhampsonia gigantea gilesi*	1
Saturniidae	*Loepa sikkima*	2
	Attacus atlas	2
Sphingidae	*Megacorma obliqua*	1
	Meganoton analis	1
	Ambulyx pryeri	1
	Daphnis hypothous	1
	Elibia dolichus	1
	Marumba juvencus	1
Uraniidae	*Lyssa zampa*	1
Total		179

Uraniidae were singletons with only a single individual collected respectively (0.56%) during this study.

As for the diversity indices, species richness and evenness were calculated and the result was shown in Table 2.

During this study, the number of moth sample collected gradually increased from first sampling night until the third sampling night, which could be influenced by the

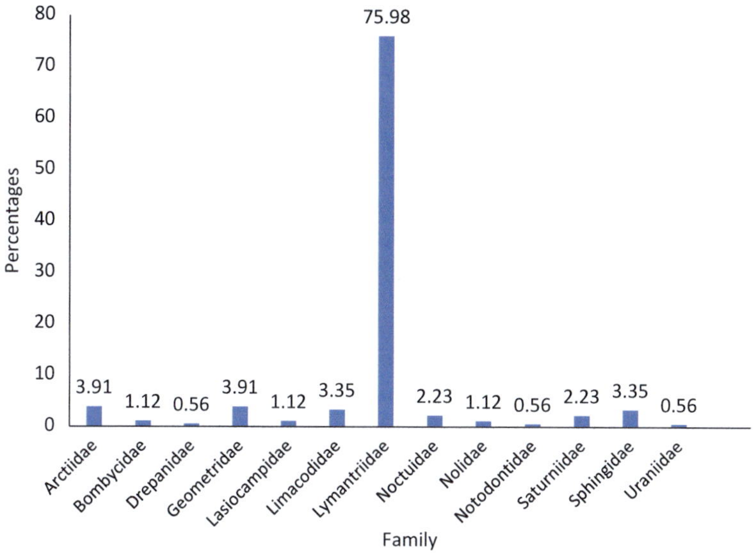

Fig. 2 Moth percentage according to family

Table 2 The diversity indices, species richness and evenness calculated for the moth community of Tasik Kenyir, Taman Negara Terengganu

Shannon index (*H′*)	1.53
Species richness(D)	2.54
Species evenness (*J*)	0.43
Fisher alpha diversity (α-diversity)	12.44

different weather conditions during the sampling. On the first and second sampling night, the weather was clear, without rain or wind during the sample collection. However, rain on the third sampling night could have caused an increase in the number of moth collected as moths usually increased in numbers when there were light rains or drizzle and little wind (Barlow 1982).

Results recorded from this study indicated that most of the species collected were characteristics of the lowland tropical forest. The Lymantriidae family (75.98%) was the most abundant family found and was recorded as having the highest proportion of lowland dipterocarp forest moths (Chong and Choo 2011; Chey 2000). The abundance of family Lymantriidae was mainly contributed by the species *Nygmia fumosa* which was the most abundant species found at the study site. *Nygmia fumosa* was found to be constantly collected during every sampling night and a large number was collected. Both females and males was collected during sampling. It was also found that most females captured in the container containing ethyl acetate will quickly lay eggs before they died.

The second highest family recorded was Arctiidae (3.91%) and Geometridae (3.91%). According to Hill and Abang (2010), Arctiidae is well-represented in South-east Asia with many species being widespread and common in the rainforest canopy. As for Geometridae, they are ecologically dominant in most forest habitats and are mostly found in forests. The most diverse family found in the study was the Sphingidae family as six genera were collected during sampling, although each genus was represented as singletons. The diversity index calculated in this study was considered as low when compared to the other studies (Choo and Chong 2011; Abang and Karim 2005; Chey 2000) with Shannon index at 1.53 and α-diversity of 12.44. This could be due to the restricted number of sampling efforts as sampling were only carried out for three nights and only using light trap. It is assumed that this forest might be diverse with the moth community and their host-plant vegetation to ensure their survival as with only three night sampling effort, 179 individuals of moth were collected, with the assumption that the number of sample collected could be increased and the number of diversity value could be greater should sampling be continued.

Conclusion

In this study, a total of 34 moth species were recorded, with 33 genuses from 13 families, representing the moth diversity and composition of Kenyir Lake. The Lymantriidae family was the most abundant, due to the abundance of one species, *N. fumosa*. Drepanidae, Notodontidae and Uraniidae were the least abundant family and were represented by singletons. The diversity index obtained in this study was considered low in lieu of other studies, which could be due to the restricted sampling period and also due to that only light trap were used in this study for sampling. As such, results of this study will form the baseline data of moth studies for conservation and monitoring changes in the biological community of Kenyir Lake and Taman Negara Terengganu.

Acknowledgements The authors would like to acknowledge Universiti Malaysia Terengganu (UMT) for financial support, and the School of Marine and Environmental Sciences and Kenyir Research Institute for the laboratory, research facilities and equipments provided.

References

Abang, F., & Karim, C. (2002). The larger moths (Lepidoptera: Heterocera) of the Crocker Range National Park, Sabah: A preliminary checklist. In G. Ismail & L. Ali (Eds.), *A scientific journey through Borneo: The Crocker Range National Park, Sabah: Natural ecosystem and species components* (pp. 147–156). London: ASEAN Academic Press.

Abang, F., & Karim, C. (2005). Diversity of macromoths (Lepidoptera: Heterracera) in the Poring Hill Dipterocarp Forest, Sabah, Borneo. *Journal Asia-Pacific Entomology, 8*(1), 69–79.

Balmer, E. (2007). *A conciseg to butterflies and moths*. Bath: Atlantic Publishing.
Barlow, H. S. (1982). *An introduction to the moths of South-East Asia*. Kuala Lumpur: The Malaysian Nature Society.
Barlow, H. S., & Woiwod, I. P. (1989). Moth diversity of a tropical forest in Peninsular Malaysia. *Journal of Tropical Ecology, 5*, 37–50.
Birkinshaw, N., & Thomas, C. D. (1999). Torch-light transect survey for moths. *Journal of Insect Conservation, 3*, 15–24.
Chey, V. K. (2000). Moth diversity in the tropical rain forest of Lanjak Entimau, Sarawak, Malaysia. *Malaysian Nature Journal, 54*, 305–318.
Chong, J. L., & Choo, S. Y. (2011). A preliminary survey of moth diversity at the Cemerung Recreational Forest, Terengganu, Malaysia, presented at Global Conference on Entomology (GCE 2011), Chiang Mai, Thailand, 2011. Chiang Mai: Global Conference.Com.
Fiedler, K., & Schulze, C. H. (2004). Forest modification affects diversity (but not dynamics) of speciose tropical pyraloid moth communities. *Biotropica, 36*(4), 615–627.
Foottit, R. G., & Adler, P. H. (2009). *Insect biodiversity: Science and society*. Oxford: Blackwell Publishing Limited.
Hill, D. S., & Abang, F. (2010). *The insects of Borneo (including South-east Asia)*. Kota Samarahan: Universiti Malaysia Sarawak.
Holloway, J. D. (1986). *The moths of Borneo part 1: Families cossidae, metarbelidae, ratardidae, dudgeoneidae, epipyropidae and limacodidae*. Kuala Lumpur: Southdene Sdn. Bhd.
Holloway, J. D. (1997). The moths of Borneo: Family Geometridae, subfamilies Sterrhinae and Larentiinae. *Malayan Nature Journal, 51*, 1–242.
Holloway, J. D. (1999). *Lymantriidae*. Retrieved from http://www.mothsofborneo.com/part5/introduction.
Ithnin, B., Mohd. Shamsudin, O., & Mohamed Mohd., S. (2008). *Serangga Perosak Buah-Buahan Tropika Malaysia*. Serdang: Institut Penyelidikan & Kemajuan Pertanian Malaysia (MARDI).
New, T. R. (2004). Moth (Insecta: Lepidoptera) and conservation: Background and perspective. *Journal of Insect Conservation, 8*, 79–94.
Pakhriazad, H. Z., Mohd. Hasmadi, I., & Aida, H. M. K. (2009). Historical and current legistration of Taman Negara National Park Peninsular Malaysia. *Journal of Politics and Law, 2*(1), 44–49.
Ramel, G. (2008). *The lepidoptera part 2 moths*. Retrieved from http://www.earthlife.net/insects/lepidop2.html.
Ricketts, T. H., Daily, G. C., & Ehrlich, P. R. (2001). Does butterfly diversity predict moth diversity? Testing a popular indicator taxon at local scales. *Biological Conservation, 103*, 361–370.
Robinson, G. S., Tuck, K. R., & Shaffer, M. (1994). *A field guide to the smaller moths of South-east Asia*. Kuala Lumpur: Malaysian Nature Society.
Shannon, C. E., & Weaver, W. (1949). *The mathematical theory of communication*. Urbana: The University of Illinois Press.
The Moths of Borneo. (2017). Retrieved from http://www.mothsofborneo.com.

Vertical Stratification of Fruit-Feeding Butterflies in Tasik Kenyir

Rosanira Mohamed, Fathihi Hakimi Rosmidi, Nur Amalina Adanan, Amirrudin Ahmad, and Mohd Tajuddin Abdullah

Abstract The diversity of fruit-feeding butterflies at different vertical stratification were studied in secondary forests in Tasik Kenyir areas using 20 baited traps at each site. This study aimed to determine the species diversity of fruit-feeding butterflies at different forest strata which included ground and canopy levels. A single rope technique was applied in 23-days sampling period and a slice of fermented pineapple was used as bait. A total of 307 individuals from 72 species and 8 subfamilies were documented. Species diversity was higher at ground level than canopy level with Shannon diversity index $H' = 3.51$ and $H' = 3.03$. Butterflies compositions at ground level have high diversity compared to canopy level. Besides that, the graph of rarefaction curve did not reach an asymptote at either level. The species richness might reach to equilibrium while the specimen of butterflies increases if the periods of sampling are extended.

Keywords Nymphalidae · Fruit-feeding butterflies · Diversity · Vertical stratification · Tasik Kenyir

R. Mohamed
School of Marine and Environmental Sciences, Universiti Malaysia Terengganu, Kuala Nerus, Terengganu, Malaysia

F. H. Rosmidi · N. A. Adanan
Institute of Tropical Biodiversity and Sustainable Development, Universiti Malaysia Terengganu, Kuala Nerus, Terengganu, Malaysia

A. Ahmad · M. T. Abdullah (✉)
Institute of Tropical Biodiversity and Sustainable Development, Universiti Malaysia Terengganu, Kuala Nerus, Terengganu, Malaysia

School of Marine and Environmental Sciences, Universiti Malaysia Terengganu, Kuala Nerus, Terengganu, Malaysia
e-mail: mohd.tajuddin@umt.edu.my

M. T. Abdullah et al. (eds.), *Greater Kenyir Landscapes*,
https://doi.org/10.1007/978-3-319-92264-5_13

Introduction

Butterflies together with moths belong to the insect order, Lepidoptera. This order has enormous diversity of species and is well distributed worldwide, but more abundant in warmer countries such as Malaysia and Indonesia (Fatimah 2006). Since time immemorial, butterflies have fascinated people. This is due to their attractive behaviour of flying that is striking to human vision because of the colourful pattern of the wings. Besides that, this insect has a high aesthetic value and one of the most collected insects and continuously available in global collections (Maryati and Nordin 1999).

Vertical stratification refers to the layering of habitat or specific strata of the forest. Canopy arthropods are the least studied taxa because of difficulty in accessibility, safety issues and they are expensive to research.

This study was conducted from July 26th, 2015 to April 4th, 2016 at four sites in Tasik Kenyir area, Hulu Terengganu, Terengganu (Fig. 1). A total of 20 baited traps were set up at each site where 10 baited traps were installed at ground level (1–1.5 m) while another 10 baited traps were established at canopy level (10–15 m). Single-rope technique used by Christharina and Abang (2014a) was applied to lift up the traps to canopy level. In this study, fermented pineapple was used as the bait. Data were analysed for species diversity using PAST (PAlaeontological Statistics) software, version 2.17 (Hammer et al. 2001) and diversity t-test was used to compare Shannon diversities in two samples of abundance data (Poole

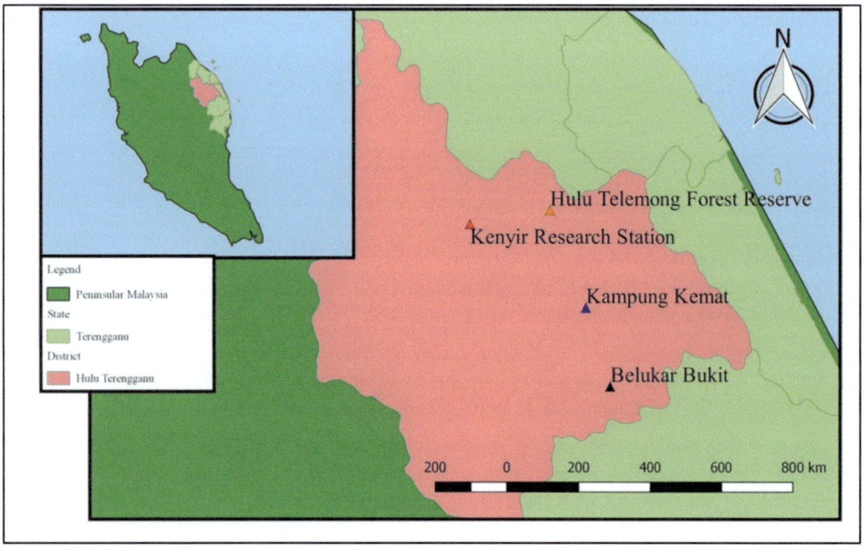

Fig. 1 Study areas in Tasik Kenyir, Hulu Terengganu, Terengganu

1974). Shannon diversity t-test was used to compare the differences of Shannon diversities between ground and canopy level.

Tasik Kenyir is the largest man-made lake in Southeast Asia and is located in Hulu Terengganu, covering over 209,199 ha. Tasik Kenyir was formed by the establishment of a dam. The area covered by Tasik Kenyir is approximately equal to the size of Singapore and is surrounded by the oldest rainforest in the world which includes Pahang and Kelantan as well as Terengganu (Shaharom 2015).

Butterflies provide many beneficial ecosystem services, for examples, as pollinator and bioindicator. Nevertheless, there are limited studies done on the vertical stratification and diversity of butterflies especially in Tasik Kenyir. Hence, this study aimed to determine the species diversity of fruit-feeding butterflies at different forest strata in Tasik Kenyir.

Species Composition of Fruit-Feeding Butterfly Community in Tasik Kenyir Forest

From the study, 72 species from 307 individuals of fruit-feeding butterflies had been recorded in Tasik Kenyir forest area, Hulu Terengganu, Terengganu (Appendix 1). Averages of 13 individuals were recorded per day throughout the sampling period. Sixty-four percent of the species were found only at ground level, 14% were found in the canopy only, and 22% were found in both strata, and total abundance at ground level was higher than canopy level (Fig. 2). At 1.0–1.5 m above the ground level eight subfamilies of butterfly were recorded with a total number of 258 individuals from 62 species. At 10–15 m above the ground level six subfamilies of butterfly were recorded with 26 species and 49 individuals. From the previous study that was carried out in eastern Ecuador, a total of 6690 individuals of butterflies were recorded comprising 44 species at ground and 25 species at canopy level (DeVries et al. 1997). In a Costa Rican rainforest, at ground level, 15 species were recorded with 24 species at canopy level and seven species occurring at both levels (Schulze et al. 2001). In Sarawak, 19 species were recorded at ground, only 5 species at canopy level and 24 species occurred at both levels (Christharina and Abang 2014b).

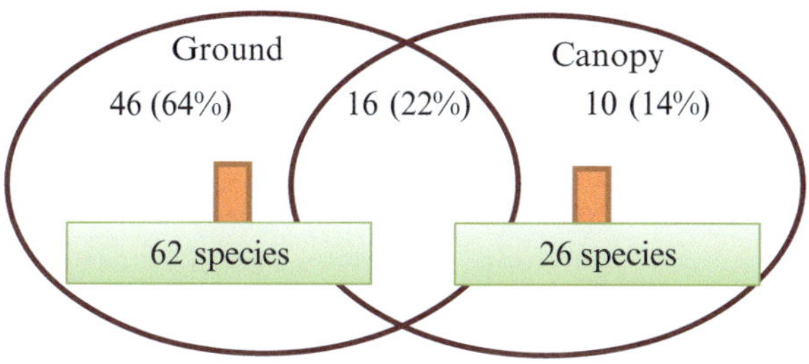

Fig. 2 Venn diagram of species overlap between ground and canopy level

Based on observation, little light intensity reached ground level, so the plants need to develop bigger leaves to achieve the required sunlight (Basset 2001). Meanwhile, it was evaluated that at canopy level 60–90% of life in the rainforest was found in the trees, making this the wealthiest natural surroundings for plant and creature life (Basset 2001) This current statement was supported by Corbet and Pendlebury (1992) who reported the butterflies highly depend on plants for food sources, oviposition and for sun basking.

Butterflies can be used as environmental bioindicators, serve as food for predators at different levels and act as pollinators (Stefanescu et al. 2004). Based on the data collected during sampling activities, only three species were recorded as least concern (LC) in IUCN (2015) Red List while other butterflies were stated as not-evaluated (NE). The species were *Euthalia merta merta*, *Tanaecia pelea pelea*, and *T. p. irenae* which are in subfamily Limenitidinae. *Tanaecia pelea* flies in the understorey, landing on leaves with wings spread and widely distributed in forests at low-to-moderate elevations (Kirton 2014). *Euthalia merta* inhabits dense primary rainforest. It is a relatively elusive butterfly that is less inclined to congregate on hilltops than other *Euthalia* species (Corbet and Pendlebury 1992). This species prefer shaded areas and they are fast flyers and very alert to movement. There were so many butterflies in this study stated as not evaluated in the IUCN Red List but are important in tropical rainforest (Momose et al. 1998).

Based on the rarefaction curve (Fig. 3), at both levels have increasing species richness due to increase in collected individuals. Rarefaction curve of both strata

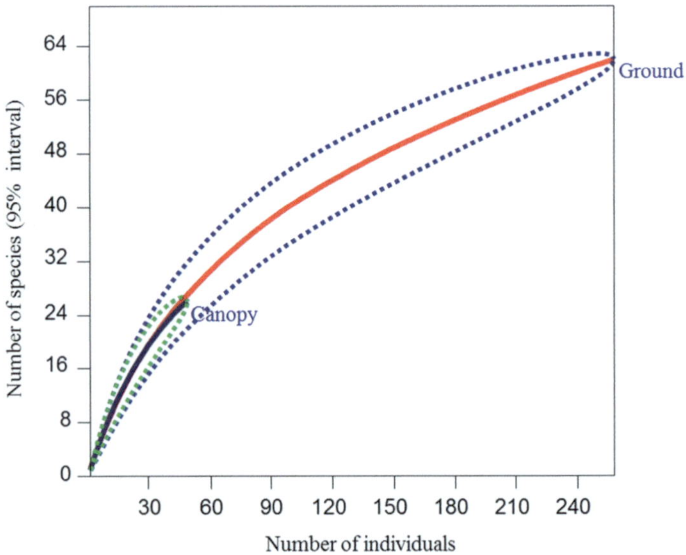

Fig. 3 Rarefaction curves comparing species diversity between ground and canopy level at Tasik Kenyir Forest, Terengganu

levels did not reach asymptote which indicates the insufficient data due to short sampling periods. The species richness might reach to equilibrium while the specimen of butterflies increases if the periods of sampling are extended (DeVries and Walla 2001). Rarefaction curve at ground and canopy level were not significantly different indicating that are certain butterfly species found at ground level can also be found at canopy level (Fig. 3). Sources such as nutrient, variation in microclimate conditions, presence of predators, and adult resources at the ground level influenced the distribution of butterflies in both strata (Basset 2001). Adult resources such as rotting fruit were mostly found scattered on the forest floor, which attract butterflies to the ground level (Christharina and Abang 2014a; Schulze et al. 2001; DeVries et al. 1997).

The abundance and diversity of the butterfly community increased with increasing food sources (Basset 2001). In this study, the result showed that ground level has availability of adult resources, and larval host plants, as well as offering avoidance from predators (Christharina and Abang 2014a; Stork and Grimbacher 2006; Tangah et al. 2004). Adult resources such as rotting fruit were mostly found scattered on the forest floor, which attracted butterflies to the ground level (Christharina and Abang 2014a; Schulze et al. 2001; DeVries et al. 1997). Besides that, species richness and abundance showed higher at ground level when habitat conditions were favourable and rich seasonal nectar and larval hosts were available (Fernandez-Hemandez 2007).

In this study, most common host-plant of butterflies at ground or understorey level were *Ixora* sp. (small tree), *Melastoma malabathricum* (shrub), *Clidemia odora* (shrub), *Asystasia intrusa* (shrub), *Mimosa pudica* (herb), *Ageratum* sp. (herb) and *Chromolaena odorata* (shrub) while at canopy level were *Dipterocarpus grandifolia*, *Shorea* sp., *Macaranga* sp., *Ficus* sp., *Saraca thaipingensis*, *Pometia pinnata*, *Monocarpia marginalis* and *Mesua ferrea* (Pesiu 2016). Moreover, on July and August 2015, the plants were fruiting and undergoing leaf-flushing, on January 2016 they were undergoing end of the fruiting and on April 2016 undergoing leaf-flushing and experiencing El-Nino. These may have effected species richness and abundance of fruit-feeding butterflies at ground and canopy level.

The Diversity Measures of Butterflies at Ground and Canopy Level

To study and measure the species diversity in each stratum, Shannon Index (H'), Richness Index (Margalef) and Evenness Index (E) were used (Ludwig and Reynolds 1988). At ground level, a higher Richness Index and Shannon Index were recorded as compared to canopy level (Table 1). This shows that species at ground level were more diverse than at canopy level. Low evenness usually indicates low diversity. Ground level was lower in the evenness, but high in the diversity (Shannon, and Richness). This indicates that a community of highly dominant species was found.

Table 1 Species diversity of fruit-feeding butterflies at two different heights at Tasik Kenyir

Compare diversity	Ground (1–1.5 m)	Canopy (10–15 m)
Taxa richness (S)	62	26
Individuals(N)	258	49
Shannon, *H′*	3.51	3.03
Evenness, *J′*	0.85	0.93
Margalef, R	10.99	6.42
Chao 1	105.9	36.11

That means, more species dominance is distributed at ground than at canopy level. Various biotic and abiotic environmental factors such as food, vegetation, and structure of habitat influence abundance of butterfly species (Schulze et al. 2001).

Conclusions

This study has successfully answered the studied objective which was to determine the species diversity of fruit-feeding butterflies at different forest strata. This study had successfully recorded additional species in Tasik Kenyir, Hulu Terengganu, Terengganu. The result from this study is important to conservation and management since controlled and ample maintaining activities done by the park rangers does not greatly reduced butterfly diversity at the areas. It can be further improved with an increase in sampling duration and the number of sampling areas covered to record more butterfly species in order to provide better information on the distribution of butterflies at ground and canopy level.

Acknowledgements We would like to thank Universiti Malaysia Terengganu (UMT), Kenyir Research Institute, and Department of Wildlife and National Park Terengganu for their support, and permission to conduct the field surveys in Tasik Kenyir. This study was supported by *Geran Galakan Penyelidikan* (GGP/68007/2014/127), Niche Research Grant Scheme (NRGS/2015/5313/2) and (NRGS/2015/5313/2), Trans Disciplinary Research Grant Scheme (TRGS/2015/59373) and *Dana Pembangunan Geopark* (KGDR/2015/53167/3) grants led by Prof Dr. Mohd Tajuddin Abdullah.

Appendices

Appendix 1: The Taxanomic of Fruit-Feeding Butterfly Recorded at Ground and Canopy in Tasik Kenyir, Hulu Terengganu, Terengganu

Species name	No. of individuals Ground (1–1.5 m)	Canopy (10–15 m)	IUCN Status (2015)
Apaturinae			
Eulacera osteria kumana	0 (0)	2 (4.08)	NE
Euripus nyctelius euploeoides	0 (0)	1 (2.04)	NE
Charaxinae			
Charaxes bernardus crepax	3(1.16)	2 (4.08)	NE
Polyura athamas athamas	0 (0)	1 (2.04)	NE
Polyura moori moori	1 (0.39)	0 (0)	NE
Prothoe franck uniformis	1 (0.39)	0 (0)	NE
Danainae			
Danaus melanippus hegesippus	1 (0.39)	0 (0)	NE
Limenitidinae			
Athyma pravara helma	0 (0)	1 (2.04)	NE
Euthalia ipona	2 (0.78)	3 (6.12)	NE
Euthalia monina monina	7 (2.71)	2 (4.08)	NE
Euthalia kanda marana	9 (3.49)	2 (4.08)	NE
Euthalia alpheda langkawica	1 (0.39)	0 (0)	NE
Euthalia mahadewa zichrina	1 (0.39)	0 (0)	NE
Euthalia adonia pinwilli	1 (0.39)	0 (0)	NE
Euthalia agnis pauper	0 (0)	1 (2.04)	NE
Euthalia eriphylae subsp.	1 (0.39)	1 (2.04)	NE
Euthalia merta merta	0 (0)	1 (2.04)	LC
Moduza procris milonia	1 (0.39)	0 (0)	NE
Tanaecia iapis puseda	5 (1.94)	1 (2.04)	NE
Tanaecia palguna consanguinea	2 (0.78)	0 (0)	NE
Tanaecia flora flora	1 (0.39)	0 (0)	NE
Tanaecia pelea pelea	1 (0.39)	0 (0)	LC
Tanaecia pelea irenae	1 (0.39)	0 (0)	LC
Tanaecia godartii asoka	2 (0.78)	0 (0)	NE
Morphinae			
Amathuxidia amythaon dilucida	4 (1.55)	0 (0)	NE
Amathusia ochraceofusca ochraceofusca	7 (2.71)	7 (14.29)	NE
Amathusia perakana perakana	1 (0.39)	0 (0)	NE
Amathusia masina malaya	1 (0.39)	0 (0)	NE

Species name	No. of individuals		IUCN Status (2015)
	Ground (1–1.5 m)	Canopy (10–15 m)	
Discophora timora perakensis	1 (0.39)	0 (0)	NE
Xanthotaenia busiris busiris	1 (0.39)	0 (0)	NE
Zeuxidia amethystus amethystus	2 (0.78)	0 (0)	NE
Zeuxidia doubledayi doubledayi	9 (3.49)	0 (0)	NE
Nymphalinae			
Athyma reta moorei	0 (0)	1 (2.04)	NE
Athyma nefte subrata	0 (0)	2 (4.08)	NE
Bassarona dunya dunya	3 (1.16)	0 (0)	NE
Bassarona teuta rayana	5 (1.94)	1 (2.04)	NE
Bassarona teuta goodrichi	3 (1.16)	0 (0)	NE
Bassarona recta monilis	1 (0.39)	0 (0)	NE
Dophla evelina compta	8 (3.10)	0 (0)	NE
Junonia atlites atlites	2 (0.78)	0 (0)	NE
Kallima limborgii amplirufa	0 (0)	1 (2.04)	NE
Rhinopalpa polynice eudoxia	5 (1.94)	1 (2.04)	NE
Pseudergolinae			
Dichorragia nesimachus deiokes	1 (0.39)	0 (0)	NE
Satyrinae			
Coelites epiminthia epiminthia	5 (1.94)	0 (0)	NE
Elymnias casiphone saueri	1 (0.39)	0 (0)	NE
Elymnias hypermnestra agina	3 (1.16)	3 (6.12)	NE
Elymnias hypermnestra tinctoria	9 (3.49)	2 (4.08)	NE
Elymnias nesaea lioneli	0 (0)	1 (2.04)	NE
Elymnias panthera panthera	10 (3.88)	6 (12.24)	NE
Elymnias penanga penanga	3 (1.16)	0 (0)	NE
Erites angularis angularis	3 (1.16)	0 (0)	NE
Lethe chandica namura	1 (0.39)	0 (0)	NE
Melanitis leda leda	6 (2.33)	2 (4.08)	NE
Melanitis phedima abdullae	7 (2.71)	0 (0)	NE
Mycalesis intermedia distanti	49 (18.99)	2 (4.08)	NE
Mycalesis oroatis ustulata	1 (0.39)	0 (0)	NE
Mycalesis orseis nautilus	5 (1.94)	0 (0)	NE
Mycalesis persoides persoides	2 (0.78)	0 (0)	NE
Mycalesis perseus Cepheus	1 (0.39)	0 (0)	NE
Mycalesis fusca fusca	4 (1.55)	0 (0)	NE
Mycalesis mineus macromalaya	10 (3.88)	0 (0)	NE
Mycalesis mineus micromalaya	1 (0.39)	0 (0)	NE
Mycalesis anapita anapita	7 (2.71)	1 (2.04)	NE
Mycalesis horsfieldi Hermana	3 (1.16)	0 (0)	NE
Mycalesis mnasicles perna	18 (6.98)	0 (0)	NE
Mycalesis janardana sagittigera	7 (2.71)	1 (2.04)	NE
Mycalesis visala pharmis	2 0.78)	0 (0)	NE

Species name	No. of individuals		IUCN Status (2015)
	Ground (1–1.5 m)	Canopy (10–15 m)	
Mycalesis patiana patiana	1 (0.39)	0 (0)	NE
Orsotriena medus cineria	1 (0.39)	0 (0)	NE
Neorina lowii neophyte	1 (0.39)	0 (0)	NE
Ypthima baldus newboldi	1 (0.39)	0 (0)	NE
Ypthima fasciata torone	1 (0.39)	0 (0)	NE
Total (72 species)	258	49	

NE Not Evaluated, *LC* Least Concern

Appendix 2: Species of Fruit-Feeding Butterflies Found at Tasik Kenyir, Hulu Terengganu, Terengganu

Moduza procris milonia

Euploea mulciber mulciber

Amathuxidia amythaon dilucida

Tanaecia aipis puseda

Dophla evelina compta

Euthalia kanda marana

Parthenos slyvia lilacinus

Rhinopalpa polynice eudoxia

Xanthotaenia busiris busiris

Tanaecia godartii osoca

Euthalia ipona

Ypthima huebneri

References

Basset, Y. (2001). Invertebrates in the canopy of tropical rain forests: How much do we really know? *Plant Ecology, 153*(1), 87–107.

Christharina, S. G., & Abang, F. (2014a). Taxonomic diversity of the fruit-feeding butterflies (Lepidoptera: Nymphalidae) in Kubah National Park, Sarawak, Southwest Borneo. *Tropical Natural History, 14*(1), 7–20.

Christharina, S. G., & Abang, F. (2014b). Diversity and abundance of the fruit-feeding butterflies (Lepidoptera: Nymphalidae) in Kubah National Park, Sarawak, Southeast Borneo. *Malayan Nature Journal, 66*(4), 390–406.

Corbet, A. S., & Pendlebury, H. W. (1992). *The butterflies of the Malay Peninsular* (4th ed.). Kuala Lumpur: Malayan Nature Society.

DeVries, P. J., & Walla, T. R. (2001). Species diversity and community structure in neotropical fruit-feeding butterflies. *Biological Journal of the Linnean Society, 74*(1), 1–15.

DeVries, P. J., Murray, D., & Lande, R. (1997). Species diversity in vertical, horizontal, and temporal dimensions of a fruit-feeding butterfly community in an Ecuadorian rainforest. *Biological Journal of the Linnean Society, 62*(3), 343–364.

Fatimah, A. (2006). *Butterflies of Malaysian Borneo: A pocket guide*. Kota Samarahan: Universiti Malaysia Sarawak.

Fernandez-Hemandez, H. M. (2007). Butterflies of the agricultural experiment station of tropical roots and tubers, and Santa Ana, Camaguey Cuba: An annotated list. *Acta Zoologica Mexicana, 23*(2), 43–47.

Hammer, Ø., Harper, D. A. T., & Rayan, P. D. (2001). PAST – PAlaeontological STatistics. ver. 1.89. *Palaoontologia Electronica, 4*(1), 1–9.

IUCN Red List. (2015). http://www.iucnredlist.org/

Kirton, L. G. (2014). *A naturalists guide to the butterflies of Peninsular Malaysia, Singapore, and Thailand*. Oxford: John Beaufoy Publishing Ltd.

Ludwig, J. A., & Reynolds, J. F. (1988). *Statistical ecology: A primer on methods and computing*. New York: Wiley.

Maryati, M., & Nordin, W. (1999). *Siri haiwan persekitaran kita kupu-kupu*. Kuala Lumpur: Percetakan Dewan Bahasa dan Pustaka.

Momose, K., Yumoto, T., Nagamitsu, T., Kato, M., Nagamasu, H., Sakai, S., Harrison, R. D., Itioka, T., Hamid, A. A., & Inoue, T. (1998). Pollination biology in a lowland dipterocarp forest Sarawak, Malaysia. I. Characteristics of the plant-pollinator community in a lowland dipterocarp forest. *American Journal of Botany, 85*(10), 1477–1501.

Pesiu, E. (2016). Personal interview.

Poole, R. W. (1974). *An introduction to quantitative ecology*. New York: McGraw Hill.

Schulze, C. H., Linsenmair, K. E., & Fiedler, K. (2001). Understory versus canopy: Patterns of vertical stratification and diversity among Lepidoptera in a Borneon rain forest. In K. E. Linsenmair, C. M. Davis, B. Fiala, & M. R. Speight (Eds.), *Tropical forest canopies: Ecology and management* (pp. 133–152). Rotterdam: Springer.

Shaharom, F. (Ed.). (2015). *Visions of Kenyir: Through the eyes of a drone*. Kuala Terengganu: Penerbit Universiti Malaysia Terengganu.

Stefanescu, C., Herrando, S., & Paramo, F. (2004). Butterfly species richness in the north-west Mediterranean Basin: The role of the natural and human-induced factors. *Journal of Biogeography, 31*(1), 905–915.

Stork, N. E., & Grimbacher, P. S. (2006). Beetle assemblages from an Australian tropical rainforest show that the canopy and the ground strata contribute equally to biodiversity. *Proceedings of the Royal Society of London B: Biological Sciences, 273*(1596), 1969–1975.

Tangah, J., Hill, J. K., Hamer, K. C., & Dawood, M. (2004). Vertical distribution of fruit-feeding butterflies in Sabah, Borneo. *Sepilok Bulletin, 1*(17), 17–27.

A Brief Survey on the Birds in Belukar Bukit, Kenyir, Terengganu, Malaysia

Gertrude David, Azuan Roslan, Elizabeth Pesiu, and Mohd Tajuddin Abdullah

Abstract There is lack of published data on avifauna in Kenyir to verify and update the bird species present in the area. Bird surveys were conducted in three sampling occasions: 15–21 May 2015, 25 September – 2 October 2015 and 31 March – 5 April 2016 at Belukar Bukit, Kenyir. The objective of the survey was to compare the bird species diversity in three occasions in relation to the forest phenology. A total of 118 birds were sampled representing 60 species from 25 families from the three sampling occasions. Pycnonotidae (14%) is the most diverse family followed by Nectariniidae (17%) and Timaliidae (7%). Among the 60 species recorded, 35 species were recorded in the first occassion, followed by 10 species from the second occasion and 19 species from the third occassion. The flowering season in May 2015 and the fruiting season from end of September to early October 2015 might be one of the factors that influence the bird diversity in three different occasions. Therefore a long-term survey using various sampling methods need to be conducted to update the bird checklist of Belukar Bukit, Kenyir.

Keywords Birds · Species diversity · Forest phenology · Tasik Kenyir

---

G. David (✉) · E. Pesiu
Kenyir Research Institute Terengganu, Universiti Malaysia Terengganu, Kuala Nerus, Terengganu, Malaysia

A. Roslan
Institute of Tropical Biodiversity and Sustainable Development, Universiti Malaysia Terengganu, Kuala Nerus, Terengganu, Malaysia

M. T. Abdullah
Institute of Tropical Biodiversity and Sustainable Development, Universiti Malaysia Terengganu, Kuala Nerus, Terengganu, Malaysia

School of Marine and Environmental Sciences, Universiti Malaysia Terengganu, Kuala Nerus, Terengganu, Malaysia
e-mail: mohd.tajuddin@umt.edu.my

M. T. Abdullah et al. (eds.), *Greater Kenyir Landscapes*,
https://doi.org/10.1007/978-3-319-92264-5_14

Fig. 1 Map showing the study site (labelled *BB* Belukar Bukit) in Kenyir, Terengganu, Malaysia

Introduction

Malaysia has one of the richest biodiversity in terms of species and habitats. The diversity of birds are particularly very rich as there are 650 species of birds recorded from Peninsular Malaysia and Singapore (Ramli et al. 2010). The lowland mixed-dipterocarp tropical rainforest habitat with abundant resources is one of the main factors towards Malaysia's rich diversity of birds. Ramli et al. (2010) also stated that the lowland rainforests of Malaysia serve as home for 311 species of birds where 274 are residents and it also provide habitats for biodiversity conservation in other tropical areas. Several studies based on species richness of birds conducted in Peninsular Malaysia have emphasized on lowland forests (Rajpar and Zakaria 2014; Nor-Hashim and Ramli 2013; Moradi et al. 2009; Rahman 2002; Wong 1986). This is because lowland rainforests contain more bird species than deforested areas due to the numerous microhabitats and diverse vegetation that provide various food resources for birds (Rajpar and Zakaria 2011; Gonzalez-Gajardo et al. 2009) which lead to the conclusion that lowland rainforests are important for bird conservation.

The objective of this study is to compare the diversity of birds at three different occasion also in relation with the forest phenology at the same study site which is at Belukar Bukit, Kenyir as in Fig. 1. Belukar Bukit was chosen because no study based on diversity of birds has been conducted in that area yet. There is only two bird surveys that has been done in Kenyir which is by Yong et al. (2011) which describe the effects of habitat fragmentation on insectivorous bird species in Tasik Kenyir and by Sulaiman et al. (2015) who conducted the survey on bird assemblages in Tanjong Mentong, Kenyir. In our study, we documented the pattern of bird diversity in three different occasions and also relate them with the forest phenology.

Located in the South West of Terengganu, Tasik Kenyir is known as the largest man-made lake in South East Asia. The 202 ha lake was surrounded by a total of 4975 ha of tropical forest and the surrounding area was flooded in 1978 to be converted into part of the Sultan Mahmud Hydro Electric Dam (Norfaizal et al. 2015). The study site is located in Belukar Bukit (N 4° 53′ 25.362″ E 102° 59′ 33.506″) which is a waterfall recreational area with a secondary forest area. The forest type of Belukar Bukit is Lowland Dipterocarp Forest and domesticated flora (plantation). The areas surrounding our study site is abundant with plant species such as *Hopea* sp. (Dipterocarpaceae), *Diespyros* sp. (Ebenaceae), *Aglaia* sp. (Meliaceae), *Knema* sp. (Myristicaceae) and *Hydrocarpus* sp. (Flacourtiaceae).

Point count method was used in this sampling and intervals between point counts are 100 m apart as the total distance for this point count is 1 km. The total numbers of point count stations are ten stations and in a radius of 25 m (Watson et al. 2004). Point count was usually for a period of 10 min (Marsden et al. 2001) and conducted twice a day which is from 0700 to 1000 h in the morning because detection rates will decrease 3 h after sunrise (Lynch 1995) and 1600 until 1900 h in the evening because birds are normally active during this hours (Ramli et al. 2009).

Mist-netting method was also used in this sampling and nets were checked every 2 h interval to reduce mortality caused by predators which can easily detect the nets whereabouts (Wong 1986). Five mist nets with four shelves (9 × 2.5 m, 36 mm mesh size) were placed at canopy level (Fig. 2) which is >10 m according to Peh et al. (2006), by using a slingshot and shoot over tall trees and five mist nets (12 × 2.5 m high, 36 mm

Fig. 2 Canopy mist net suspended in the canopy level

mesh size) were placed understorey based on their accessibility, bird activities and the bird flyways. The forest phenology data were obtained from Pesiu (2018).

Checklist

This study provides the first checklist of birds for Belukar Bukit, Kenyir, since there has been no published record for this area. A total of 118 individuals were collected from the survey which belongs to 60 species from 25 families (Table 1). Family Pycnonotidae is the most diverse family comprising of 10 species, followed by five species for family Nectariniidae, Timaliidae and Picidae respectively.

The abundance of family Pycnonotidae especially in the second sampling occasion (Fig. 3) is due to its feeding guild which is frugivorous that mainly feeds on berries (Phillipps and Phillipps 2014) and depend widely on fruiting seasons. The berries that are mainly observed eaten by these bulbuls are *Melastoma malabathricum* berries which blooms all year round (Engel and Phummai 2011). The abundance of family Nectariniidae are influenced by its nectarivorous feeding guild where there are known as specialist nectar feeders that also feeds on insects and small berries (Phillipps and Phillipps 2014). Ginger and wild banana are commonly found in the study site which leads to the abundance of the Nectariniidae family. Timaliidae and Picidae are both insectivorous where they were caught and observed mostly in the forest interior. Woodpeckers from the family Picidae are considered to be sensitive towards the forest disturbance because they need to nest in the cavities of large trees and feed on bark-boring insects of both living and dead in large trees (Sodhi and Brook 2006). Timaliidae mostly hunts insects on the ground, sensitive towards the changes in habitat and difficult to adapt in disturbed environments (Wells 2007a, b, 1999) and Picidae mostly hunts ants on tree trunks (Phillipps and Phillipps 2014).

Throughout the three sampling occasions, 34 species (46%) were collected in 15th–21st May 2015, 10 species (14%) were collected in 25th September – 2nd October 2015, 30 species (41%) from 31st March to 5th April 2016 and a total of 60 species for all sampling occasions. Based on Fig. 3, Pycnonotidae is the highest number of individuals according to family in the first sampling survey while Pycnonotidae, Nectariniidae and Columbidae families has the highest number of individuals (21%) in the second sampling survey and Nectariniidae is the highest family in the third sampling survey with a total of 22%. However, the diversity of species is low during the second sampling survey period due to the unfavourable weather which was heavy thunderstorm. The highest family in all three sampling occasions is Pycnonotidae with a total of 14% based on Fig. 4.

The first sampling (15th–21st May 2015) in Belukar Bukit has the most diverse species with a total of 34 species respectively followed by the third sampling (31st March – 5th April 2016) with a total of 30 species and the second sampling (25th September – 2nd October 2015) with the least species which is 10 species (Fig. 5). The abundance in the two sampling periods is due to flowering seasons (Pesiu 2018).

Understorey insectivores such as babblers does not favour to cross open areas (Develey and Stouffer 2001; Borges and Stouffer 1999) causing the increased canopy open areas close to the edge (Haugaasen et al. 2003; Barlow et al. 2002) to limit

Table 1 Checklist of birds collected from three occasions

Bil	Family/species Name	Common name	No. of individuals	1st occasion	2nd occasion	3rd occasion	Method of collection	Conservation status (IUCN 2016)
	Muscicapidae							
1	*Copsychus malabaricus*	White-rumped shama	2	✓		✓	MN	LC
2	*Copsychus saularis*	Oriental magpie robin	4	✓		✓	PC	LC
	Alcedinidae							
3	*Alcedo meninting*	Blue-eared kingfisher	1	✓			MN	LC
4	*Halcyon smyrnensis*	White-throated kingfisher	1		✓		MN	LC
	Bucerotidae							
5	*Buceros rhinoceros*	Rhinoceros hornbill	1	✓			PC	NT
6	*Brenicornis comatus*	White-crowned hornbill	5			✓	PC	NT
	Ardeidae							
7	*Ixobrychus cinnamomeus*	Cinnamon bittern	1			✓	MN	LC
	Pycnonotidae							
8	*Alophoixus phaeocephalus*	Yellow-bellied bulbul	14	✓		✓	MN and PC	LC
9	*Criniger finschy*	Finsch's bulbul	1	✓			MN	NT
10	*Tricholestes criniger*	Hairy-backed bulbul	1	✓			MN	LC
11	*Pycnonotus erythropthalmos*	Spectacled bulbul	1	✓			MN	LC
12	*Pycnonotus simplex*	Cream-vented bulbul	1	✓			MN	LC
13	*Iole olivacea*	Buff-vented bulbul	1	✓			MN	NT
14	*Hemixos castanonotus*	Chestnut bulbul	1			✓	PC	LC
15	*Pycnonotus cyaniventris*	Grey-bellied bulbul	2	✓	✓		MN	NT
16	*Pycnonotus brunneus*	Red-eyed bulbul	1		✓		MN	LC
17	*Pycnonotus goiavier*	Yellow vented bulbul	2			✓	PC	LC

(continued)

Table 1 (continued)

Bil	Family/species Name	Common name	No. of individuals	1st occasion	2nd occasion	3rd occasion	Method of collection	Conservation status (IUCN 2016)
	Timaliidae							
18	*Stachyris poliocephala*	Grey-headed babbler	1	✓			MN	LC
19	*Stachyris erythroptera*	Chestnut winged babbler	3	✓		✓	MN	LC
20	*Trichastoma malaccense*	Short-tailed babbler	1	✓			MN	NT
21	*Pellorreum capistratum*	Black-capped babbler	2	✓		✓	MN	LC
22	*Mixornis gularis*	Pin-striped tit babbler	1			✓	MN	LC
	Nectariniidae							
23	*Arachnothera longirostra*	Little spiderhunter	11	✓	✓	✓	MN	LC
24	*Arachnothera crassirostris*	Thick-billed spiderhunter	1	✓			MN	LC
25	*Arachnothera robusta*	Long-billed spiderhunter	2	✓		✓	MN	LC
26	*Arachnothera modesta*	Grey-breasted spiderhunter	1			✓	MN	LC
27	*Hypogramma hypogrammicum*	Purple-naped sunbird	5	✓	✓	✓	MN	LC
	Dicaeidae							
28	*Prionichilus maculatus*	Yellow-breasted flowerpecker	2	✓		✓	MN	LC
29	*Dicaeum trigonostigma*	Orange-bellied flowerpecker	1	✓			MN	LC
30	*Prionochilus thoracicus*	Scarlet-Backed flowerpecker	1	✓			PC	LC
31	*Dicaeum chrysorrheum*	Yellow vented flowerpecker	1			✓	PC	LC
	Dicruridae							
32	*Dicrurus remifer*	Lesser-racket tailed drongo	1			✓	PC	LC
33	*Dicrurus paradiseus*	Greater-racket tailed drongo	2			✓	PC	LC
	Picidae							
34	*Meglyptes tukki*	Buff-necked woodpecker	1	✓			MN	NT
35	*Sasia abnormis*	Rufous piculet	1	✓			MN	LC
36	*Picus miniaceus*	Banded woodpecker	1	✓			PC	LC
37	*Chryscophlegma humii*	Chequer-throated yellownape	1	✓			MN	NT
38	*Picus chlorolophus*	Lesser yellownape	1			✓	PC	LC

	Monarchidae							
39	*Philentoma pyrhopterum*	Rufous-winged philentoma	3	✓			MN	LC
40	*Hypothymis azurea*	Black-naped monarch	1	✓			MN	LC
	Hirundinidae							
41	*Hirundo tahitica*	Pacific swallow	1			✓	MN	LC
	Irenidae							
42	*Chloropsis cyanopogon*	Lesser-green leafbird	3	✓			MN and PC	NT
	Columbidae							
43	*Chalcophaps indica*	Emerald dove	4	✓	✓	✓	MN	LC
44	*Geopelia strata*	Zebra dove	1		✓		MN	LC
	Calyptomeni dae							
45	*Calyptomena viridis*	Green broadbill	1	✓			MN	NT
	Cettidae							
46	*Abroscopus superciliaris*	Yellow bellied warbler	1			✓	MN	LC
	Cisticolidae							
47	*Orthotomus sericeus*	Rufous tailed tailorbird	1			✓	PC	LC
	Corvidae							
48	*Platysmurus leucopterus*	Black magpie	2			✓	PC	NT
	Cuculidae							
49	*Cacomantis merulinus*	Plaintive cuckoo	1			✓	MN	LC
50	*Rhinortha chlorophaea*	Raffles's Malkoha	1			✓	PC	LC
	Eurylaimidae							
51	*Cymbirhynchus macrorhynchos*	Black and red broadbill	4			✓	MN and PC	LC
	Megalaimidae							
52	*Magalaima mystocophanos*	Red-throated barbet	1	✓			MN	NT
53	*Psilopogon rafflesii*	Red-crowned barbet	1			✓	PC	NT

(continued)

Table 1 (continued)

Bil	Family/species Name	Common name	No. of individuals	1st occasion	2nd occasion	3rd occasion	Method of collection	Conservation status (IUCN 2016)
	Accipitridae							
54	*Accipiter soloensis*	Chinese sparrowhawk	1		✓		MN	LC
55	*Spilornis cheela*	Crested serpent eagle	3	✓			PC	LC
56	*Nisaetus cirrhatus*	Changeable-hawk eagle	1			✓		LC
	Meropidae							
57	*Nyctyornis amictus*	Red-bearded bee-eater	3		✓		MN	LC
58	*Merops viridis*	Blue-throated bee-eater	1	✓			PC	LC
	Motacillidae							
59	*Motacilla cinerea*	Grey wagtail	1		✓		MN	LC
	Oriolidae							
60	*Oriolus xanthonotus*	Dark-throated oriole	1	✓			PC	NT
		Total	118					

IUCN status following The IUCN Red List of Threatened Species (Version 2016-2)
MN Mist Net, *PC* Point Count, *LC* Least Concerned, *NT* Near Threatened

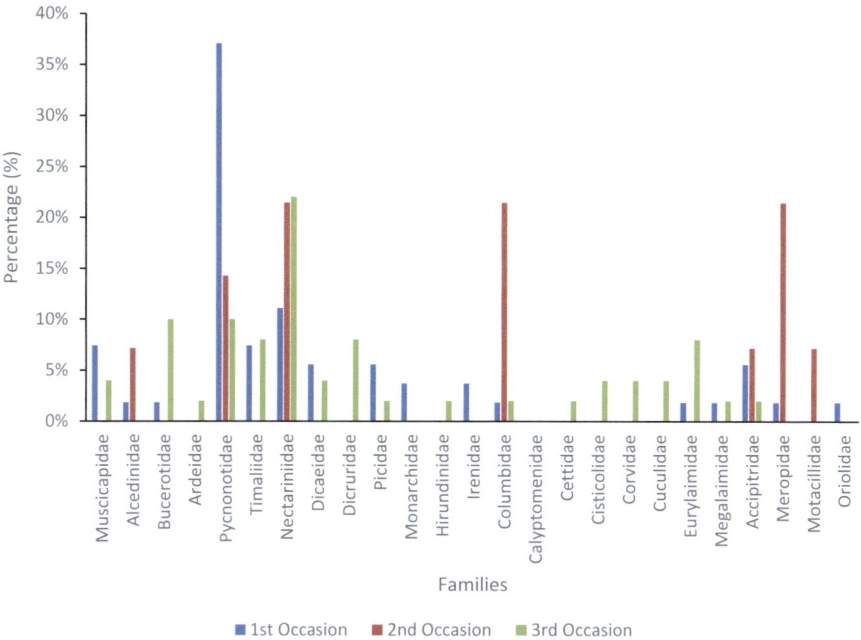

Fig. 3 Percentage of birds recorded according to families in three sampling occasions in Belukar Bukit, Kenyir

the movement and dispersal of most insectivorous understorey birds associated with forest interior, especially terrestrial insectivores such as babblers. This explains the low number of species under the family Timaliidae (Fig. 4). Food resources for birds are different with fragment size as a result of edge effects, because invertebrates may not survive well in edge habitat, which is often warmer and drier than the forest interior as they are easily dehydrated (Zanette et al. 2000). However, bulbuls from the family Pycnonotidae are arboreal frugivorous birds that are known as colonizer or secondary species and are common in disturbed areas (Zakaria et al. 2005; Nordin and Zakaria 1997). The bulbuls that were abundant at the edge are capable of enduring the high temperature and light intensity (Zakaria et al. 2002). There are a total of ten species of bulbul from three sampling occasions with *Criniger phaecephalus* (Yellow Bellied Bulbul) (Fig. 6) being the most abundant species in this survey with a total of 14 individuals *Criniger phaecephalus* is the most abundant species in this survey as it is a common bird in the lowland rainforest and often forages on low fruiting shrubs solitarilyy or sometimes together with other birds (Strange and Jeyarajasingam 1993). Bulbuls from the family Pycnonotidae which are insectivore-frugivore species is a group of birds that can adapt to the seasonal availability of fruits (Azman et al. 2011) and also capable of switching their diet (Gray et al. 2007). This explains why Pycnonotidae is the family with the most number of species. Some species however showed no evidence for an edge effect

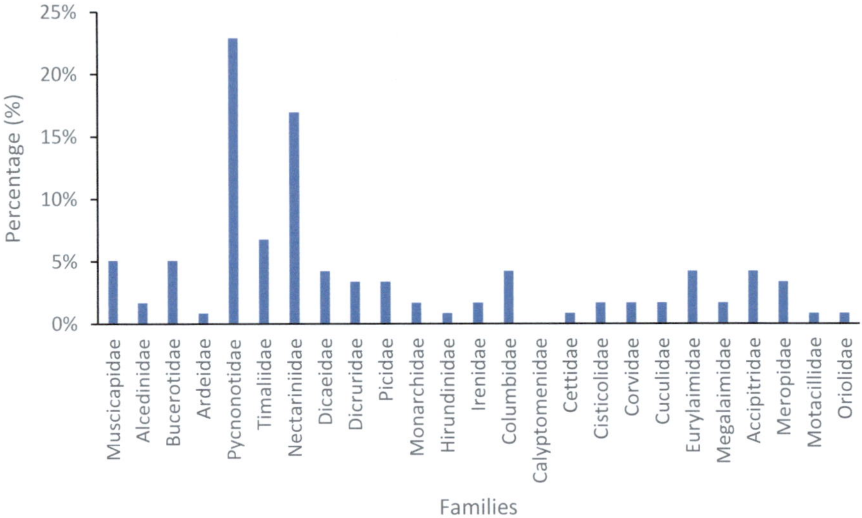

Fig. 4 Total Percentage of birds recorded according to family in three sampling occasions in Belukar Bukit, Kenyir

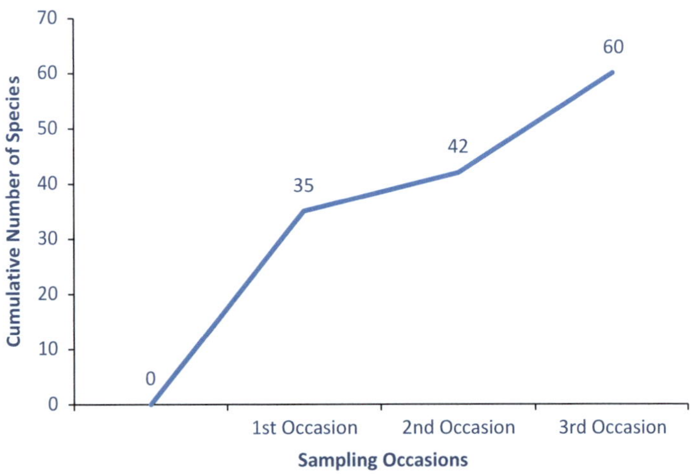

Fig. 5 Species accumulation curve for all three occasions

due to microhabitat variables, sufficient justifications of habitat variables at either edge or interior sites.

Arachnothera longirostra (Little Spiderhunter) (Fig. 7) from the family Nectariniidae is the second most abundant species for these surveys with a total of 11 individuals because it is a common resident in both primary and secondary lowland rainforest (Phillipps and Phillipps 2014). Being both nectarivorous and insec-

Fig. 6 *Criniger phaeocephalus* (yellow bellied bulbul). (Photographed by Ahmad Fadhli Adzemi)

Fig. 7 *Arachnothera longirostra* (little spiderhunter)

tivorous, it is especially found in areas abundant with wild bananas and ginger because it is a specialist feeder on this plants. It also feeds on insects occasionally (Strange and Jeyarajasingam 1993). It feeds frequently at banana inflorescences and is common in disturbed forest.

Hypogramma hypogrammicum (Purple-naped Sunbird) (Fig. 8) from the family Nectariniidae and *Brenicornis comatus* (White-crowned Hornbill) from the family Bucerotidae is the third most abundant species with a total of five individu-

Fig. 8 *Hypogramma hypogrammicum* (purple-naped sunbird, male). (Photographed by Mohd. Abid Kamaruzzaman)

Table 2 Comparison of results with Sulaiman et al. (2015)

	Sulaiman et al. (2015)	Our study
Total number of individuals	21	118
Total number of families	10	25
Total number of species	12	60
Methods used	Mist-netting	Mist-netting and point count
Duration	49 net-days	18 sampling days
Study site	Tanjung Mentong, Tasik Kenyir	Belukar Bukit, Kenyir

als respectively. *Hypogramma hypogrammicum* is common in closed-canopy forest (Davison and Fook 2003). It is common in lower storey and usually feeds on nectar of wild banana and ginger, sometimes forage on insects and fruits (Shi 2012). *Brenicornis comatus* prefers dense and shrubby vegetation nearby rivers in lowland and hill forest and usually hunts in small parties for insects and small animals (Phillipps and Phillipps 2014). It is known as the most carnivorous hornbill and has a varied diet on fruit, insects and small animals (Phillipps and Phillipps 2014).

Table 2 showed the comparison of bird species with Sulaiman et al. (2015). Even though this survey recorded higher diversity of birds, but the species accumulation curve has not reached asymptote yet and did not provide a complete checklist of species available here due to the brief sampling period, no data of rainfall and habitat changes. Hence, a longer sampling period covering a wider area would likely provide a better representation of the bird richness in Belukar Bukit particularly.

Conclusion

This study shows that there are high diversity of birds in Belukar Bukit, which is greatly influenced by the flowering and fruiting period which is the food resources for this birds. However, the present list of bird is not complete, hence prolonged surveys using various sampling methods on more favourable weathers and surveys in more interior part of the forest is recommended to update the bird checklist.

Acknowledgements We would like to thank Kenyir Research Institute for providing us with the field equipments, accommodation, aid in logistic, and administrative supports. We would also like to thank Universiti Malaysia Terengganu (UMT) for the transportation provided during the field. Special thanks to Muhd Tarmizi Abdullah for the assistance during the survey. We acknowledged financial support from the Trans-disciplinary Grant Scheme (TRGS/2014/59373), *Geran Galakan Penyelidikan*, UMT (GGP/68007/2014/127), and Niche Research Grant Scheme (NRGS/2015/5313/2) led by M. T. Abdullah.

References

Azman, N. M., Latip, N. S. A., Sah, S. A. M., Akil, M. A. M. M., Shafie, N. J., & Khairuddin, N. L. (2011). Avian diversity and feeding guilds in a secondary forest, an oil palm plantation and a paddy field in riparian areas of the Kerian River Basin, Perak, Malaysia. *Tropical Life Sciences Research, 22*(2), 45–64.

Barlow, J., Haugaasen, T., & Peres, C. A. (2002). Effects of surface wildfires on understorey bird assemblages in Amazonian forests. *Biological Conservation, 105*, 157–169.

Borges, S. H., & Stouffer, P. C. (1999). Bird communities in two types of anthropogenic successional vegetation in central Amazonia. *The Condor, 101*, 529–536.

Davison, G. W. H., & Fook, C. Y. (2003). *A photographic guide to birds of Peninsular Malaysia and Singapore*. London: New Holland Publishers Ltd..

Develey, P. F., & Stouffer, P. C. (2001). Effects of roads on movements by understorey birds in mixed-species flocks in central Amazonian Brazil. *Conservation Biology, 15*, 1416–1422.

Engel, D. H., & Phummai, S. (2011). *A field guide to tropical plants of Asia*. Singapore: Marshall Cavendish Edition.

Gonzalez-Gajardo, A., Sepulveda, P. V., & Schlatter, R. (2009). Waterbird assemblages and habitat characteristics in wetlands; Influence of temporal variability on species-habitat relationships. *Waterbirds, 32*(2), 225–233.

Gray, M. A., Baldauf, S. L., Mayhew, P. J., & Hill, J. K. (2007). The response of avian feeding guilds to tropical forest disturbance. *Conservation Biology, 21*(1), 133–141.

Haugaasen, T., Barlow, J., & Peres, C. A. (2003). Surface wildfires in central Amazonia: Short-term impact on forest structure and carbon loss. *Forest Ecology and Management, 179*, 321–331.

IUCN. (2016). The IUCN red list of threatened species. Version 2016-2. Retrieved from http://www.iucnredlist.org.

Lynch, J. F. (1995). Effects of point count duration, time-of-day, and aural stimuli on detectability of migratory and resident bird species in Quintana Roo, Mexico. In C. J. Ralph, J. R. Sauer, & S. Droege (Eds.), *Monitoring bird populations by point counts* (pp. 1–6). Albany: Forest Service General Technical Report PSW-GTR-149.

Marsden, S. J., Whiffin, M., & Galetti, M. (2001). Bird diversity and abundance in forest fragments and *Eucalyptus* plantations around an Atlantic forest reserve, Brazil. *Biodiversity and Conservation, 10*, 737–751.

Moradi, H. V., Zakaria, M., & Yusof, A. B. M. E. (2009). Insectivorous birds and environmental factors across and edge-interior gradient in tropical rainforest of Malaysia. *International Journal of Zoological Research, 5*(1), 27–41.

Nordin, M., & Zakaria, M. (1997). Some effects of logging in mixed lowland dipterocarp forests on birds. In O. B. Gaik (Ed.), *State of the Malaysian environment* (pp. 161–166). Penang: Consumers' Association of Penang Press.

Nor-Hashim, E., & Ramli, R. (2013). Comparative study of understorey birds diversity inhabiting lowland rainforest virgin jungle reserve and regenerated forest. *The Scientific World Journal*. https://doi.org/10.1155/2013/676507.

Norfaizal, G. M., Masrom, H., & Muhammad Radzali, M. (2015). Flora diversity of Pulau Tekak Besar, Tasik Kenyir, Hulu Terengganu, Malaysia. *International Journal of Current Research in Biosciences and Plant Biology, 2*(5), 179–183.

Peh, K. S. H., Sodhi, N. S., De Jong, J., Sekercioglu, C. H., Yap, C. A. M., & Lim, S. L. H. (2006). Conservation value of degraded habitats for forest birds in southern Peninsular Malaysia. *Diversity and Distributions, 12*(5), 572–581.

Pesiu, E. (2018). Tree community structure, composition and above ground biomass in off coasts islands, setiu wetlands and Hulu Terengganu, Terengganu. (Unpublished master's thesis). Universiti Malaysia Terengganu, Kuala Nerus, Terengganu.

Phillipps, Q., & Phillipps, K. (2014). *Phillipps field guide to the birds of Borneo* (3rd ed.). Oxford: John Beaufoy Publishing.

Rahman, M. A. (2002). Using mist-nets on canopy walkways in Malaysia to study canopy avifauna. *The Raffles Bulletin of Zoology, 50*(2), 499–506.

Rajpar, M. N., & Zakaria, M. (2011). Bird species abundance and their correlationship with microclimate and habitat variables at natural wetland reserve, Peninsular Malaysia. *International Journal of Zoology, 758573*, 17. https://doi.org/10.1155/2011/758573.

Rajpar, M. N., & Zakaria, M. (2014). Effects of habitat characteristics on waterbird distribution and richness in wetland ecosystem of Malaysia. *Journal of Wildlife and Parks, 28*, 105–120.

Ramli, R., Ya'cob, Z., & Hashim, R. (2009). Diversity of birds in Kenaboi forest reserve, Jelebu, Negeri Sembilan, Malaysia. *Malaysian Journal of Science, 28*(4), 465–480.

Ramli, R., Ya'cob, Z., Aimi, F., & Ezyan, N. H. (2010). A survey of avifauna in Bachok District, Kelantan, Peninsular Malaysia. *Malaysia Journal of Science, 29*(Special Issue), 121–130.

Shi, W. T. (2012). *A naturalist's guide to the birds of Borneo (Sabah, Sarawak, Brunei and Kalimantan)*. Oxford: John Beaufoy Publishing Limited.

Sodhi, N. S., & Brook, B. W. (2006). *Southeast Asian biodiversity in crisis*. New York: Cambridge University Press.

Strange, M., & Jeyarajasingam, A. (1993). *Birds: A photographic guide to the birds of Peninsular Malaysia and Singapore*. Singapore: Sun Tree Publishing Pte Ltd..

Sulaiman, M. H., Embong, M., Mamat, M. A., Tahir, N. F. D. A., Latip, N. A., Murni, R., & Azhar, I. M. (2015). Preliminary rurvey of the bird assemblage at Tanjong Mentong, Lake Kenyir, Hulu Terengganu, Malaysia. *Tropical Natural History, 15*(1), 87–90.

Watson, J. E. M., Whittaker, R. J., & Dawson, T. P. (2004). Habitat structure and proximity to forest edge affect the abundance and distribution of forest-dependent birds in tropical coastal forests of south-eastern Madagascar. *Biological Conservation, 120*, 311–327.

Wells, D. R. (2007a). *The birds of Thai-Malay peninsula: Volume I: Non-passerines*. London: Christopher Helm Publishers Ltd.

Wells, D. R. (2007b). *The birds of the Thai-Malay peninsula. Volume II: Passerines*. London: Christopher Helm Publishers Ltd..

Wells, D. R. (1999). *The birds of Thai-Malay Peninsula: Non-passerines*. London: Academic Press.

Wong, M. (1986). Trophic organization of understory birds in a Malaysian dipterocarp forest. *The Auk, 103*, 100–116.

Yong, D. L., Qie, L., Sodhi, N. S., Poh, L. P., Peh, K. S. H., Lee, T. M., Lim, H. C., & Lim, S. L. H. (2011). Do insectivorous bird communities decline on land-bridge forest islands in Peninsular Malaysia? *Journal of Tropical Ecology, 27*, 1–14.

Zakaria, M., Khairul, A., & Jamalun, N. (2002). Comparison of understorey bird species composition in a primary and logged mixed hill dipterocarp forest in Peninsular Malaysia. *Malaysian Nature Journal, 19*, 74–85.

Zakaria, M., Leong, P. C., & Ezhar, Y. M. (2005). Comparison of species composition in three forest types: Toward using bird as indicator of forest ecosystem heath. *Journal of Biological Sciences, 5*(6), 734–737.

Zanette, L., Doyle, P., & Tremont, S. M. (2000). Food shortage in small fragments: Evidence from an area-sensitive passerine. *Ecology, 81*, 1654–1666.

Species Richness Estimation of Reptiles in Selected Sites of Tasik Kenyir, Hulu Terengganu, Malaysia

Amirah Azizah Zakaria, Noor Aisyah A. Rahim, Amirrudin Ahmad, and Mohd Tajuddin Abdullah

Abstract The knowledge on the composition of reptile species at Tasik Kenyir and its vicinity is still lacking as there is no comprehensive checklist available to date. This study aimed to provide a checklist and to estimate the species richness of reptiles in recreational forests located at and around Tasik Kenyir, namely Sungai Buweh, Belukar Bukit, Sekayu, and Saok. Visual Encounter Survey (VES), pitfall trapping, and opportunistic survey methods were used in this study. From 46 sampling days between April 2015 and February 2016, a total of 118 individuals from 26 species, 19 genera and nine families of reptiles were successfully recorded. Almost half of the reptile species belongs to family Gekkonidae (42%), followed by Agamidae (23%) and Scincidae (11%). The most abundant species observed were *Eutropis multifasciata* (24%), *Hemidactylus frenatus* (16%) and *Gekko monarchus* (13%). Eleven of the recorded species are protected under Wildlife Conservation Act 2010 and nine species are listed as Least Concern by IUCN 2016. The number of species estimated by non-parametric estimators, Chao 1 and Chao 2 were 35 and 39 species respectively. The results from this study could be useful to monitor the impact of habitat changes and human disturbances at Tasik Kenyir and the neighbouring areas.

Keywords Estimation · Reptiles · Richness · Tasik Kenyir

A. A. Zakaria (✉) · N. A. A. Rahim
Institute of Tropical Biodiversity and Sustainable Development,
Universiti Malaysia Terengganu, Kuala Nerus, Terengganu, Malaysia

A. Ahmad · M. T. Abdullah (✉)
Institute of Tropical Biodiversity and Sustainable Development,
Universiti Malaysia Terengganu, Kuala Nerus, Terengganu, Malaysia

School of Marine and Environmental Sciences, Universiti Malaysia Terengganu,
Kuala Nerus, Terengganu, Malaysia
e-mail: mohd.tajuddin@umt.edu.my

M. T. Abdullah et al. (eds.), *Greater Kenyir Landscapes*,
https://doi.org/10.1007/978-3-319-92264-5_15

Introduction

Reptiles play important roles in ecosystems, as prey and predators, grazers, agents for seed dispersal, commensal species, and their often specific microhabitat associations may be used to illustrate evolutionary processes of speciation (Raxworthy et al. 2008; Read 1998). Due to its ecological relevance, the declines of reptile populations may cause negative impact on the natural systems. In order to maintain healthy ecosystems, biodiversity studies are crucial as they provide the knowledge to prioritize conservation areas and habitats, and to model sustainable use of resources (Sohdi et al. 2004). Species richness data for instance are vital so that the responsible authorities are aware of the wildlife present in an area. Mitigation measures that need to be included in the development plans considering the plight of the animals can be developed using the data, thus ensuring the continued existence of the species in their natural habitat. Without substantial amount of biodiversity publications, species will be prone to extermination as no action taken on threats faced by a particular species.

In Malaysia, most of the reptile studies were focusing on species richness (Sumarli et al. 2015, 2016; Voris 2015; Shahriza and Ibrahim 2014; Nurulhuda et al. 2014; Amirah et al. 2013; Quah et al. 2013; Shahriza et al. 2012; Chan et al. 2010a, b; Ibrahim et al. 2008, 2012, 2013; Grismer et al. 2004, 2006, 2014a, b, c) and only a single study have been published on the estimation of species richness (van Rooijen et al. 2011). It has been reported that 397 species of reptiles have been recorded around Malaysia and among them, an approximate of 256 species (65%) were identified to be present in Peninsular Malaysia (Das and Norsham 2007). Since then, many new species have been discovered along with the rediscovery of rare species (Grismer et al. 2008a, b, c, 2009, 2010a, b, 2011, 2014a, b, c, 2016a, b, c, d; Sumarli et al. 2016; Grismer and Quah 2015; Johnson et al. 2012; Grismer 2008, 2011; Chan et al. 2010a; Wood et al. 2009). These discoveries clearly show that Malaysian forests are very rich in reptile assemblage, which underlines the need for more studies and surveys to be carried out.

Reptiles in Tasik Kenyir, Hulu Terengganu are still understudied although this place is a home to a significant number of flora and fauna (Faizah et al. 2015). Therefore, this study was conducted to examine the species richness of reptiles in selected sites surrounding Tasik Kenyir, Terengganu. It is hoped that this study can be useful for future reptile studies, and for the conservation and management purposes at Tasik Kenyir and other similar habitats particularly in Terengganu.

The survey was conducted from April 2015 to February 2016 at four sampling sites located at and around Tasik Kenyir, Terengganu, namely Sungai Buweh, Belukar Bukit, Sekayu and Saok, (Table 1, Fig. 1). These four sampling sites are recreational areas surrounded by lowland dipterocarp forests. They were chosen as the study sites because there were no or very limited information on reptilian fauna of these areas which may harbour a diverse set of species.

In this study, pitfall trapping method was used to collect samples at all sites but not in all occasions. In order to increase the probability of reptiles to fall into the traps, drift fences from aluminium sheets, wire nets or fallen logs were used. The

Table 1 List of sampling sites, dates and methods used

Site	Date	Methods
Sungai Buweh	20.4.2015 – 26.4.2015	VES; 14 PT
(5° 08′ 56.0″ N 102° 46′ 20.2″ E)	28.4.2015	OS
	26.8.2015	OS
	23.1.2016	OS
Belukar Bukit	15.5.2015 – 21.5.2015	VES; 18 PT
(4° 53′ 27.0″ N 102° 59′ 36.1″ E)	25.9.2015 – 1.10.2015	VES; 18 PT
Sekayu	21.6.2015 – 27.6.2015	VES; 4 PT
(4° 57′ 50.1″ N 102° 57′ 28.1″ E)	17.10.2015 – 24.10.2015	VES; 30 PT
Saok	22.2.2016 – 28.2.2016	VES; 30 PT
(5° 00′ 35.8″ N 102° 39′ 01.2″ E)		

VES Visual Encounter Survey, *PT* pitfall traps, *OS* opportunistic sampling (opportunistic samplings were done by other members of our zoological survey team)

Fig. 1 Map showing the four sampling sites around Tasik Kenyir, Hulu Terengganu. (Source: Google Maps)

numbers of pitfall traps used were unequal between sampling sites due to geographical constraint and manpower limitations. Visual Encounter Surveys (VES) was also conducted to search and to collect reptiles. Active searching were carried out for three sessions per day (in the morning, evening and at night time) along an approximately 1 km transect, for 2–3 h per session. All observed reptile species were recorded and captured, when possible, using sweep nets, hand-picking device, or simply by hand grabbing.

The specimens were placed inside an individual plastic or cloth bag and were taken back to the laboratory for sample processing. Voucher specimens were preserved in 10% formalin and stored in 70% ethanol and deposited in laboratory situated in Centre of Kenyir Ecosystems Research, Kenyir Research Institute, Universiti Malaysia Terengganu (UMT) for future reference. Taxonomy of reptiles or species identification was based on Das (2004, 2007, 2010). The conservation status of the reptile species was assessed based on IUCN Red List of Threatened Species (IUCN 2016) and Wildlife Conservation Act 2010.

To estimate the species richness of reptiles in the study area, the data were analysed using non-parametric richness estimators calculated in EstimateS version 9.1 (Colwell 2013) Chao 1 and Chao 2 (Chao 1984). Chao 1 uses abundance-based data while Chao 2 uses incidence-based data. These estimators use information on the frequency of rare species in a sample to estimate the number of undetected species in an assemblage (Gotelli and Chao 2013; Gotelli and Colwell 2011).

Checklist of Reptiles

From 46 sampling days, 118 individuals of reptiles from nine families, 19 genera and 26 species were recorded from study sites at and around Tasik Kenyir, Terengganu (Table 2). The most abundant family was Gekkonidae (42%) followed by Agamidae (23%), Scincidae (11%), and the other six families, Geoemydidae, Trionychidae, Varanidae, Colubridae, Elapidae and Viperidae each contributed 4% for the total number of species (Fig. 2). The most diverse reptile genus recorded were *Hemidactylus* with three species followed by two species for *Draco, Gonocephalus, Cyrtodactylus, Gekko* and *Eutropis* each. Only one species was recorded for the genus *Cyclemys, Dogania, Aphaniotis, Calotes, Cnemaspis, Gehyra, Hemiphyllodactylus, Luperosaurus, Lipinia, Varanus, Calamaria, Naja* and *Tropidolaemus.* The proportion of species that have been assessed for their conservation status was 54%. Based on IUCN 2016, only nine out of the 26 species were categorised as Least Concern and the rest were not yet assessed. In Wildlife Conservation Act 2010, 11 of the total species recorded were protected, mostly in the Agamidae family.

The numbers of reptile species estimated by Chao 1 and Chao 2 estimators were all higher than the number of observed species, 26 (Table 3, Fig. 3). The percentage of observed richness based on the estimated richness of Chao 1 and Chao 2 was 74% and 67% respectively.

In this study, Belukar Bukit and Sekayu have the highest number of reptile species, mainly because the sampling duration in these study sites were longer, 14 and 15 days respectively, compared to Sungai Buweh (10 days) and Saok (seven days). Restricted time allows limited coverage area for sampling activities, thus contributes to limited finding of species (Ibrahim et al. 2012; Shahriza et al. 2012). Herpetological surveys especially require longer periods of sampling due to the elusive and secretive nature of amphibians and reptiles (Ibrahim et al. 2013) therefore

Table 2 Checklist of reptiles in selected sites of Tasik Kenyir, Terengganu

Order family species	Common name	SB	BB	SK	SO	WCA 2010	IUCN 2016
Chelonia							
Geoemydidae							
Cyclemys sp.	Leaf turtle	0	2	0	0	NA	NA
Trionychidae							
Dogania subplana	Malayan softshell turtle	0	0	0	1	P	LC
Squamata							
Agamidae							
Aphaniotis fusca	Brown shrub lizard	0	1	0	0	P	LC
Calotes versicolor	Garden lizard	0	0	1	0	P	NA
Draco melanopogon	Black-bearded flying lizard	0	1	1	0	P	NA
Draco sumatranus	Common flying lizard	0	0	1	0	NP	NA
Gonocephalus bellii	Blue-necked angle-headed lizard	2	1	0	1	P	NA
Gonocephalus grandis	Giant angle-headed lizard	0	1	1	0	P	LC
Gekkonidae							
Cnemaspis peninsularis	Peninsular rock gecko	3	1	2	0	NP	NA
Cyrtodactylus consobrinus	Peter's bent-toed gecko	3	0	1	0	P	NA
Cyrtodactylus quadrivirgatus	Four-striped bent-toed gecko	2	0	0	1	P	NA
Gehyra mutilata	Common four-clawed gecko	2	0	0	0	NP	NA
Gekko monarchus	Warty house gecko	2	3	7	3	NP	NA
Gekko smithii	Smith's giant gecko	0	3	0	0	NP	LC
Hemidactylus craspedotus	Frilly forest gecko	0	1	0	0	NP	NA
Hemidactylus frenatus	Asian house gecko	0	14	4	1	NP	LC
Hemidactylus garnotii	Garnot's house gecko	0	2	1	0	NP	NA
Hemiphyllodactylus typus	Common worm gecko	0	0	0	1	NP	NA
Luperosaurus browni	Brown's camouflage gecko	0	0	0	1	NP	NA
Scincidae							
Eutropis longicaudata	Long-tailed ground skink	1	0	0	0	NP	NA
Eutropis multifasciata	Common sun skink	3	12	13	0	NP	NA
Lipinia vittigera	Common striped skink	2	0	0	0	NP	NA
Varanidae							
Varanus salvator	Water monitor lizard	0	6	5	1	P	LC
Colubridae							
Calamaria pavimentata	Brown reed snake	1	0	0	0	NP	LC
Elapidae							
Naja kaouthia	Monocled cobra	0	0	1	0	P	LC
Viperidae							

(continued)

Table 2 (continued)

Order family species	Common name	SB	BB	SK	SO	WCA 2010	IUCN 2016
Tropidolaemus wagleri	Wagler's keeled green pit viper	0	0	1	0	P	LC
	No. of individual	21	48	39	10		
	No. of species	10	13	13	8		
	No. of genus	8	9	11	8		
	No. of family	4	5	6	4		
	No. of individual	118					
	No of genus	19					
	No. of species	26					
	No. of family	9					

Study sites: *SB* Sungai Buweh, *BB* Belukar Bukit, *SK* Sekayu, *SO* Saok; Conservation status: *WCA* Wildlife Conservation Act, *IUCN* International Union for Conservation of Nature

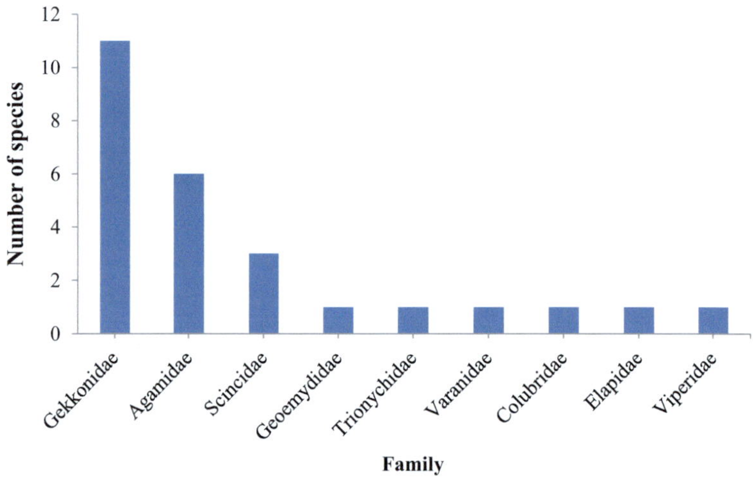

Fig. 2 Composition of reptile in each family in selected sites of Tasik Kenyir, Terengganu

additional surveys in the study sites are necessary to increase the chances of species detection. Gekkonidae was found to be the most dominant family at Tasik Kenyir with almost half of the total species recorded (42%) were from this family. This is because most of the species in this family such as *Gehyra mutilata*, *Gekko monarchus, G. smithii, Hemidactylus frenatus* and *H. garnotii* were considered as commensal species and were easily spotted around human residences. Some of these species may come from nearby forested habitats to forage in human habitations where the food source is available.

Eutropis multifasciata from family Scincidae (Fig. 4) was the most abundant species (23.7%) recorded in this study, found at all study sites except Saok. This

Table 3 Number of estimated species from abundance- and incidence-based richness estimators, and the percentage of predicted richness actually observed of reptiles in the study area

Estimators	Mean±SD	Percentage observed
Chao 1	35.09 ± 7.33	74%
Chao 2	39.13 ± 9.29	67%

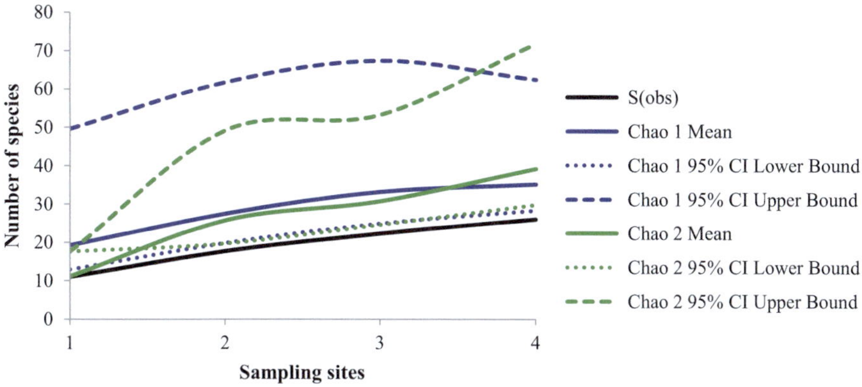

Fig. 3 Observed species, S(obs), Chao 1 and Chao 2 estimator curves for number of reptile species

Fig. 4 *Eutropis multifasciata* was observed basking on a rock at Taman Pertanian Sekayu, Kenyir, Terengganu. (Photograph by Fathihi Hakimi Rosmidi)

Fig. 5 *Tropidolaemus wagleri*, Wagler's keeled green pit viper (female)

commensal species is known to be common in forest edges and around human settlements. The specimens of this species were found around buildings, along forest trails, and on rocky outcrops near water bodies. Two individuals of *E. multifasciata*, one in Sekayu and one in Belukar Bukit were caught in cage traps that were used by other zoological team members to catch small mammals. Two individuals of *Cyclemys* sp. were also caught in cage traps placed along the river banks in Belukar Bukit.

In this study, only three snake species managed to be recorded. *Calamaria pavimentata* was observed on the ground near rocky area of Sungai Buweh waterfall while *Naja kaouthia* and *Tropidolaemus wagleri* (Fig. 5) were both found along the river at Sekayu. The low number of snakes spotted may be due to their highly elusive and cryptic behaviour. Longer sampling duration may not necessarily produce a higher number of snake species encountered. For instance, only three species of snakes were documented by Ibrahim et al. (2012) in Bukit Perangin Forest Reserve, Kedah from January 2009 till end of May 2010.

One of the interesting species encountered in the study area was *Luperosaurus browni* (Fig. 6), which was found tangled in a mist nest deployed for bird survey at Saok. This species is highly elusive as it dwells in the canopy, and the grey-brown colouration of this small gecko enables it to camouflage well on tree trunks.

The Chao 1 and Chao 2 species richness estimator showed a completeness of 74% and 67% respectively. Although this result revealed an acceptable estimate based on the sampling efforts (Shazali et al. 2016), additional surveys with standardised sampling methods are highly recommended to better estimate the reptile species composition at Tasik Kenyir. The duration of sampling of this study ranged from seven to 15 days per study site, which was not enough to cover the entire forests, rivers and waterfalls, especially with limited manpower.

Fig. 6 *Luperosaurus browni* (Brown camouflage gecko) found in Saok, Tasik Kenyir, Terengganu

Conclusion

Based on this study, Tasik Kenyir is expected to hold a higher number of reptile species, as suggested by the estimators (up to 39 species), and the opportunities to discover rare and new species are infinite. Extensive surveys covering areas deep in the forest that are more pristine should be conducted to establish a better picture of its reptile diversity. All of the study sites are always frequented by visitors from time to time therefore it is hoped that this study would be useful for the management purposes of Tasik Kenyir. The composition of reptiles and other taxa should be taken into account before any changes to the habitat is made in order to avoid species loss especially for species that are sensitive to environmental changes.

Acknowledgements We would like to thank Department of Wildlife and National Park (DWNP) for granting permission to conduct studies at Kenyir and Setiu (Special Permit: T-00370-16-16). Kenyir Research Institute, Universiti Malaysia Terengganu (UMT) and Trans-disciplinary Research Grant Scheme (TRGS/2015/59373) under Prof. Dr. Mohd Tajuddin Abdullah are also acknowledged for the administrative and logistic supports throughout the survey. For assistance in the field we are grateful to Nur Amalina Adanan, Gertrude David, Elizabeth Pesiu, Nur Izzah Izzati, Fathihi Hakimi Rosmidi and Azuan Roslan.

References

Amirah, H., Mohd Azmeer, A. B., Sharma, D., Nurolhuda, N., Sharma, R., Ahmad Ridzuan, Y. A., & Ibrahim, J. (2013). An updated checklist of the herpetofauna of the Belum-Temengor forest reserves, Hulu Perak, Peninsular Malaysia. Proceedings of the 3rd Annual International Conference Syiah Kuala University (AIC Unsyiah) in conjunction with the 2nd International Conference on Multidisciplinary Research (ICMR). Banda Achec, 2012. Banda Acheh: Syiah Kuala University Press.

Chan, K. O., Grismer, L. L., Shahrul, A., Quah, E., Muin, M. A., Savage, A. E., Grismer, J. L., Norhayati, A., Remegio, A. C., & Greer, L. F. (2010a). A new endemic Rock Gecko *Cnemaspis* Strauch 1887 (Squamata: Gekkonidae) from Gunung Jerai, Kedah, northwestern Peninsular Malaysia. *Zootaxa, 2576*, 59–68.

Chan, K. O., van Rooijen, J., Grismer, L. L., Daicus, B., Mohd. Abdul Muin, M. A., Hamidi, J., Gregory, R., & Norhayati, A. (2010b). First report on the herpetofauna of Pulau Pangkor, Perak, Malaysia. *Russian Journal of Herpetology, 17*(2), 139–146.

Chao, A. (1984). Nonparametric estimation of the number of classes in a population. *Scandinavian Journal of Statistics, 11*, 265–270.

Colwell, R. K. (2013). Estimates: Statistical estimation of species richness and shared species from samples. Version 9. Retrieved from http://purl.oclc.org/estimates.

Das, I. (2004). *Lizards of Borneo*. Kota Kinabalu: Natural History Publications (Borneo).

Das, I. (2007). *Amphibians and reptiles of Brunei: A pocket guide*. Kota Kinabalu: Natural History Publications (Borneo).

Das, I. (2010). *A pocket guide to the reptiles of South-east Asia*. London: New Holland Publishers.

Das, I., & Norsham, Y. (2007). Status of knowledge of the Malaysian Herpetofauna. In L. S. L. Chua, L. G. Kirton, & L. G. Saw (Eds.), *Status of biological diversity in Malaysia and threat assessment of plant species in Malaysia: Proceedings of the Seminar and Workshop 28–30 June 2005* (pp. 31–38). Kepong: Forest Research Institute Malaysia.

Faizah, M. S., Abdullah, M. T., Che Aziz, A., & Ismail, R. (2015). *Geopark Tasik Kenyir*. Kuala Terengganu: Penerbit Universiti Malaysia Terengganu.

Gotelli, N. J., & Chao, A. (2013). Measuring and estimating species richness, species diversity, and biotic similarity from sampling data. In S. A. Levin (Ed.), *Encyclopedia of Biodiversity* (2nd ed., pp. 195–211). Cambridge, MA: Academic.

Gotelli, N. J., & Colwell, R. K. (2011). Estimating species richness. In A. E. Magurran & B. J. McGill (Eds.), *Frontiers in measuring biodiversity* (pp. 39–54). New York: Oxford University Press.

Grismer, L. L. (2008). A new species of insular skink (Genus *Sphenomorphus* Fitzinger 1843) from the Langkawi Archipelago, Kedah, West Malaysia with the first report of the herpetofauna of Pulau Singa Besar and an updated checklist of the herpetofauna of Pulau Langkawi. *Zootaxa, 1691*, 53–66.

Grismer, L. L. (2011). *Lizards of Peninsular Malaysia, Singapore and their adjacent archipelagos*. Frankfurt: Edition Chimaira.

Grismer, L. L., & Quah, S. H. E. (2015). The rediscovery of *Sphenomorphus malayanus* Doria, 1888 (Squamata: Scincidae) from the Titiwangsa Mountain Range of Peninsular Malaysia and its re-description as *S. senja* sp. nov. *Zootaxa, 3931*(1), 63–70.

Grismer, J. L., Grismer, L. L., Das, I., Norsham, S. Y., Lim, B. L., Tzi, M. L., Youmans, T. M., & Kaiser, H. (2004). Species diversity and checklist of the herpetofauna of Pulau Tioman, Peninsular Malaysia, with a preliminary overview of habitat utilization. *Asiatic Herpetological Research, 10*, 247–279.

Grismer, L. L., Youmans, T. M., Wood, P. L., Jr., & Grismer, J. L. (2006). Checklist of the herpetofauna of the Seribuat Archipelago, West Malaysia with comments on biogeography, natural history and adaptive types. *The Raffles Bulletin of Zoology, 54*(1), 157–180.

Grismer, L. L., Chan, K. O., Nurulhuda, N., & Sumontha, M. (2008a). A new species of karst dwelling gecko (Genus Cnemaspis Stauch 1887) from the border region of Thailand and Peninsular Malaysia. *Zootaxa, 1875*, 51–68.

Grismer, L. L., Grismer, J. L., Wood, P. L., Jr., & Chan, K. O. (2008b). The distribution, taxonomy, and redescription of the geckos *Cnemaspis affinis* (Stoliczka 1887) and *C. flavolineata* (Nicholls 1949) with descriptions of a new montane species and two new lowland, karst-dwelling species from Peninsular Malaysia. *Zootaxa, 1931*, 1–24.

Grismer, L. L., Norhayati, A., Chan, K. O., Daicus, B., Muin, M. A., Grismer, J. L., & Wood, P. L., Jr. (2008c). Two new diminutive species of *Cnemaspis* Strauch 1887 (Squamata: Gekkonidae) from Peninsular Malaysia. *Zootaxa, 2019*, 40–56.

Grismer, L. L., Wood, P. L., Jr., & Grismer, J. L. (2009). A new insular skink of the genus *Sphenomorphus* Strauch 1887 (Squamata: Scincidae) from Pulau Perhentian Besar, Terengganu, Peninsular Malaysia. *Tropical Life Sciences Research, 20*, 51–69.

Grismer, L. L., Chan, K. O., Quah, E., Muin, M. A., Savage, A. E., Grismer, J. L., Norhayati, A., Greer, L. F., & Remegio, A. C. (2010a). Another new, diminutive Rock Gecko (*Cnemaspis* Strauch) from Peninsular Malaysia and a discussion of resource partitioning in sympatric species pairs. *Zootaxa, 2569*, 55–66.

Grismer, L. L., Shahrul, A., Quah, E., Muin, M. A., Chan, K. O., Grismer, J. L., & Norhayati, A. (2010b). A new spiny, prehensile-tailed species of *Cyrtodactylus* (Squamata: Gekkonidae) from Peninsular Malaysia with a preliminary hypothesis of relationships based on morphology. *Zootaxa, 2625*, 40–52.

Grismer, L. L., Quah, E. S. H., Siler, C. D., Chan, K. O., Wood, P. L., Jr., Grismer, J. L., Shahrul Anuar, M. S., & Norhayati, A. (2011). Peninsular Malaysia's first limbless lizard: A new species of skink of the genus *Larutia* (Bohme) from Pulau Pinang with a phylogeny of the genus. *Zootaxa, 2799*, 29–40.

Grismer, L. L., Lukman, H. B. I., Awang, M. T., Syed, A. R., & Ahmad, B. A. (2014a). A new species of lowland skink (genus *Lipinia* Gray, 1845) from northestern Peninsular Malaysia. *Zootaxa, 3821*(4), 457–464.

Grismer, L. L., Wood, P. L., Jr., Anuar, S., Riyanto, A., Norhayati, A., Muin, M. A., Sumontha, M., Grismer, J. L., Chan, K. O., Quah, E. S. H., & Pauwels, O. S. A. (2014b). Systematics and natural history of Southeast Asian Rock Geckos (genus *Cnemaspis* Strauch, 1887) with descriptions of eight new species from Malaysia, Thailand, and Indonesia. *Zootaxa, 3880*(1), 1–147.

Grismer, L. L., Wood, P. L., Jr., Ahmad, A., Sumarli, A. S., Vasquez, J. J., Ismail, L. H., Nance, R., Mohd-Amin, M. A., Othman, M. N., Rizaijessika, S. A., Kuss, M., Murdoch, M., & Cobos, A. (2014c). A new species of insular Rock Gecko (Genus *Cnemaspis* Strauch, 1887) from the Bidong Archipelago, Terengganu, Peninsular Malaysia. *Zootaxa, 3755*(5), 447–456.

Grismer, L. L., Wood, P. L., Jr., Shahrul, A., Davis, H. R., Cobos, A. J., & Murdoch, M. L. (2016a). A new species of karst forest Bent-toed Gecko (genus *Cyrtodactylus* Gray) not yet threatened by foreign cement companies and a summary of Peninsular Malaysia's endemic karst forest herpetofauna and the need for its conservation. *Zootaxa, 4061*(1), 1–17.

Grismer, L. L., Quah, E. S. H., Wood, P. L., Jr., Shahrul, A., Muin, A., Davis, H. R., Murdoch, M. L., Grismer, J. L., Cota, M., & Cobos, A. J. (2016b). Dragons in the mist: Three new species of *Pseudocalotes* Fitzinger (Squamata: Agamidae) from sky island archipelago of Peninsular Malaysia. *Zootaxa, 4136*(3), 461–490.

Grismer, L. L., Wood, P. L., Jr., Shahrul, A., Grismer, M. S., Quah, E. S. H., Murdoch, M. T., Muin, M. A., Davis, H. R., Aguilar, C., Klabacka, R., Cobos, J., Aowphol, A., & Sites, J. W., Jr. (2016c). Two new Bent-toed Geckos of the *Cyrtodactylus pulchellus* complex from Peninsular Malaysia and multiple instances of convergent adaptation to limestone forest ecosystems. *Zootaxa, 4105*(5), 401–429.

Grismer, L. L., Wood, P. L., Jr., Muhammad, F. S., Baizul, H. B., Rizal, S. A., Ahmad, B. A., & Quah, E. S. H. (2016d). On the taxonomy and phylogeny of the skinks *Lipinia sekayuensis* and *Lipinia surda* Boulenger from Peninsular Malaysia. *Zootaxa, 4147*(1), 59–66.

Ibrahim, J., Shahrul Anuar, M. S., Norhayati, A., Chan, K. O., & Muin, M. A. (2008). *The common amphibians and reptiles of Penang Island*. Penang: Penang State Forestry Department.

Ibrahim, J., Zalina, A., Shahriza, S., Shahrul Anuar, M. S., Nur Hafizah, I., Amirah, H., Nurul Dalila, A. R., Min, M. A., & Amirudin, I. (2012). Checklist of the herpetofauna of Bukit Perangin Forest Reserve, Kedah, Malaysia. *Sains Malaysiana, 41*(6), 691–696.

Ibrahim, J., Amirah, H., Shahriza, S., Nur Hafizah, I., Zalina, A., Yap, C. H., Nurliza, A. M., & Nur Hafizah, C. Z. (2013). Additions to the herpetofauna of Jerejak Island, Penang, Peninsular Malaysia. *Malayan Nature Journal, 64*(4), 213–232.

IUCN Red List of Threatened Species. (2016). Retrieved from http://www.iucnredlist.org/.

Johnson, C. B., Quah, S. H. E., Anuar, S., Muin, M. A., Wood, P. L., Jr., Grismer, J. L., Greer, L. F., Chan, K. O., Norhayati, A., Bauer, A. M., & Grismer, L. L. (2012). Phylogeography, geographic variation, and taxonomy of the Bent-toed Gecko *Cyrtodactylus quadrivirgatus* Taylor, 1962 from Peninsular Malaysia with the description of a new swamp dwelling species. *Zootaxa, 3406*, 39–58.

Nurulhuda, Z., Juliana, S., Fakhrul Hatta, M., Daicus, B., Chan, K. O., Shukor, M. N., & Norhayati, A. (2014). Species composition of amphibians and reptiles in Krau Wildlife Reserve, Pahang, Peninsular Malaysia. *Check List, 10*(2), 335–343.

Quah, E. S. H., Shahrul Anuar, M. S., Muin, M. A., Nur Amira, A. R., Fatim Syakirah, M., & Grismer, L. L. (2013). Species diversity of herpetofauna of Bukit Panchor State Park, Penang, Peninsular Malaysia. *Malayan Nature Journal, 64*(4), 193–211.

Raxworthy, C. J., Pearson, R. G., Zimkus, B. M., Reddy, S., Deo, A. J., Nussbaum, R. A., & Ingram, C. M. (2008). Continental speciation in the tropics: Contrasting biogeographic patterns of divergence in the *Uroplatus* leaf-tailed gecko radiation of Madagascar. *Journal of Zoology, 275*, 423–440.

Read, J. L. (1998). Are geckos useful bioindicators of air pollution? *Oecologia, 114*, 180–187.

Shahriza, S., & Ibrahim, J. (2014). Reptiles of Lata Bukit Hijau, Kedah, Malaysia. *Songklanakarin Journal of Science and Technology, 36*(1), 37–44.

Shahriza, S., Ibrahim, J., Shahrul Anuar, M. S., & Abdul Muin, M. A. (2012). Herpetofauna of Peta area of Endau-Rompin National Park, Johor, Malaysia. *Pertanika Journal of Tropical Agricultural Science, 35*(3), 553–567.

Shazali, N., Sultana, P. H. R., Nurul, F. D. A. T., Rafik, M., Nurshilawati, A. L., Nur Mukminah, N., & Tingga, R. C. T. (2016). Small mammals from Miri, northeastern region of Sarawak, Malaysian Borneo: Note on new locality records. *Check List, 12*(2), 1863.

Sohdi, N. S., Koh, L. P., Brook, B. W., & Ng, P. K. L. (2004). Southeast Asian biodiversity: An impending disaster. *Trends in Ecology & Evolution, 19*(12), 654–660.

Sumarli, A. X., Grismer, L. L., Shahrul, A., Muin, M. A., & Quah, E. S. H. (2015). First report on the amphibians and reptiles of a remote mountain, Gunung Tebu in northeastern Peninsular Malaysia. *Check List, 11*(4), 1679.

Sumarli, A., Grismer, L. L., Wood, P. L., Jr., Ahmad, B. A., Syed, R., Ismail, L. H., Nur Amalina, M. I., Norhayati, A., & Linkem, C. W. (2016). The first riparian skink (Genus: *Sphenomorphus* Strauch, 1887) from Peninsular Malaysia and its relationship to other Indochinese and Sundaic species. *Zootaxa, 4173*(1), 29–44.

van Rooijen, J., Chan, K. O., Grismer, L. L., & Norhayati, A. (2011). Estimating the herpetofaunal species richness of Pangkor Island, Peninsular Malaysia. *Bonn Zoological Bulletin, 60*(1), 3–8.

Voris, H. K. (2015). Marine snake diversity in the mouth of the Muar River, Malaysia. *Tropical Natural History, 15*(1), 1–21.

Wood, P. L., Jr., Grismer, J. L., Grismer, L. L., Norhayati, A., Chan, K. O., & Bauer, A. M. (2009). Two new montane species of *Acanthosaura* Gray, 1831 (Squamata: Agamidae) from Peninsular Malaysia. *Zootaxa, 2012*, 28–46.

Note on Pollen and Seeds Dispersed by Pteropodid Bats in Kenyir Forest Area, Terengganu

Nor Nadhirah Ahmad Yazid, Sri Handayani Gunawan, Azuan Roslan, Mohamed Nor Zalipah, and Mohd Tajuddin Abdullah

Abstract A study on the role of bats as pollinating and seed dispersal agents was conducted at Kenyir forest area, Terengganu. The objectives of this study were to identify the species of bats that potentially act as pollinating and seed dispersal agents at Kenyir forest area and to determine the plant species visited by this bats from the pollen types and seeds collected from the bats. In this study, six pollen types were recorded including *Ceiba pentandra*, *Melaleuca* sp., *Musa* sp., *Oroxylum indicum*, *Parkia* sp. and *Sonneratia* sp. From the total of eight species of bats caught at the study area, only *Balionycteris maculata*, *Cynopterus brachyotis* and *Eonycteris spelaea* were identified as pollinating agents. *Cynopterus brachyotis* carried the highest number of pollen types which was six, with the highest number of pollen grains was from *Musa* sp. *Balionycteris maculata and E. spelaea* on the other hand carried five and four pollen types respectively, with *Parkia* sp. pollen grains collected the most from both species. *Cynopterus brachyotis*, *C. horsefieldi* and *E. spelaea* were the fruit bats that act as seed dispersal agents for five different *Ficus* sp. Not only the fruits were ingested, four fruits of three *Ficus* sp. were found to be carried by the bats in flight. Our study indicates the important of pteropodid bats particularly *C. brachyotis* as pollinator and seed dispersal at the study area.

Keywords Pteropodids · Fruit-eating bats · Pollinators · Seed dispersal · Tasik Kenyir

---

N. N. A. Yazid · S. H. Gunawan · M. Nor Zalipah (✉)
School of Marine and Environmental Sciences, Universiti Malaysia Terengganu, Kuala Nerus, Terengganu, Malaysia
e-mail: zalipah@umt.edu.my

A. Roslan
Institute of Tropical Biodiversity and Sustainable Development, Universiti Malaysia Terengganu, Kuala Nerus, Terengganu, Malaysia

M. T. Abdullah
Institute of Tropical Biodiversity and Sustainable Development, Universiti Malaysia Terengganu, Kuala Nerus, Terengganu, Malaysia

School of Marine and Environmental Sciences, Universiti Malaysia Terengganu, Kuala Nerus, Terengganu, Malaysia

M. T. Abdullah et al. (eds.), *Greater Kenyir Landscapes*,
https://doi.org/10.1007/978-3-319-92264-5_16

Introduction

About one third of bat species rely upon plants for food either partially or wholly (Marshall 1983). These bats feed mainly on nectar, pollen, fruits and other edible plant tissues such as sweet bracts, petals and leaves. Bats in return are used by these plants to promote pollination and seed dispersal. Because pollen and seeds are the two ways by which plants transfer their genes to the next generation, their dispersal also affects the genetic structure of the plant population, and hence has important evolutionary consequences (Fleming and Kress 2013). As plant-visiting bats commute long distances each night to search for food (Horner et al. 1998; Marshall 1983), they potentially play an important role in the maintenance and regeneration of the forests by moving pollen and seeds over large distances (Andriafidison et al. 2006; Arias-Coyotl et al. 2006; Law and Lean 1999). By moving pollen and seeds away from the parental trees, not only do the bats assist with gene flow among isolated populations (Dick et al. 2008), but they also prevent inbreeding especially in plants that are capable of selfing (Lobo et al. 2005).

In Peninsular Malaysia, there are more than 50 plant species that depend on bats for pollination and seed dispersal including *Ceiba pentandra*, *Ficus* spp., *Durio zibethinus*, *Musa* sp., *Oroxylum indicum*, *Palaquium* sp., *Parkia* sp. and *Sonneratia* spp. (Nor Zalipah et al. 2016; Muscarella and Fleming 2007; Hodgkison et al. 2003; Tan et al. 1999; Start and Marshall 1976). Many of these plants pollinated by bats are endemics (Fleming and Muchhala 2008) and many are of considerable economic value (Bumrungsri et al. 2013; Kunz et al. 2011; Fujita and Tuttle 1991). Therefore bats are not only important for ecological functions, but also benefit humans with their pollination behaviour. For example, bats are important pollinators of bananas (*Musa* sp.), one of the most widely consumed aseasonal fruits with numerous varieties. In Malaysia, the revenue of banana cultivation increased from RM3.4 million in 2009 to RM8.1 million in 2012 (Zazali 2013). The value of durian (*Durio* sp.) exports from Malaysia to China in 2013 was reported to have yielded about RM4 million (Noor Azura 2013). Bitter bean (*Parkia* sp.), silk cotton (*Ceiba pentandra*) and Indian trumpet (*Oroxylum indicum*) are of great local and regional value. The bitter bean seeds are consumed mainly in Malaysia, Indonesia and India, while the fruits and shoots of the Indian trumpet are eaten by humans particularly in the east coast of Peninsular Malaysia and the southern part of Thailand. The whitish fibre extracted from the silk cotton fruits is used as stuffing material in mattresses, pillows and toys. Other than that, the oil extracted from the seeds of the silk cotton fruits was found to have biodiesel potential (Ong et al. 2013; Berry 1979).

Bats are among the most common and significant pollinators and seed dispersal agents in tropical region (i.e. Nor Zalipah et al. 2016; Kaiser-Bunbury et al. 2010; Fleming et al. 2009; Fleming and Muchhala 2008; Sekercioglu 2006), which also are among the most susceptible to disturbance by human. Extinction rates due to habitat destruction and over exploitation of natural resources however are acute in tropics particularly Southeast Asia, given the high species richness of these habitats (Bradshaw et al. 2009). The loss and degradation of essential ecosystem functions

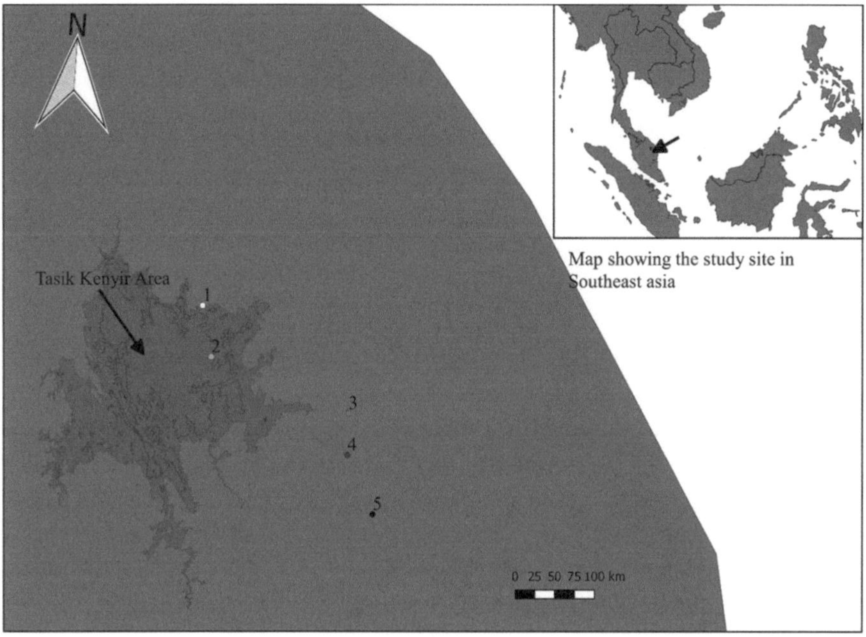

Fig. 1 Location of sampling sites at Kenyir forest area. 1 = UMT Research Station, 2 = Sauk Waterfall, 3 = Kampung Kemat, 4 = Taman Pertanian Sekayu, 5 = Belukar Bukit

(pollination, seed dispersal, carbon sequestration, water cycling etc.) and services (flood mitigation, topsoil retention, non-timber products) are threatening billions of people living in tropical countries. Therefore, better understanding in the significant role of bats as pollinators and seed dispersal agents in their environment will certainly enhance the effectiveness of tropical forest conservation practices, which critically needed at time when these forests are seriously threatened by human actions. Here, we report the plant species with high potential to be pollinated and dispersed by the pteropodid bats in Kenyir forest area, based on pollen grains attached at their bodies, and seeds they consume and carried while in flight.

Nettings were conducted at various sites of Kenyir forest area (Fig. 1) in between August and September 2015, and between January and February 2016, to collect pollen attached to the bats' bodies and seeds from their faeces. Pollen grains adhering to the bat's bodies were sampled by carefully rubbing their bodies with a small block (2 mm^3) of glycerin jelly to which pollen grains adhere (Wooller et al. 1983). This block is then placed on a microscope slide with a coverslip resting on top and warmed gently by holding the slide above cigarette lighter until the gel melts. The slides were then keep in slide box for further pollen identification in the lab. Collection of pollen grains from the bats however was not conducted at Sauk Waterfall. Identifications of the pollen grains collected were conducted following Nor Zalipah et al. (2015). Netted bats were kept in cloth bag individually for an hour, in which most individual defecated to get their faeces (Corlett 1998). Faeces collected were kept in individual vial containing 75% ethanol for further examinations in the lab. The seeds were identified based on its morphology and colour.

Fruitbats as Pollinators and Seed Dispersers

A total of 43 pteropodid bats from eight species were recorded throughout this study. *Cynopterus brachyotis* recorded the highest number of individuals caught, representing 74.4% of the total captures, while single individual were recorded for *Rousettus amplexicaudatus*, *Chironax melanocephalus*, *Megaerops ecaudatus* and *Macroglossus sobrinus* (Table 1). However, from the eight species, only three bats species recorded pollen grains on their bodies namely *C. brachyotis, Balionycteris maculata* and *Eonycteris spelaea*, while three species recorded with faecal containing seeds, which were *C. brachyotis, C. horsfieldi* and *E. spelaea* (Fig. 2).

We recorded six pollen types from the three bat species which include *Ceiba pentandra*, *Melaleuca* sp., *Musa* sp., *Oroxylum indicum*, *Parkia* sp. and *Sonneratia* sp. *Cynopterus brachyotis* (N = 18) collected all six pollen types, with the most pollen grains belonged to *Musa* sp. Each bat individual carried two to five pollen types at the time of captured. *Balionycteris maculata* (N = 3) collected five types of pollen grains (except *Melaleuca* sp.) with individuals carried four to five pollen types while *E. spelaea* (N = 2) carried four species (except *C. pentandra* and *Melaleuca* sp.). The two individuals of *E. spelaea* recorded three and four pollen types on their bodies respectively. For both *B. maculata* and *E. spelaea,* pollen grains of *Parkia* sp. was the most collected attached at their bodies.

Table 1 The list of pteropodid bats captured in Kenyir forest area, Terengganu, with number of individuals with seed content in faecal samples and pollen grains on their bodies

Bat species	Common name	Number of individuals	Number of samples with seeds in faeces	Number of individuals carrying pollen grains
Rousettus amplexicaudatus	Geoffroy's rousette bat	1	0	0
Cynopterus brachyotis	Lesser short-nosed fruit bat	32	16	18
Cynopeterus horsfieldii	Horsfield's fruit bat	3	1	0
Balionycteris maculata	Spotted winged fruit bat	2	0	3
Chironax melanocephalus	Black-capped fruit bat	1	0	0
Megaerops ecaudatus	Temminck's tailless fruit bat	1	0	0
Eonycteris spelaea	Lesser dawn bat	2	1	2
Macroglossus sobrinus	Greater long-tongued nectar bat	1	0	0

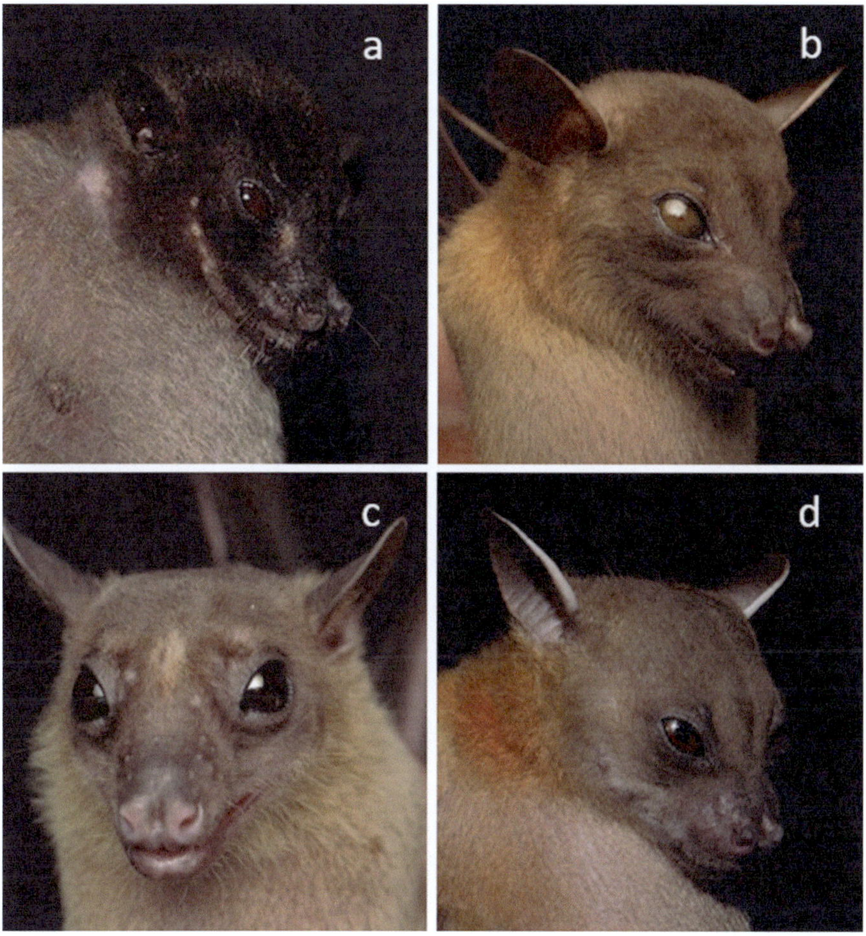

Fig. 2 The bat species captured at Kenyir forest areas which carried pollen grains on their bodies and seed/fruit to be dispersed. (**a**) *Balionycteris maculata,* (**b**) *Cynopterus brachyotis,* (**c**) *Cynopterus horsefieldi,* and (**d**) *Eonycteris spelaea.* Photographs by Azuan Roslan

The pollen types recorded in our study was only half of the species listed by Nor Zalipah et al. (2015) for pteropodid bats in mangrove habitat, due to short sampling duration of this study. It is however comparable to four plant species (*Parkia javanica, P. speciosa, Palaquium bispidum* and *P. obovatum*) reported to be visited by the bats in lowland rainforest elsewhere in Peninsular Malaysia from the pollen grains observed on the bats' bodies (Hodgkison et al. 2003). It is interesting to note that we recorded *Sonneratia* sp. pollen grains in this study as it is a genus of mangrove species. This confirms that bats travel long distance to feed as Kenyir forest area located in inland side and more than 100 km away from the mangrove forest along Terengganu coastline.

Fig. 3 The seeds of four different *Ficus* species recorded to be dispersed by bats in Kenyir forest area. (**a**–**c**) are the seeds observed from fecal samples, (**c**) is also recorded from partially eaten fruit carried by the bats in flight. (**d**) is a seed extracted from uneaten fruit collected in the mist nets together with the an entangled *C. brachyotis*

Meanwhile, the seeds observed in the faeces were from three different *Ficus* species based on the seed size and seed coat morphology (Fig. 3). Apart from that, four individuals of *C. brachyotis* were netted with partially eaten and uneaten *Ficus* fruits. Examination of the seeds showed that the fruits were of three different *Ficus* species, with seeds of one of the *Ficus* fruits were also found in the faeces. Therefore, all together, the bats recorded to feed on five different *Ficus* sp. *Cynopterus brachyotis* (N = 16) recorded all five *Ficus* species while *C. horsfieldi* (N = 1) and *E. spelaea* (N = 1) recorded with one different *Ficus* species respectively in their faeces.

The pteropodid bats in Peninsular Malaysia particularly *C. brachyotis* were reported to have good predilection towards *Ficus* spp. by Tan et al. (1999), therefore help in dispersing the seeds. The fruits are usually consumed in situ by the bats, but are sometimes carried to feeding roosts before consumption to prevent disturbance from other bats. The flesh is bitten off or rasped away by the bats to extract juice, and the bats often ingest only the juice and pulp. Larger seeds are often dropped, while smaller seeds are usually ingested or rejected in the form of solid pellets (Hodgkison et al. 2003; Phua and Corlett 1989; Marshall 1983). Pteropodid bats therefore act as important seed dispersal agent by moving the seeds far away from the parent tree (Seltzer et al. 2013; Tang et al. 2012).

Conclusion

This study recorded six plant species visited by pteropodid bats to feed on flowers, and five *Ficus* sp. were visited for fruits. These bats therefore potentially help in pollination and seed dispersal of the plant species. *Cynopterus brachyotis* in particularly is important for the survival of *Musa* sp., *Sonneratia* sp., *Oroxylum indicum*, *Parkia* sp., *Ceiba pentandra*, *Melaleuca* sp. and *Ficus* spp. in Terengganu. Pollen and seeds are the two ways by which plants transfer their genes to the next generation, and their dispersal influence the genetic structure of the plant population. Bats as pollinating and seed dispersal agents are therefore important ecosystem engineers in which they determine ecosystem structure of the habitat they live in. As forest ecosystem are responsible for the regulation of water drainage system, and Kenyir Dam play important role in flood mitigation in east coast of Peninsular Malaysia particularly Terengganu and Kelantan, this study provides an important information for resources managers and planners of Kenyir as this ecosystem is where the Kenyir Dam is located.

Acknowledgements We would like to thank School of Marine and Environmental Sciences and Kenyir Research Institute, UMT for equipment and logistical support to conduct this research. This study was supported by various research grants, *Geran Galakan Penyelidikan* (GGP/68007/2014/127) and Trans Disciplinary Research Grant Scheme (TRGS/2015/59373).

References

Andriafidison, D., Andrianaivoarivelo, R. A., Ramilijaona, O. R., Razanahoera, M. R., MacKinnon, J., Jenkins, R. K. B., & Racey, P. A. (2006). Nectarivory by endemic Malagasy fruit bats during the dry season. *Biotropica, 38*, 85–90.

Arias-Coyotl, E., Stoner, K. E., & Casas, A. (2006). Effectiveness of bats as pollinators of *Stenocereus stellatus* (Cactaceae) in wild, managed in situ, and cultivated populations in La Mixteca Baja, central Mexico. *American Journal of Botany, 93*, 1675–1683.

Berry, S. K. (1979). The characteristics of the kapok (*Ceiba petandra*, Gaertn.) seed oil. *Pertanika, 2*(1), 1–4.

Bradshaw, C. J. A., Sodhi, N. S., & Brook, B. W. (2009). Tropical turmoil: A biodiversity tragedy in progress. *Frontiers in Ecology and Evolution, 7*, 79–87.

Bumrungsri, S., Lang, D., Harrower, C., Sripaoraya, E., Kitpipit, K., & Racey, P. A. (2013). The dawn bat, *Eonycteris spelaea* Dobson (Chriptera: Pteropodidae) feeds mainly on pollen of economically important food plants in Thailand. *Acta Chiropterologica, 15*, 95–104.

Corlett, R. T. (1998). Frugivory and seed dispersal by vertebrates in the Oriental (Indomalayan) region. *Biological Reviews, 73*, 413–448.

Dick, C. W., Hardy, O. J., Jones, A. F., & Petit, R. J. (2008). Spatial scales of pollen and seed-mediated gene flow in tropical rain forest trees. *Tropical Plant Biology, 1*, 20–33.

Fleming, T. H., & Kress, W. J. (2013). *The ornaments of life, coevolution and conservation in the tropics*. Chicago: The University of Chicago Press.

Fleming, T. H., & Muchhala, N. (2008). Nectar-feeding bird and bat niches in two worlds: Pantropical comparisons of vertebrate pollination systems. *Journal of Biogeography, 35*, 764–780.

Fleming, T. H., Geiselman, C., & Kress, W. J. (2009). The evolution of bat pollination: A phylogenetic perspective. *Annals of Botany, 104*, 1017–1043.

Fujita, M. S., & Tuttle, M. D. (1991). Flying foxes (Chiroptera: Pteropodidae): Threatened animals of key ecological and economic importance. *Conservation Biology, 5*, 455–463.

Hodgkison, R., Balding, S., Zubaid, A., & Kunz, T. (2003). Fruit bats (Chiroptera: Pteropodidae) as seed dispersers and pollinators in a lowland Malaysian rain forest. *Biotropica, 35*(4), 491–502.

Horner, M. A., Fleming, T. H., & Sahley, C. T. (1998). Foraging behaviour and energetics of a nectar-feeding bat, *Leptonycteris curasoae* (Chiroptera: Phyllostomidae). *Journal of Zoology, 244*, 575–586.

Kaiser-Bunbury, C. N., Traveset, A., & Hansen, D. M. (2010). Conservation and restoration of plant-animal mutualism on oceanic islands. *Perspectives in Plant, Ecology and Systematics, 12*, 131–143.

Kunz, T. H., de Torrez, E. B., Bauer, D., Lobova, T., & Fleming, T. H. (2011). Ecosystem services provided by bats. *Annals of the New York Academy of Sciences, 1223*, 1–38.

Law, B. S., & Lean, M. (1999). Common blossom bats (*Syconycteris australis*) as pollinators in fragmented Australian tropical rainforest. *Biological Conservation, 91*, 201–212.

Lobo, J. A., Quesada, M., & Stoner, K. E. (2005). Effects of pollination by bats on the mating system of *Ceiba pentandra* (Bombacaceae) populations in two tropical life zones in Costa Rica. *American Journal of Botany, 92*, 370–376.

Marshall, A. G. (1983). Bats, flowers and fruit: Evolutionary relationships in the Old World. *Biological Journal of the Linnean Society, 20*, 115–135.

Muscarella, R., & Fleming, T. H. (2007). The role of frugivorous bats in tropical forest succession. *Biological Reviews, 82*(4), 573–590.

Noor Azura, S. (2013). *Eksport durian beku meningkat RM24 juta*. Retrieved from http://www.doa.gov.my/c/document_library/get_file?uuid=b3ff5ea1-8df3-4241-a1f0-204ac24c9dac&groupId=38371.

Nor Zalipah, M., Mohd Lokman, H., & Sulong. (2015). Pollination importance of bats in mangrove communities od Setiu Wetland, Terengganu. In M. Faridah, M. S. Jamilah, M. J. Jarina, & S. Rohani (Eds.), *Setiu Wetlands: Species, ecosystems and livelihoods* (pp. 73–84). Kuala Terengganu: Penerbit Universiti Malaysia Terengganu.

Nor Zalipah, M., Shahrul Anuar, M. S., & Jones, G. (2016). Potential significance of nectar-feeding bats as pollinators in mangrove habitats of Peninsular Malaysia. *Biotropica, 48*, 425–428.

Ong, H. C., Silitonga, A. S., Masjuki, H. H., Mahlia, T. M. I., Chong, W. T., & Boosroh, M. H. (2013). Production and comparative fuel properties of biodiesel from non-edible oils: *Jatropha curcas, Sterculia foetida and Ceiba petandra. Energy Conservation and Management, 73*, 245–255.

Phua, P. B., & Corlett, R. (1989). Seed dispersal by the lesser short-nosed fruit bat (*Cynopterus brachyotis*, Pteropodidae, Megachiroptera). *Malayan Nature Journal, 42*, 251–256.

Sekercioglu, C. H. (2006). Increasing awareness of avian ecological function. *Trends in Ecology & Evolution, 21*, 464–471.

Seltzer, C. E., Ndangalasi, H. J., & Cordeiro, N. J. (2013). Seed dispersal in the dark: Shedding light on the role of fruit bats in Africa. *Biotropica, 45*, 450–456.

Start, A. N., & Marshall, A. N. (1976). Nectarivorous bats as pollinators of trees in west Malaysia. In J. Burley & B. Styles (Eds.), *Tropical trees: Variation, breeding, and conservation* (pp. 141–150). London: Academic.

Tan, K. H., Zubaid, A., & Kunz, T. H. (1999). Fruit dispersal by the lesser dog-faced fruit bat *Cynopterus brachyotis* (Muller) (Chiroptera: Pteropodidae). *Malayan Nature Journal, 54*(1), 51–62.

Tang, Z. H., Xu, J. L., Flanders, J., Ding, X. M., Ma, X. F., Sheng, L., & Cao, M. (2012). Seed dispersal of *Syzygium oblatum* (Myrtaceae) by two species of fruit bat (*Cynopterus sphinx* and *Rousettus leschenaulti*) in south west China. *Journal of Tropical Ecology, 28*, 255–261.

Wooller, R., Russell, E., & Renfree, M. (1983). A technique for sampling pollen carried by vertebrates. *Technical Note, 10*, 433–434.

Zazali, M. (2013). Banana business stays bountiful. Retrieved from http://www.thestar.com.my/News/Community/2013/02/22/Banana-business-stays-bountiful.aspx.

Taxonomic Composition of Non-volant Small Mammal Assemblages in Tasik Kenyir, Hulu Terengganu, Terengganu

Nur Ainnurq Mohammad Noor, Noor Aisyah A. Rahim, Nur Izzah Izzati Ahmad, and Mohd Tajuddin Abdullah

Abstract A study was carried out in the area of Tasik Kenyir, Hulu Terengganu from July 2015 to April 2016. The aims of the study were to access the taxonomic composition of non-volant small mammal assemblages in Tasik Kenyir area and to investigate if the types of ecosystems influence the species richness and diversity of non-volant small mammal assemblages in that area. A total of 56 individuals comprising five orders, six families and 12 species of non-volant small mammals were recorded. The Common Treeshrew (*Tupaia glis*) was the most abundant species of non-volant small mammals recorded in Tasik Kenyir area. The highlights of this study included the Grey-cheeked Flying Squirrel (*Hylopetes platyurus*) and the world's smallest mammal, Savi's Pigmy Shrew (*Suncus etruscus*). In conclusion, Kampung Kemat recorded the highest diversity of small mammals with 13 individuals belonging to three orders, four families and six species. The types of habitat may influence the species richness and diversity of non-volant small mammals by their tendency to adapt to those types of habitat due to the fact that certain species are more adaptable to many kinds of habitats than the other species.

Keywords Composition · Habitat · Non-volant small mammals · Tasik Kenyir

---

N. A. Mohammad Noor (✉)
School of Marine and Environmental Sciences, Universiti Malaysia Terengganu,
Kuala Nerus, Terengganu, Malaysia

N. A. A. Rahim · N. I. I. Ahmad
Institute of Tropical Biodiversity and Sustainable Development,
Universiti Malaysia Terengganu, Kuala Nerus, Terengganu, Malaysia

M. T. Abdullah (✉)
Institute of Tropical Biodiversity and Sustainable Development, Universiti Malaysia Terengganu, Kuala Nerus, Terengganu, Malaysia

School of Marine and Environmental Sciences, Universiti Malaysia Terengganu,
Kuala Nerus, Terengganu, Malaysia
e-mail: mohd.tajuddin@umt.edu.my

M. T. Abdullah et al. (eds.), *Greater Kenyir Landscapes*,
https://doi.org/10.1007/978-3-319-92264-5_17

Introduction

Merritt (2010) stated that small mammals are mammalians weighing 5 kg or less. They are further categorized into volant and non-volant small mammals. Small mammals are not surveyed thoroughly in many areas in Southeast Asia (Francis 2001). The studies of the diversity and abundance of small mammals in Malaysia have been scarce (Ruppert et al. 2015). Small mammals are often caught at emergent, forest litter, rotting logs, seedlings or rough barks as these kinds of habitats are suitable for breeding and nesting sites for them as well as plenty of food resources being available (Zakaria et al. 2001; Zakaria and Nordin 1998; Medway 1983).

Tasik Kenyir is located at Hulu Terengganu and covers over 209,199 ha. It is the largest man-made lake in Southeast Asia. This lake is also surrounded by one of the oldest rainforests in the world, ranging from Kelantan to Pahang (Shaharom 2015). There was only one study carried out for studying the diversity of small mammals of Tasik Kenyir. The study only recorded 14 species of volant small mammals from four families of chiropteran and no non-volant small mammals captured due to the unattractiveness of the baits and inadequate sampling efforts (Mazlan et al. 2015).

The objectives of the study were to assess the taxonomic composition of non-volant small mammal assemblages in Tasik Kenyir and to determine if the types of habitat influence the species richness and diversity of non-volant small mammals in Tasik Kenyir.

The study was carried out at in Hulu Telemong Forest Reserve (5° 13′48.0″N, 102° 50′08.9″E), Kenyir Research Station, Universiti Malaysia Terengganu (5° 08′59.5″N, 102° 45′48.9″E), Kampung Kemat (5° 00′52.2″N, 102° 57′10.4″E) and Belukar Bukit (4° 53′37.4″N, 102° 59′23.4″E) (Fig. 1). All sampling sites mainly comprised lowland dipterocarp forests. Hulu Telemong Forest Reserve was also a riparian forest as there was a river present in the sampling site. Furthermore, it was disturbed by ongoing human activities. The forest located at Kenyir Research Station was a regenerated forest as the forest was logged previously. The forest located at Kampung Kemat was disturbed as there were human settlements nearby as well as ongoing human activities. Belukar Bukit was a riparian forest and recently became a disturbed forest due to human intrusion.

A total of 100 cage traps were set up for each study site in Tasik Kenyir area and its vicinity. Selecting suitable multiple microhabitats for trap placement at each station is important to increase the documentation of non-volant small mammals (Jayaraj et al. 2013). Bananas were used as baits as proposed by Bernard (2003) to be the most suitable baits used to capture small mammals. Other live trapping methods used were pitfall trapping and mist netting. Thirty pitfall traps and ten mist nets were set up during this study. The traps were checked daily between 8.00 a.m. until 10.00 a.m. and 3.00 p.m. until 5.00 p.m.

A line transect along 1 km trail was used to conduct visual encounter survey (VES) (Ramli and Hashim 2009; Crump and Scott 1994). Diurnal and nocturnal VES were carried out as frequently as possible during the sampling period. The Bushnell Powerview 20 × 50 Optics Green Film binoculars were used for diurnal VES and widebeam spotlight was used during nocturnal VES.

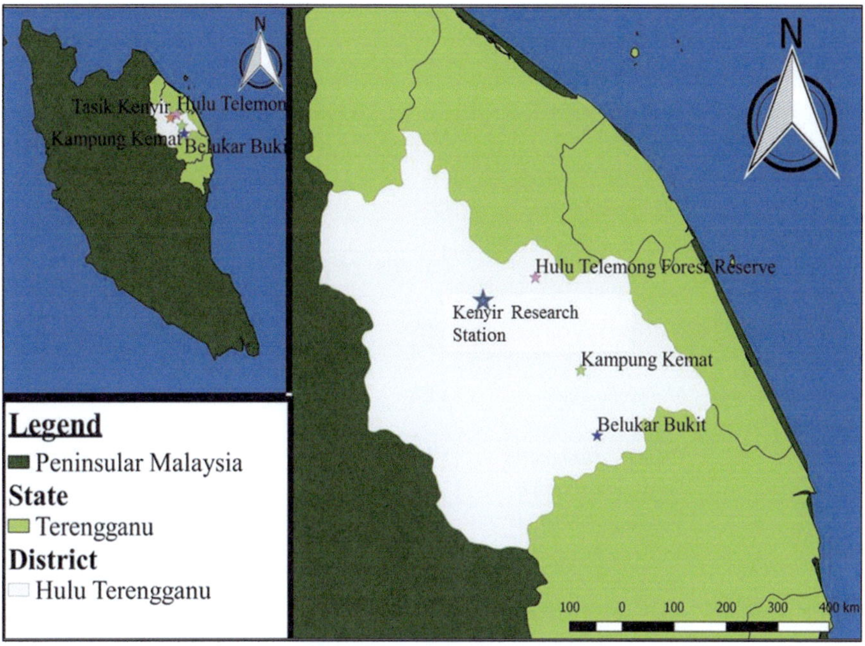

Fig. 1 Map showing the study area in Tasik Kenyir area, Hulu Terengganu

All individuals captured and observed were identified following Francis (2013), Shepherd and Shepherd (2012), Francis (2001, 2008), Payne and Francis (1998), and Medway (1983). All standard measurements of captured individuals were recorded as well. Individuals for museum vouchers were euthanized by using chloroform before being tagged and preserved in 70% ethanol. The species diversity and richness were calculated and compared using t-test and a dendrogram constructed by using PAST software, version 2.17 (Caceres et al. 2011; Hammer et al. 2001).

Species Diversity and Richness of Non-volant Small Mammals in Tasik Kenyir

A total of 56 individuals comprising five orders, six families and 12 species were recorded in Tasik Kenyir area, Hulu Terengganu (Table 1). The most abundant species was the Common Treeshrew (*Tupaia glis*) from family Tupaiidae of order Scandentia. Most species recorded were from order Rodentia consisting of five species of sciurids and three species of murids.

The cluster analysis of non-volant small mammal assemblages separated Kenyir Research Station (KRS) which comprises lowland dipterocarp forest and regenerated forest with the other three sites. The disturbed forest of Kampung Kemat (KK) was closely clustered with Hulu Telemong Forest Reserve (HTFR) with

Table 1 Taxonomic list of non-volant small mammal species recorded with different types of habitat in Tasik Kenyir area, Hulu Terengganu

			Sampling site			
			HTFR	BB	KRS	KK
ORDER family	Species	Method	n (RA)	n (RA)	n (RA)	n (RA)
Insectivora						
Soricidae	*Suncus etruscus*	PF	0 (0)	0 (0)	1 (5)	1 (8)
Scandentia						
Tupaiidae	*Tupaia glis*	CT/VES	6 (67)	4 (29)	9 (45)	3 (23)
Dermoptera						
Cynocephalidae	*Galeopterus variegatus*	VES	0 (0)	0 (0)	1 (5)	0 (0)
Ecarnivora						
Mustelidae	*Martes flavigula*	VES	0 (0)	0 (0)	1 (5)	0 (0)
Rodentia						
Sciuridae	*Callosciurus notatus*	VES	1 (11)	3 (21)	0 (0)	2 (15)
	Sundasciurus lowii	VES	0 (0)	5 (36)	3 (15)	0 (0)
	Sundasciurus tenuis	VES	2 (22)	1 (7)	0 (0)	2 (15)
	Lariscus insignis	VES	0 (0)	1 (7)	0 (0)	0 (0)
	Hylopetes platyurus	MN	0 (0)	0 (0)	0 (0)	1 (8)
Muridae	*Rattus rattus*	CT	0 (0)	0 (0)	0 (0)	4 (31)
	Sundamys muelleri	CT	0 (0)	0 (0)	1 (5)	0 (0)
	Maxomys surifer	CT	0 (0)	0 (0)	4 (20)	0 (0)

Sampling site – *HTFR* Hulu Telemong Forest Reserve, *BB* Belukar Bukit, *KRS* Kenyir Research Station (UMT), *KK* Kampung Kemat
Method – *CT* cage trapping, *VES* visual encounter survey, *PF* pitfall trapping, *MN* mist netting
n number of individuals, *RA* relative abundance

approximately 90% similarity and Belukar Bukit (BB) which consists of lowland dipterocarp forest and riparian forest was secondly clustered with both sites with about 73% similarity (Fig. 2).

Kampung Kemat (KK) has recorded the highest diversity and richness ($H' = 1.67$, $R_1 = 1.95$, $J' = 0.93$) with 13 individuals belong to three orders, four families six species while the lowest species diversity and richness was recorded for Hulu Telemong Forest Reserve (HTFR) ($H' = 0.85$, $R_1 = 0.91$, $J' = 0.77$) (Table 2) a result of the lowest recorded number of total individuals of only nine individuals from two orders, two families and three species among the four sampling sites.

The species cumulative curves showed that the total of non-volant small mammals species recorded in all sites have yet to reach asymptote indicated that there were still additional species to be found in the future (Fig. 3).

The Common Treeshrew (*T. glis*) was the dominant species in Tasik Kenyir area because it was found in all the study sites and proved that this species is a generalist that could occupy all types of habitat present in this study area. Moreover, the high abundance of this ground-dwelling species may result from its ability to breed anytime throughout the year, short gestation period and the species is not restricted to breeding season (Francis 2013; Medway 1983). Bananas were used in this study as

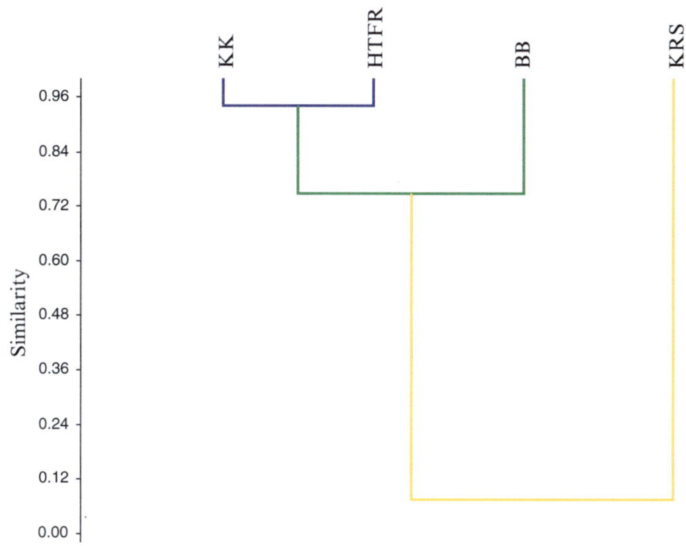

Fig. 2 Dendrogram shows the similarity between non-volant small mammal assemblages for each site

Table 2 Diversity measures of non-volant small mammal assemblages in Tasik Kenyir area

	HTFR	BB	KRS	KK
Shannon diversity index (*H'*)	0.85	1.43	1.57	1.67
Margalef richness index (R_1)	0.91	1.52	2.00	1.95
Evenness (*J'*)	0.77	0.89	0.80	0.93

HTFR Hulu Telemong Forest Reserve, *BB* Belukar Bukit, *KRS* Kenyir Research Station (UMT), *KK* Kampung Kemat

the species is easily attracted to any conventional baits (Bernard 2003; Medway 1983). Previous studies in Selangor, Kelantan, Pahang, Kelantan and Perak had recorded this species (Ruppert et al. 2015; Jayaraj et al. 2012, 2013; Tingga et al. 2012; Zakaria et al. 2001).

The non-volant small mammal assemblages in Kampung Kemat (KK) and Hulu Telemong Forest Reserve (HTFR) were clustered together with more than 90% similarity of species richness or assemblages in both sampling sites (Fig. 2). Subsequently, the two sites were followed by Belukar Bukit (BB) with approximately 73% of similarity. These three sites were found to be distantly clustered to Kenyir Research Station (KRS) with less than 12% similarity signified that most species found in the regenerated forest were different from disturbed and riparian forests.

The presence of variety of disturbed habitats such as human settlements, oil palm plantations and orchards in Kampung Kemat showed that the high abundance of non-volant small mammals which included the pests and commensal species, House Rat (*Rattus rattus*) that mainly found in correlation with human and the rarely

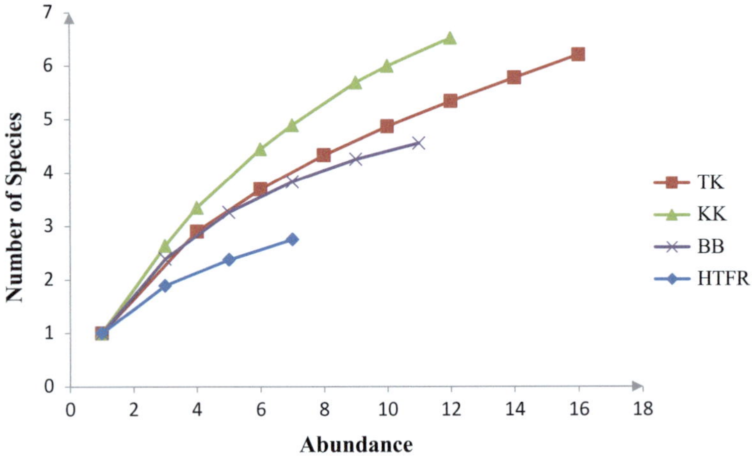

Fig. 3 Species accumulation curves for non-volant small mammals in Tasik Kenyir area, Hulu Terengganu

captured flying squirrel, Grey-cheeked Flying Squirrel (*Hylopetes platyurus*). The non-volant small mammals may rely on the abundance of resources available in these kinds of habitat (Wells et al. 2004). The high diversity of non-volant small mammal assemblage in Kampung Kemat may also be influenced by the presence of microhabitats such as tree logs, burrows and tree holes where these microhabitats could provide shelters and nesting places. Zakaria and Nordin (1998) suggested that variability of habitats is affecting the species assemblages in an area aside from other factors such as distribution and abundance of food supplies. In contrast, Hulu Telemong Forest Reserve recorded the lowest species diversity and richness because of human encroachment and disturbance are affecting the non-volant small mammal community. Zakaria et al. (2001) and Corlett (1992) stated that the Common Treeshrew (*T. glis*) and the Plantain Squirrel (*Callosciurus notatus*) have high tolerance towards habitat disturbance.

The species accumulation curves have shown that 23 sampling nights were inadequate to record all non-volant small mammal assemblages in Tasik Kenyir area as the curves were increasing exponentially. Other species may be added to the species checklist of Tasik Kenyir if the sampling duration is extended.

Conclusion

The variety of microhabitats could sustain a high diversity of non-volant small mammals. Although the disturbed forest of Kampung Kemat recorded the highest diversity of non-volant small mammals with 13 individuals belong to three orders, four families and six species, most of the species such as the Common Treeshrew (*T. glis*) and Plantain Squirrel (*C. notatus*) are very adaptable in any kind of habitat.

Therefore, types of habitat may influence the species richness and diversity of non-volant small mammals by their tendency to adapt to those types of habitat. Therefore, it is suggested that a proper forest inventory and stricter laws be established for the protection and conservation of endangered species in Tasik Kenyir area, Hulu Terengganu to conserve the non-volant small mammal community in the area.

Acknowledgements The authors would like to thank Universiti Malaysia Terengganu, Kenyir Research Institute, local authorities, villagers, RELA officers at Pusat Latihan RELA Wilayah Timur and Department of Wildlife and National Parks of Peninsular Malaysia (DWNP) for granted the support, permission and permit, reference no. T-00372-16-16 to examine and collect samples. The funding was partially supported by Geopark grant (GEOPARK/2015/53167), Trans Disciplinary Research Grant Scheme (TRGS/2014/59373), *Geran Galakan Penyelidikan* (GGP/68007/2014/127) and Niche Research Grant Scheme (NRGS/2015/5313/2) led by Prof Dr. Mohd Tajuddin Abdullah.

Appendix

Some photos of the non-volant small mammals recorded at Tasik Kenyir, Hulu Terengganu, Terengganu (a) Savi's Pigmy Shrew (*Suncus etruscus*), (b) Red Spiny Maxomys (*Maxomys surifer*) and (c) Müller's Rat (*Sundamys muelleri*).

References

Bernard, H. (2003). Bait preferences of some small mammal species of North Borneo based on line-trapping with wire-mesh cage traps. *Sabah Parks Nature Journal, 6*, 27–44.

Caceres, N. C., Nápoli, R. P., & Hannibal, W. (2011). Differential trapping success for small mammals using pitfall and standard cage traps in a woodland savannah region of Southwestern Brazil. *Mammalia, 75*(1), 45–52.

Corlett, R. T. (1992). The ecological transformation of Singapore, 1819–1990. *Journal of Biogeography, 19*(4), 411–420.

Crump, M. L., & Scott, N. J. (1994). Visual encounter surveys. In W. R. Heyer, M. A. Donnelly, R. W. McDiarmid, L. C. Hayek, & M. S. Foster (Eds.), *Measuring and monitoring biological diversity: Standard methods for amphibians* (pp. 84–92). Washington, DC: Smithsonian Institution Press.

Francis, C. M. (2001). *A photographic guide to mammals of South-East Asia: Including Thailand, Malaysia, Singapore, Myanmar, Laos, Vietnam, Cambodia, Java, Sumatra, Bali and Borneo*. Sanibel Island: Ralph Curtis Books.

Francis, C. M. (2008). *A field guide to the mammals of South-east Asia*. London: New Holland Publisher (UK) Ltd.

Francis, C. M. (2013). *A photographic guide to Mammals of South-east Asia: Including Thailand, Malaysia, Singapore, Myanmar, Laos, Cambodia, Vietnam, Java, Sumatra, Bali and Borneo*. London: New Holland Publishers (UK) Ltd.

Hammer, Ø., Harper, D. A. T., & Ryan, P. D. (2001). PAST-PAlaeontological STatistics, ver. 1.89. *Palaeontologia Electronica, 4*(1), 1–9.

Jayaraj, V. K., Tahir, N. F. D. A., Udin, N. A., Baharin, N. F. K., Ismail, S. K., & Zakaria, S. N. A. (2012). Species diversity of small mammals at Gunung Stong State Park, Kelantan, Malaysia. *Journal of Threatened Taxa, 4*(6), 2617–2628.

Jayaraj, V. K., Daud, S. H. M., Azhar, M., Sah, S. A. M., Mokhtar, S. I., & Abdullah, M. T. (2013). Diversity and conservation status of mammals in Wang Kelian State Park, Perlis, Malaysia. *Check List, 9*(6), 1439–1448.

Mazlan, N., Tan, C., Kamaruzzaman, M. A., Adrus, M., & Abdullah, M. T. (2015). Survey of small mammals in Bukit Taat, Tasik Kenyir, Hulu Terengganu, Malaysia. *Borneo Journal of Resource Science and Technology, 5*(2), 79–83.

Medway, L. (1983). *The wild mammals of Malaya (Peninsular Malaysia) and Singapore* (2nd ed.). Kuala Lumpur: Oxford University Press.

Merritt, J. F. (2010). *The biology of small mammals*. Baltimore: John Hopkins University Press.

Payne, J., & Francis, C. M. (1998). *A field guide to the mammals of Borneo*. Kota Kinabalu: Sabah Society.

Ramli, R., & Hashim, R. (2009). Diversity of small mammals inhabiting disturbed forest: A case study on Kenaboi Forest Reserve, Jelebu, Negeri Sembilan, Malaysia. *Malaysian Journal of Science, 28*(4), 481–490.

Ruppert, N., Mansor, A., & Anuar, S. M. S. (2015). Diversity and biomass of terrestrial small mammals at a primary rainforest reserve (Segari Melintang Forest Reserve, Peninsular Malaysia). *Journal of Tropical Life Science, 5*(1), 35–44.

Shaharom, F. (Ed.). (2015). *Visions of Kenyir: Through the eyes of a drone*. Kuala Terengganu: Penerbit UMT.

Shepherd, C. R., & Shepherd, L. A. (2012). *A naturalist's guide to the mammals of South-east Asia: Brunei, Cambodia, Indonesia, Laos, Malaysia, Myanmar, the Philippines, Singapore, Thailand and Vietnam*. Oxford: John Beaufoy Publishing.

Tingga, R. C. T., Anwarali, F. A., Mohd Ridwan, A. R., Senawi, J., & Abdullah, M. T. (2012). Small mammals from Kuala Atok, Taman Negara Pahang, Malaysia. *Sains Malaysiana, 41*(6), 659–669.

Wells, K., Pfeiffer, M., Lakim, M. B., & Linsenmair, K. E. (2004). Use of arboreal and terrestrial space by a small mammal community in a tropical rain forest in Borneo, Malaysia. *Journal of Biogeography, 31*, 641–652.

Zakaria, M., & Nordin, M. (1998). Comparison of visitation rates of frugivorous birds in primary and logged forest in Sabah lowland dipterocarp forest. *Tropical Biodiversity, 5*(1), 1–9.

Zakaria, M., Silang, S., & Mudin, R. (2001). Species composition of small mammals at the Ayer Hitam Forest Reserve, Puchong, Selangor. *Pertanika Journal of Tropical Agricultural Science, 24*(1), 19–22.

Checklist of Small Mammals of Hulu Terengganu, Terengganu

Mohamed Nor Zalipah, Azuan Roslan, Juliana Senawi, Vijaya Kumaran Jayaraj, Mohd Isham Azhar, Mohd Tajuddin Abdullah, and Boo Liat Lim

Abstract Despite their ecological importance, small mammals record in east-coast of Peninsular Malaysia is particularly rare and scattered, as compared to records made in west of Peninsular Malaysia. In this paper, we assembled data from various trapping occasions conducted at surrounding areas of Tasik Kenyir, Hulu Terengganu, including Taman Negara Terengganu, from 1990 to present. We documented 55 spe-

---

M. Nor Zalipah (✉)
School of Marine and Environmental Science, Universiti Malaysia Terengganu, Kuala Nerus, Terengganu, Malaysia

Department of Wildlife and National Parks, Institute for Biological Diversity, Bukit Rengit, Lanchang, Pahang, Malaysia
e-mail: zalipah@umt.edu.my

A. Roslan
Institute of Tropical Biodiversity and Sustainable Development, Universiti Malaysia Terengganu, Kuala Nerus, Terengganu, Malaysia

J. Senawi
Department of Wildlife and National Parks, Institute for Biological Diversity, Bukit Rengit, Lanchang, Pahang, Malaysia

Institute for Environment and Development (LESTARI), Universiti Kebangsaan Malaysia, Bangi, Selangor, Malaysia

V. K. Jayaraj
Faculty of Earth Science, Universiti Malaysia Kelantan, Jeli, Kelantan, Malaysia

M. I. Azhar
Faculty of Natural Science and Sustainability, University College Sabah Foundation, Kota Kinabalu, Sabah, Malaysia

M. T. Abdullah
Institute of Tropical Biodiversity and Sustainable Development, Universiti Malaysia Terengganu, Kuala Nerus, Terengganu, Malaysia

School of Marine and Environmental Sciences, Universiti Malaysia Terengganu, Kuala Nerus, Terengganu, Malaysia

B. L. Lim
Department of Wildlife and National Parks Peninsular Malaysia, Lanchang, Pahang, Malaysia

M. T. Abdullah et al. (eds.), *Greater Kenyir Landscapes*,
https://doi.org/10.1007/978-3-319-92264-5_18

cies of bats and 19 species of non-volant small mammals, mainly of rats and squirrels. This represent approximately 30% of the total non-marine mammal species reported for Peninsular Malaysia, indicating the important of Hulu Kenyir forest areas as conservation area for small mammals.

Keywords Bats · Rats · Squirrels · Tasik Kenyir

Introduction

Hulu Terengganu is the largest district in the state of Terengganu with approximately 383,000 ha area. More than 25% of the area is covered by primary forest, which includes Terengganu National Park (TNP) and the adjacent forest reserves (Terengganu Forestry 2016). TNP is accessible through Tasik Kenyir, the largest artificial lake in Southeast Asia, also famous for the Sultan Mahmud Hydroelectric Power Station. The breath-taking terrain of Tasik Kenyir's natural habitat and its diverse forms of life at present makes it a renown ecotourism destination (KETENGAH 2016). Among the noteworthy life forms that further commended its reputation is the presence of carnivores and ungulates deemed as umbrella species (Hassan 1993). Moreover, the scenic landscape complimented by the unique topography of Tasik Kenyir has great potential to become one of the noteworthy geoparks in Malaysia for the future (Faizah et al. 2015).

Small mammals are very diverse throughout the country as well as in Hulu Terengganu. Over the years, brief surveys ranging from 4 to 10 days were carried out for inventory collections at TNP, Hutan Simpan Tembat, Hutan Simpan Hulu Terengganu and Hutan Simpan Hulu Terengganu Tambahan (Fig. 1) (Norfarhana et al. 2015; UMK: 2012 unpubl. data; UMT: 2008–2011, 2015--2016 unpubl. data; Shahrul Anuar and Mohd Hifni 2001; Lim B.L: 1990–2002 unpubl. data; DWNP: 1990–2002, 2006 unpubl. data). The results revealed that bats are one of the most abundant animals found in the area, in which the trapping success were three times greater compared to the non-volant small mammals.

Checklist

The present checklist consists of 74 species of small mammals, comprising of 55 species of bats whereas the remaining 19 species were non-volant small mammals (Table 1). Hence small mammals of Hulu Terengganu represent approximately 30% of the total mammal species recorded in Peninsular Malaysia (Abdullah 2016). This figure is relatively high and comparable to the data from Kelantan and Krau, with 84 species of small mammals (Jayaraj et al. 2016) and 69 species of bats (Kingston et al. 2006), respectively.

Bats are divided into two main groups, the fruit-eating and the insect-eating bats. The 11 species of fruit-eating bats recorded are divided into two major diet groups: the true nectar feeders which feed mainly on nectar, pollen and flowers parts and the

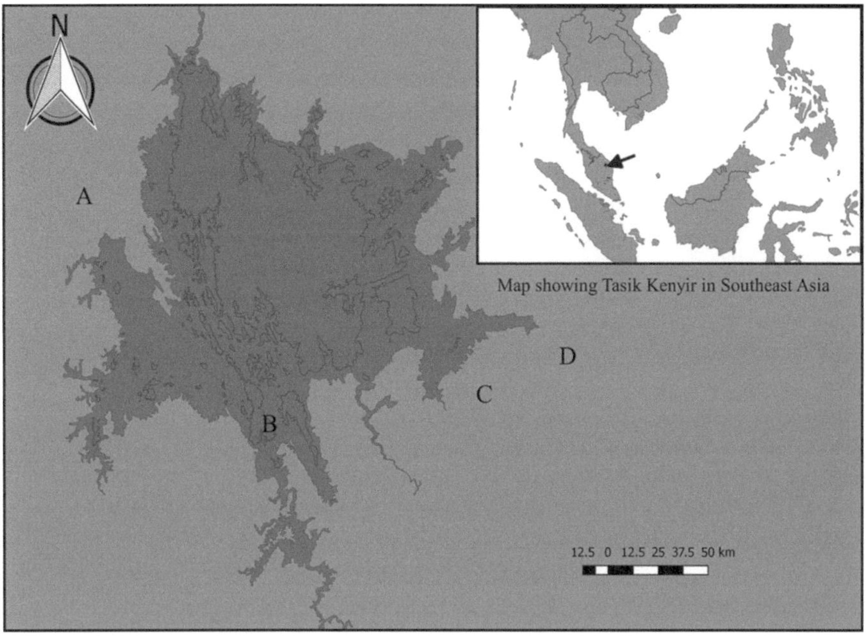

Fig. 1 Location of sampling sites in Hulu Terengganu, Terengganu. A = Hutan Simpan Tembat, B = Taman Negara Terengganu, C = Hutan Simpan Hulu Terengganu, D = Hutan Simpan Hulu Terengganu Tambahan

fruit feeders which consume fruit flesh, juice, seeds and leaves. In Hulu Terengganu, the nectar feeding bats recorded were *Eonycteris spelaea*, *Macroglossus minimus* and *M. sobrinus* (Fig. 2). These bats showed specialised morphological adaptations for nectar feeding such as elongated snouts and long specialised tongues to extract the nectar, which are usually hidden within the flower. Other pteropodid bats are fruit feeders with shorter and broader snouts, giving them a dog-like appearance. Their feeding behaviour makes the pteropodid bats significant as seed dispersing and pollinating agents for the plants that they feed on (Hodgkison et al. 2003; Start and Marshall 1976).

Meanwhile, there are 44 species of insect-eating (insectivorous) bats comprising of 7 families (Emballonuridae, Megadermatidae, Nycteridae, Hipposideridae, Rhinolophidae, Vespertitionidae and Molossidae) identified. These include the capture of few rare and poorly recorded species such as *Ceolops robinsoni* and *Chaerephon johorensis*, which also contribute to new locality records for species such as *Kerivoula krauensis* and *Rhinolophus chiewkweeae* (Fig. 3). Both *C. robinsoni* and *C. johorensis* are listed as vulnerable (VU) by IUCN (2016), due to the diminishing lowland forest areas (Francis 2008). In 2010, the once thought endemic Krau woolly bat, *K. krauensis*, was recorded at Sekayu Recreational Forest in Hulu Terengganu Tambahan Forest Reserve, hence extending the distribution of this species from Krau, Pahang right up to Hulu Terengganu in Peninsular Malaysia

Table 1 The checklist of small mammal species recorded for Hulu Terengganu and their conservation status

Family species	Common name	IUCN (2016)	Wildlife Act (2010)
Cynocephalidae			
Galeopterus (Cynocephalus) variegatus	Flying lemur	LC	TP
Pteropodidae			
Rousettus amplexicaudatus	Geoffroy's rousette bat	LC	NP
Cynopterus brachyotis	Lesser short-nosed fruit bat	LC	NP
Cynopeterus horsfieldii	Horsfield's fruit bat	LC	NP
Cynopterus sphinx	Greater short-nosed fruit bat	LC	NP
Penthetor lucasi	Lucas's short-nosed fruit bat	LC	NP
Balionycteris maculata	Spotted winged fruit bat	LC	NP
Chironax melanocephalus	Black-capped fruit bat	LC	NP
Megaerops ecaudatus	Temminck's tailless fruit bat	LC	NP
Eonycteris spelaea	Lesser dawn bat	LC	NP
Macroglossus minimus	Lesser long-tongued nectar	LC	NP
Macroglossus sobrinus	Greater long-tongued nectar bat	LC	NP
Emballonuridae			
Emballonura monticola	Lesser sheath-tailed bat	LC	NP
Nycteridae			
Nycteris tragata	Malayan slit-faced bat	NT	NP
Megadermatidae			
Megaderma spasma	Lesser false vampire	LC	NP
Megaderma lyra	Greater false vampire	LC	NP
Rhinolophidae			
Rhinolophus affinis	Intermediate horseshoe bat	LC	NP
Rhinolophus stheno	Lesser brown horseshoe bat	LC	NP
Rhinolophus robinsoni	Peninsular horseshoe bat	LC	NP
Rhinolophus lepidus	Blyth's horseshoe bat	LC	NP
Rhinolophus pusillus	Least horseshoe bat	LC	NP
Rhinolophus sedulus	Lesser woolly horseshoe bat	NT	NP
Rhinolophus luctus	Woolly horseshoe bar	LC	NP
Rhinolophus chiewkweeae	Chiewkwee's horseshoe bat	NT	NP
Hipposideridae			
Hipposideros bicolor	Temminck bicoloured roundleaf horseshoe bat	LC	NP
Hipposideros cineraceus	Least roundleaf horseshoe bat	LC	NP
Hipposideros ridleyi	Ridley's roundleaf horseshoe bat	VU	NP
Hipposideros galeritus	Cantor's roundleaf horseshoe bat	LC	NP
Hipposideros armiger	Great roundleaf horseshoe bat	LC	NP
Hipposideros larvatus	Large roundleaf horseshoe bat	LC	NP
Hipposideros diadema	Diadem roundleaf bat	LC	NP

(continued)

Table 1 (continued)

Family species	Common name	IUCN (2016)	Wildlife Act (2010)
Hipposideros cervinus	Fawn-coloured roundleaf bat	LC	NP
Hipposideros dyacorum	Dayak roundleaf bat	LC	NP
Ceolops robinsoni	Malaysian tailless horseshoe bat	VU	NP
Vespertilionidae			
Miniopterus screibersii	Schreiber's long-fingered bat	NT	NP
Miniopterus ridleyi	Ridley's bat	NT	NP
Murina cyclotis	Dobson's round-eared tube-nosed bat	LC	NP
Murina suilla	Temminck brown tube-nosed bat	LC	NP
Murina rozendalii	Gilded tube-nosed bat	VU	NP
Kerivoula papillosa	Papillose bat	LC	NP
Kerivoula intermedia	Small woolly bat	NT	NP
Kerivoula pellucida	Clear-winged bat	NT	NP
Kerivoula minuta	Least woolly bat	NT	NP
Kerivoula hardwickii	Hardwicke's woolly bat	LC	NP
Kerivoula krauensis	Krau woolly bat	DD	NP
Phoniscus jagorii	Peter's trumpet eared bat	LC	NP
Tylonycteris robustula	Greater bamboo bat	LC	NP
Tylonycteris pachypus	Lesser flat-headed bat	LC	NP
Glischropus tylopus	Common thick-thumbed bat	LC	NP
Myotis ridleyi	Ridley's myotis	NT	NP
Myotis muricola	Napalese whiskered bat	LC	NP
Myotis ater	Peters's myotis	LC	NP
Myotis siligorensis	Himalayan whiskered bat	LC	NP
Myotis horsfieldii	Horsfield's myotis	LC	NP
Myotis mystacinus	Eurasian whiskered myotis	LC	NP
Scotophilus kuhlii	Lesser asiatic yellow house bat	LC	NP
Philetor brachypterus	Narrow-winged brown bat	LC	NP
Molossidae			
Chaerephon johorensis	Northern free-tailed bat	VU	NP
Tupaiidae			
Tupaia glis	Common treeshrew	LC	P
Sciuridae			
Ratufa bicolor	Black giant squirrel	NT	TP
Callosciurus notatus	Plantain squirrel	LC	NP
Callosciurus caniceps	Grey-bellied squirrel	LC	NP
Callosciurus prevostii	Prevost's squirrel	LC	TP
Sundasciurus tenuis	Slender squirrel	LC	NP
Lariscus insignis	Three-striped ground squirrel	LC	NP

(continued)

Table 1 (continued)

Family species	Common name	IUCN (2016)	Wildlife Act (2010)
Muridae			
Rattus tiomanicus	Malayan wood rat	LC	NP
Rattus exulans	Polynesian rat	LC	NP
Rattus annandalei	Annandale's rat	LC	NP
Sundamys muelleri	Mueller's rat	LC	NP
Niviventer cremoriventer	Dark-tailed tree rat	VU	NP
Maxomys surifer	Red spiny rat	LC	NP
Maxomys rajah	Brown spiny rat	VU	NP
Maxomys whiteheadi	Whitehead's rat	VU	NP
Leopoldamys sabanus	Long-tailed giant rat	LC	NP
Viverridae			
Viverra tangalunga	Malay civet	LC	P
Paradoxurus hermaphroditus	Common palm civet	LC	P

IUCN status following The IUCN Red List of Threatened Species (Version 2016-2). Wildlife Act status according to Laws of Malaysia Act 716, Wildlife Conservation Act (2010)
LC Least concern, *NP* Not protected, *NT* Near threatened, *P* Protected, *VU* Vulnerable, *TP* Totally protected

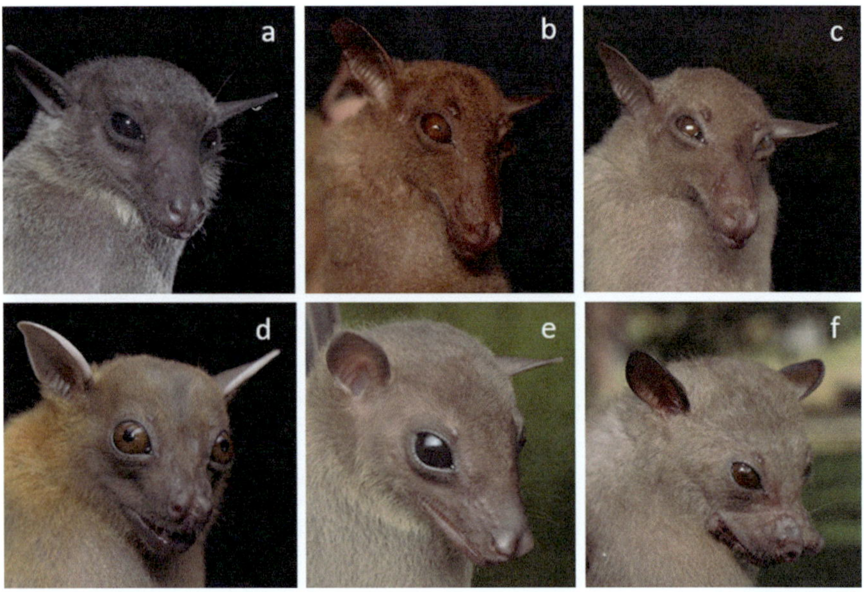

Fig. 2 The nectar-feeding bats (**a**–**c**) and the fruit-feeding bats (**d**–**f**) recorded at Hulu Terengganu forest areas. (**a**) *Eonycteris spelaea*, (**b**) *Macroglossus minimus*, (**c**) *Macroglossus sobrinus*, (**d**) *Cynopterus brachyotis*, (**e**) *Rousettus amplexicaudatus*, (**f**) *Megaerops ecadatus* (Photographs by Mohd Isham Mohd Azhar and Azuan Roslan)

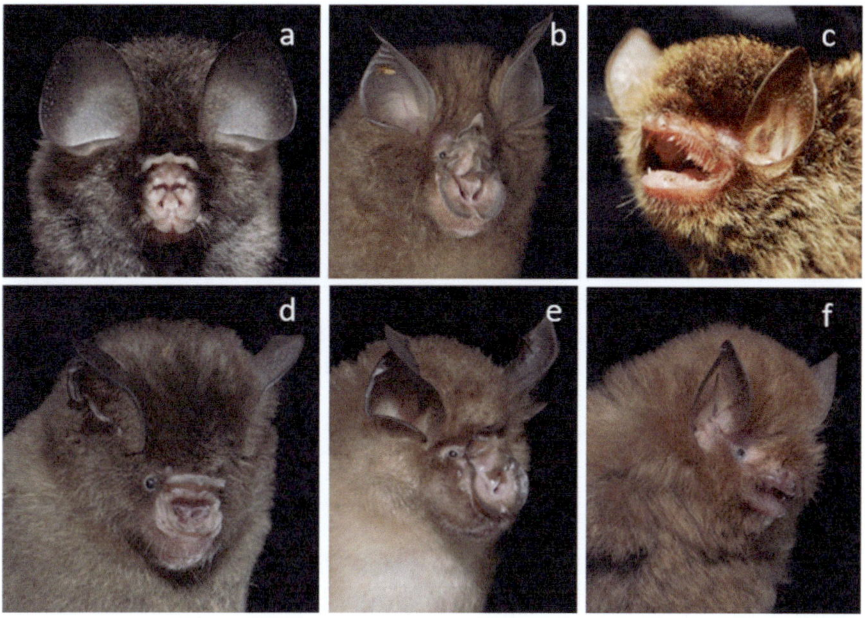

Fig. 3 The insect-feeding bats recorded at Hulu Terengganu forest areas. (**a**) *Ceolops robinsoni*, (**b**) *Rhinolophus chiewkweeae*, (**c**) *Kerivoula krauensis*, (**d**) *Hipposideros cervinus*, (**e**) *Rhinolophus stheno*, (**f**) *Kerivoula intermedia*. (Photographs by Mohd Isham Mohd Azhar, Azuan Roslan and Syed Ahmad Rizal Tuan Nek)

(Struebig et al. 2016; Francis et al. 2007). For *R. chiewkweeae*, Hulu Terengganu is the seventh locality known to harbor this species in Malaysia (Muhd Amsyari et al. 2016). This species was first described by Yoshiyuki and Lim (2005) from specimens collected at the west-coast of Peninsular Malaysia (Lubok Semilan and Weng Sub-catchment area, Kedah, Wang Kelian State Park, Perlis, Asahan Forest Reserve, Melaka, Gunung Ledang and Labis Forest Reserve, Johore).

Kingston et al. (2003) stated that *Hipposideros cervinus, R. stheno, R. lepidus refulgens* and *Kerivoula intermedia* are common insectivorous bats caught in lowland rainforest of Peninsular Malaysia. Insectivorous bats prey heavily on nocturnal insects which are mainly pests and of health importance; therefore, these bats are important in providing ecological services and warrant conservation efforts (Kunz et al. 2013). In contrast to the fruit bats, nine species of insectivorous bats were classified as near threatened (NT) and four species were vulnerable (VU), mainly due to decreasing population trend from habitat loss (Francis 2008).

Contrary to bats, only 17 species of the non-volant mammals from 5 families (Cynocephalidae, Muridae, Scruridae, Tupaiidae and Viviverridae) were recorded. Most species recorded in Hulu Terengganu were common mammals found in lowland dipterocarp forest of Peninsular Malaysia (Ruppert et al. 2015) and are important seed dispersal agents (Corlett 1998). This is particularly true for many species from the families of Muridae, Sciuridae and Tupaiidae, as some studies found that their faeces have fig seeds (*Ficus* sp.), indicating their significant roles as seed dispersers in lowland rainforests of Borneo (Wells et al. 2009; Wells and Bagchi 2005). *Niviventer cremoriventer, Maxomys rajah* and *M. whiteheadii* from the family Muridae are listed as vulnerable (VU) in the IUCN Red List of Threatened Species (IUCN 2016). Although these three species are common in Southeast Asia, the population of these rats are thought to be at risk due to habitat destruction in some parts of the region (Francis 2008). The enigmatic giant squirrel, *Ratufa affinis*, is categorised as Near Threatened (NT) in the IUCN (2016) and is totally protected in Peninsular Malaysia under the Wildlife Conservation Act (2010). On the other hand, the Prevost's squirrel, *Callosciurus prevostii*, is also protected in Peninsular Malaysia under the Wildlife Conservation Act (2010), this species is listed as of Least Concern (LC) in the IUCN Red List of Threatened Species (IUCN 2016).

Conclusion

Hulu Terengganu harbours a total of 74 small mammal species, which is about 30% of the total mammals species (including large mammals) recorded for Peninsular Malaysia. This shows that Hulu Terengganu is fairly rich in small mammal diversity. Besides, it is the last remaining large block of forest on the east coast of Peninsular Malaysia and therefore is an important area for conservation. The disparity between bats and non-volant mammals recorded (74% of the total species are bats and only 26% are tree shrews, rodents and civets) is due to trapping success rate; mist nets and harp traps are highly efficient methods in trapping bats and could trap more than one individual at a time as compared to a wire cage trap, which can only trap one individual at a time. There is a greater potential to find more species in the area if regular and systematic assessments are conducted.

Acknowledgements We would like to thank DWNP personnel who were involved in samplings in 1990–2002 (L.B.L) and in 2006 (N.Z.M and J.S.) particularly in TNTNP area. We also would like to thank UMT for logistical and transportation support for N.Z.M to conduct samplings from 2008 to 2011, and for A.R. and A.M.T in 2015–2016. J.V.K would like to acknowledge the following grants awarded to him and colleagues for the sampling work in Tasik Kenyir, Terengganu (UMK-SGJP R/SGJP/001/2010/000037, MTSF R/MTSF/003/2012/000071 and MOHE RACE R/RACE/002/2012/000085). This study was partly supported by various research grants, Geran Galakan Penyelidikan (GGP/68007/2014/127) and Trans Disciplinary Research Grant Scheme (TRGS/2015/59373) for A.R and M.T.A. Photos in this article are provided by M.I.M.A and A.R, except for *K. krauensis*, by Syed Ahmad Rizal B Tuan Nek, senior lab assistant, School of Marine and Environmental Sciences, UMT.

References

Abdullah, M. T. (2016). *Evolusi dan Biogeografi Mamalia Malaysia (Syarahan Inaugural Universiti Malaysia Terengganu)*. Kuala Terengganu: Penerbit Universiti Malaysia Terengganu.

Corlett, R. T. (1998). Frugivory and seed dispersal by vertebrates in the Oriental (Indomalayan) region. *Biological Reviews, 73*, 413–448.

Faizah, M. S., Abdullah, M. T., Che Aziz, A., & Roslina, I. (2015). *Geopark Tasik Kenyir*. Kuala Terengganu: Penerbit Universiti Malaysia Terengganu.

Francis, C. M. (2008). *A field guide to the mammals of South-East Asia*. London: New Holland Publishers Ltd..

Francis, C. M., Kingston, T., & Zubaid, A. (2007). A new species of *Kerivoula* (Chiroptera: Vespertilionidae) from peninsula Malaysia. *Acta Chiropterologica, 9*, 1–12.

Hassan, K. (1993). Distribution of large mammal in Tasek Kenyir, Taman Negara Terengganu. *Journal of Wildlife and Parks, 12*, 22–24.

Hodgkison, R., Balding, S. T., Zubaid, A., & Kunz, T. H. (2003). Fruit bats (Chiroptera: Pteropodidae) as seed dispersers and pollinators in a lowland Malaysia rain forest. *Biotopica, 35*, 491–502.

IUCN. (2016). The IUCN red list of threatened species. Version 2016-2. Retrieved from http://www.iucnredlist.org.

Jayaraj, V. K., Khan, F. A. A., Mohd Isham, M. A., Ean, W. C., Mohd Rohanif, M. A., Ahmad, A., & Mohd Yusoff, A. (2016). Diversity and conservation status of small mammals in Kelantan, Malaysia. *Songklanarin Journal of Science and Technology, 38*(2), 213–220.

KETENGAH. (2016). Tasik Kenyir. Retrieved from http://www.ketengah.gov.my/lakekenyir.

Kingston, T., Francis, C. M., Zubaid, A., & Kunz, T. H. (2003). Species richness in an insectivorous bat assemblage from Malaysia. *Journal of Tropical Ecology, 19*, 67–79.

Kingston, T., Lim, B. L., & Akbar, Z. (2006). *Bats of Krau Wildlife Reserve*. Bangi: Penerbit Universiti Kebangsaan Malaysia.

Kunz, T. H., de Torrez, E. B., Bauer, D., Lobova, T., & Fleming, T. H. (2013). Ecosystem services provided by bats. *Annals of the New York Academy of Sciences*. https://doi.org/10.1111/j.1749-6632.2011.06004.x.

Muhd Amsyari, M., Nurul Farah Diyana, A. T., Qhairil Shyamri, R., Julius, W. D., Isham, A., Azuan, R., Muhamad Aidil, Z., Abdullah, M. T., & Faisal Ali, A. K. (2016). New record of *Rhinolophus chiewkweeae* (Chiroptera: Rhinolophidae) from the east coast of Peninsular Malaysia with new information on their echolocation calls, genetics and their taxonomy. *The Raffles Bulletin of Zoology, 64*, 242–249.

Norfarhana, M., Tan, C.-H., Kamaruzzaman, M. A., Madinah, A., & Abdullah, M. T. (2015). Survey of small mammals in Bukit Taat, Tasik Kenyir, Hulu Terengganu, Malaysia. *Borneo Journal of Resource Science and Technology, 5*, 79–83.

Ruppert, N. B., Asyraf, M., & Shahrul Anuar, M. S. (2015). Diversity and biomass of terrestrial small mammals at a Malaysian primary rainforest (Segari Melintang Forest Reserve, Peninsular Malaysia). *Journal of Tropical Life Science, 5*, 35–44.

Shahrul Anuar, M. S., & Mohd Hifni, B. (2001). *Conservation of mammal and bird fauna at Hulu Kenyir, Trengganu, International Conference on In-situ and Ex-situ Biodiversity Conservation in the New Millennium. Kota Kinabalu, 2001*. Kota Kinabalu: Jabatan Muzium Sabah.

Start, A. N., & Marshall, A. G. (1976). Nectarivorous bats as pollinators of trees in west Malaysia. In J. Burley & B. T. Styles (Eds.), *Variation, breeding and conservation of tropical forest trees* (pp. 141–150). London: Academic.

Struebig, M. J., Huang, J. C. C., Nor Zalipah, M., Noerfahmy, S., Schoner, C. R., Schoner, M. G., & Francis, C. M. (2016). Forest surveys extend the range of the Krau Wolly Bat (*Kerivoula krauensis*) in the Malay-Thai Peninsula, Borneo and Sumatra. *Mammalia*. https://doi.org/10.1515/mammalia-2015-2014.

Terengganu Forestry. (2016). Maklumat Perhutanan: Hutan Simpan Kekal. Retrieved from http://trgforestry.terengganu.gov.my/index.php?option=com_content&view=article&id=79&Itemid=321&lang=my.

Wells, K., & Bagchi, R. (2005). Eat it or take away-seed predation and removal by rats (Muridae) during a fruiting event in a dipterocarp rainforest. *The Raffles Bulletin of Zoology, 53*, 281–286.

Wells, K., Corlett, R. T., Lakim, M. B., Kalko, E. K. V., & Pfeiffer, M. (2009). Seed consumption by small mammals from Borneo. *Journal of Tropical Ecology, 25*, 555–558.

Yoshiyuki, M., & Lim, B. L. (2005). A new horseshoe bat, *Rhinolophus chiewkweeae* (Chiroptera, Rhinolophidae), from Malaysia. *Bulletin of the National Science Museum Tokyo, Series A, 31*, 29–36.

Rapid Assessment and Taxonomic Checklist of Vertebrates at the Foot of Gunung Tebu Forest Reserve, Terengganu

Nur Iema Omar, Muazzah Abd Latif, Nursamiyah Shamsul, Munirah Izzati Sharif Katullah, Hasrulzaman Hassan Basri, Amirul Asyraf Mazlan, Nur Farhana Azmi, Romanrio Ering, Salmi Abdullah, Habibah Anuar, Nurul Ahlam Ismail, Muhammad Hafiz Ahmad, Mohammad Naufal Mohammad Shah, Khairul Bariah Mohd Johan, and Mohd Tajuddin Abdullah

Abstract The vertebrate diversity at the foot of Gunung Tebu Forest Reserve (GTFR) is not well documented therefore a study was conducted at lowland dipterocarp forest of Gunung Tebu Forest Reserve, Terengganu from 7th to 13th August 2016 and 27th August 2016 to 2nd September 2016 to record and construct a checklist of vertebrates from selected taxonomic group i.e.; mammals, aves and amphibians. A total of 470 individuals were observe and recorded consisting of 16 species of mammals (8 families and 4 orders), 64 species of aves (25 families and 10 orders) and 20 species of herpetofauna (6 families and a single order). The highest species abundance for bats recorded was Bicoloured Roundleaf Bat (*Hipposideros bicolor*)

N. I. Omar · M. Abd Latif · N. Shamsul · M. I. Sharif Katullah
Department of Zoology, Faculty of Resource Science and Technology, Universiti Malaysia Sarawak, Kota Samarahan, Sarawak, Malaysia

Centre for Kenyir Ecosystems Research, Institute of Tropical Biodiversity and Sustainable Development, Universiti Malaysia Terengganu, Kuala Nerus, Terengganu, Malaysia

H. H. Basri (✉)
Centre for Kenyir Ecosystems Research, Institute of Tropical Biodiversity and Sustainable Development, Universiti Malaysia Terengganu, Kuala Nerus, Terengganu, Malaysia

A. A. Mazlan · N. F. Azmi · R. Ering · S. Abdullah · H. Anuar · N. A. Ismail · M. H. Ahmad · M. N. Mohammad Shah · K. B. Mohd Johan
School of Marine and Environmental Sciences, Universiti Malaysia Terengganu, Kuala Nerus, Terengganu, Malaysia

M. T. Abdullah
Institute of Tropical Biodiversity and Sustainable Development, Universiti Malaysia Terengganu, Kuala Nerus, Terengganu, Malaysia

School of Marine and Environmental Sciences, Universiti Malaysia Terengganu, Kuala Nerus, Terengganu, Malaysia

M. T. Abdullah et al. (eds.), *Greater Kenyir Landscapes*,
https://doi.org/10.1007/978-3-319-92264-5_19

whereas for birds, the Greater Racket-tailed Drongo (*Dicrurus paradiseus)* shown highest species observed meanwhile for anurans, the highest species abundance documented was Spotted Litter Frog (*Leptobrachium hendricksoni).* An increase of sampling effort, number of transects and survey period could increase the number of species recorded in the area.

Keywords Amphibians · Aves · Mammals · Gunung Tebu

Introduction

Gunung Tebu Forest Reserve (GTFR) is located on northeast of the Banjaran Timur and its summit reaching 1039 m. It consists of dipterocarp forest containing a significant variety of flora and fauna of conservation interest which indicates that the forest reserve is in healthy condition (Sulaiman et al. 2014). Our study was conducted at the surrounding of lowland region which is at Peladang Setiu Agro Resort (PSAR), Kampung Bintang situated at the foothill of GTFR. Although the forest reserve is surrounded by oil palm plantation, PSAR attracts many visitors for recreational outdoor activities featuring a landscape of rainforest and beautiful lakes. Moreover, the vegetation in this surrounding lowland region is characterized as low hill dipterocarp forest (Sumarli et al. 2015). From our observations in PSAR, there are small flowing streams, large granite rocks and man-made pool which are suitable for herpetological, ornithology and mammalogy studies.

The information on the biodiversity of fauna such as amphibian, aves and mammals in lowland region is deficient but many were discovered in uplands of GTFR (Sumarli et al. 2015; Grismer et al. 2013) making the lowland region virtually unexplored and documented. Hence, this study was aimed to document the species diversity and to construct an updated checklist of vertebrates from selected taxa (mammals, aves and amphibians) found at the foot of GTFR. In addition, Adrus et al. (2013) pointed out that the diverse type of forest in Malaysia give access to different taxonomic composition of vertebrate fauna. Thus, we are also comparing the biodiversity assessment from previous studies in Peninsular Malaysia by Adrus et al. (2013).

The survey was conducted in PSAR which is located at the foot of GTFR (5.5914° N, 102.6122° E) (Fig. 1) situated 90 km away from Kuala Terengganu on the 7th to 13th August 2016 and 27th August 2016 to 2nd September 2016. The sampling events mostly took place at the PSAR headquarters, camp site, nearby palm oil plantation and in the forest trails.

Anurans Sampling

The sampling method used was active searching where the anurans were caught by hand and small scoop net and placed in plastic bags. The details such as time caught and microhabitats of the captured anurans were recorded and written on the plastic bag using marker pen.

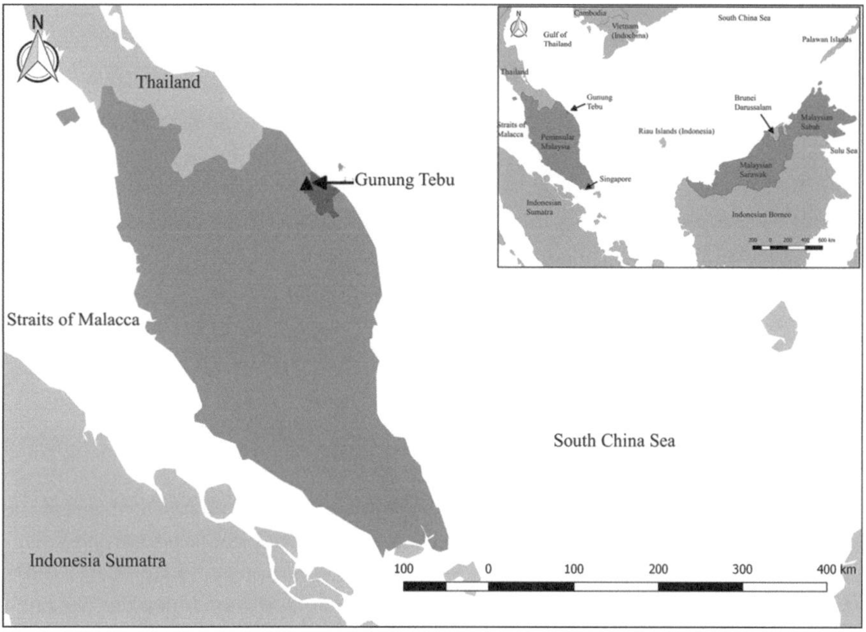

Fig. 1 Map showing the sampling site in Gunung Tebu Forest Reserve, Terengganu

As most amphibians were nocturnal and easily detected by their calls, catching anurans with active searching method were done at night around swampy areas, in leaves litter, decaying logs (Ingram et al. 1975) and low vegetation. The amphibians collected were identified based on Norhayati (2016) and Inger (2005).

Aves Sampling

Mist Net

Ten understory mist nets with 2–3 m above ground and four canopy mist nets with more than 15 m above ground were set up. Mist nets were checked thrice daylight, at 0700, 1200, and 1600 h. Vernier callipers, rulers, cloth bags, gloves, spring and electronic balance were also used to record measurements of bill length (BL), bill depth (BD), bill width (BW), tarsus length (TR), tail length (T), wing length (WL), wing span (WS), total length (TL) and weight (W). All birds were identified and measured with the guidance from Wong (2012), Craig (2000), Davison and Chew (1996, 2003), and Morten and Jeyarajasingam (1993). Nail polish was used to mark captured individuals to recognize any recaptured individuals during the sampling period.

Point Count

Point count method involves the observer to stay at a certain point for a fixed time period to do the visual and audio observation (Fuller and Langslow 1984). Hence, every 50 m of 1 km of point count distance, we stopped for 7 min for bird observation. Point count distance from the subject, method of visualisation and type of habitat were recorded and observed. Bushnell binocular 7 × 42, roller or measuring tape were also used to estimate length from each point.

Line Transect

Line transect method is a type of observation method which is widely used in term of estimating any biological abundance (Burnham et al. 1980). During sampling, we did 1 km transect line from the forest edge to the middle of the forest to increase the probability of bird detection along the transect line. There were two session of sampling, morning session started at 0700 h and evening session started at 1630 h. GPS Garmin was used to record the coordinate of the observation and binocular Bushnell 7 × 24 was used to help observing and identifying birds from afar.

Mammals Sampling

Ten understory mist nets were placed 2–3 m above ground and four canopy mist nets with more than 15 m above ground were used during this sampling to capture volant small mammals. All mist nets were positioned along small streams, area which is near to big rocks and along the forest trails. As for harp traps, the location selected were between trees that have slight pathways between them as stated in Tingga et al. (2012). Mist nets with low capture rate were relocated to obtain more bats species and results.

For chiropterans, mist nets were checked every 2 h interval starting from 1800 to 2300 and final check at 0600 h on the next day. The captured bats were put in cloth bag and brought to station for the process of measurement and identification. Identification is based on Francis (2008), Payne et al. (2007), and Kingston et al. (2006). The captured bats were weighted using digital weight balance, and digital calliper was used to measure forearm length (FA), tail length (T), ear length (E), and tibia length (TB), head body (HB) and hindfoot (HF). As for non-volant small mammals, observations were conducted randomly either by vocalisation or visual encountered (Adrus et al. 2013). The mammals were identified based on Francis (2008) and Medway (1978). All recorded species of anurans, aves and mammals were cross checked with the IUCN (2016) red list and Wildlife Conservation Act (2010) for their population status in the wild.

Taxonomic Checklist

Anuran

A total of 208 individuals of anuran comprising of 20 species from the family Bufonidae, Dicroglossidae, Megophyridae, Microhylidae, Ranidae and Rhacophoridae were recorded (Table 1). The highest number of individuals recorded was *Leptobrachium hendricksoni* (Spotted Litter Frog) from family of Megophyridae with 61 individuals followed by *Hylarana erythraea* (Common Green Frog) from family of Ranidae with 42 individuals.

Table 1 Taxonomy checklist of anuran fauna at the foot of Gunung Tebu forest reserve

				Conservation status	
Family	Species	Common name	Total	WCA (2010)	IUCN (2016)
Bufonidae	*Duttaphrynus melanostictus*	Common sunda toad	7	NP	LC
	Ingerophrynus parvus	Lesser stream toad	26	NP	LC
	Phrynoidis asper	Asian giant toad	4	NP	LC
Dicroglossidae	*Fejervarya limnocharis*	Asian grass frog	7	NP	LC
	Limnonectes malesianus	Malesian frog	6	P	NT
	Limnonectes pamacrodon	Lesser swamp frog	1	P	LC
	Limnonectes plicatellus	Penang wart frog	1	NP	LC
	Occidozyga laevis	Common puddle frog	1	NP	LC
Megophyridae	*Leptobrachium hendricksoni*	Spotted litter frog	61	NP	LC
Microhylidae	*Kalophrynus palmatissimus*	Lowland grainy frog	1	P	EN
	Kalophrynus limbooliati	Lim Boo Liat sticky frog	1	NP	LC
	Kaloula pulchra	Common bullfrog	6	NP	LC
	Microhyla butleri	Butler's rice frog	6	NP	LC
	Microhyla fissipes	Ornate narrow-mouth frog	5	NP	LC
	Microhyla heymonsi	Dark sided Chorus frog	4	NP	LC
Ranidae	*Chalcorana labialis*	White lipped frog	15	NP	LC
	Hylarana erythraea	Common green frog	42	NP	LC
	Hylarana glandulosa	Rough-sided frog	1	NP	LC
	Hylarana laterimaculata	Lesser swamp frog	5	NP	LC
Rhacophoridae	*Polypedates leucomystax*	Common tree frog	8	NP	LC
Total order			1		
Total family			6		
Total species			20		
Total individuals			208		

WCA Wildlife Conservation Act, *NP* Not Protected, *P* Protected, *TP* Totally Protected, *LC* Least Concern, *NT* Near Threatened, *EN* Endangered

Table 2 The updated checklist of anuran fauna of Gunung Tebu forest reserve

Order Family	Species	Common name	This study	Sumarli et al. (2015)
Anuran				
Bufonidae	*Duttaphrynus melanostictus*	Common Sunda toad	+	+
	Ingerophrynus parvus	Lesser stream toad	+	+
	Phrynoidis asper	Asian giant toad	+	+
Dicroglossidae	*Fejervarya limnocharis*	Asian grass frog	+	+
	Limnonectes malesianus	Malesian frog	+	–
	Limnonectes pamacrodon	Lesser swamp frog	+	–
	Limnonectes plicatellus	Penang wart frog	+	–
	Occidozyga laevis	Common puddle frog	+	–
Megophyridae	*Leptobrachium hendricksoni*	Spotted litter frog	+	+
Microhylidae	*Kalophrynus palmatissimus*	Lowland grainy frog	+	–
	Kalophrynus limbooliati	Lim Boo Liat sticky frog	+	–
	Kaloula pulchra	Common bullfrog	+	–
	Microhyla butleri	Butler's rice frog	+	–
	Microhyla fissipes	Ornate narrow-mouth frog	+	+
	Microhyla heymonsi	Dark sided chorus frog	+	+
Ranidae	*Chalcorana labialis*	White lipped frog	+	+
	Hylarana erythraea	Common green frog	+	–
	Hylarana glandulosa	Rough-sided frog	+	–
	Hylarana laterimaculata	Lesser swamp frog	+	+
Rhacophoridae	*Polypedates leucomystax*	Common tree frog	+	+

According to Sumarli et al. (2015), anuran species were usually found on the leaf litter on forest floor, at the side of the stream, puddle near the stream, rocky stream in a swampy area along the stream, human residential area and on the leaves of low vegetation. Most of the species captured were human communal (Adrus et al. 2013). *Kalophrynus palmatissimus* was listed as endangered, *L. malesianus* was listed as near threatened while others are listed as least concern in IUCN (2016) red list. There were three species namely *L. malesianus, L. pamacrodon* and *K. palmatissimus* that were listed as protected in Wildlife Conservation Act (2010).

Comparison of anurans record of GTFR with previous study conducted by Sumarli et al. (2015) is as shown in Table 2. The later study was conducted at the lowland and upland of GTFR while this study covered only the lowland area. There were not only anurans included in this previous research but also caecilians, lizards, snakes and turtles. Based on Table 2, species that were not recorded previously but were documented from this study were *L. malesianus, L. pamacrodon, L. plicatellus, Occidozyga laevis, K. palmatissimus, K. limbooliati, Kaloula pulchra, Microhyla butleri, H. erythraea* and *H. glandulosa*.

Aves

During our 14 days of sampling, we observed 217 individuals of birds from 64 species to 25 families (Table 3). Order Passeriformes recorded the highest number of families (n = 10). Family Nectariniidae was the most diverse family with seven different species, followed by family Cuculidae and Muscicapidae with six species while Bucerotidae and Megalaimidae each documented four species. *Dicrurus paradiseus* (Greater Racket-tailed Drongo) from family Dicruridae showed the highest abundance with 16 individuals. *Copsychus saularis* (Oriental Magpie Robin) from family Muscicapidae recorded a total of 15 individuals, followed by *Passer montanus* (Eurasian Tree Sparrow) with 12 individuals.

Dicrurus paradiseus's vocal signal could be the factor that makes it easily detectable since they were mostly documented from point count and line transects method. Only one individual caught in mist net during our sampling. *Copsychus saularis* is a common insectivorous birds which can adapt to eat various type of food such as insects (Bhattacharya et al. 2007) and invertebrates (Sreekar 2010). This features allowed *C. saularis* to adapt with human disturbance from the tourism activities at lowland region of GTFR. According to IUCN (2016) red list, 14 out of 64 species are listed under near threatened, while 45 species captured are categorised as totally protected animals and 2 species are under protected animals.

Mammals

A total of 16 species of mammals were recorded in this study (Table 4), 11 species were captured by traps, while another 5 species were recorded through observation. These 11 species of chiropterans were mostly caught by mist nets comprising of four families, Pteropodidae, Rhinolophidae, Hipposideridae and Vespertilionidae. The most abundance species recorded was the insectivorous bats which were *Hipposideros bicolor* (Bicoloured Roundleaf Bat) with 19 individuals followed by *Hipposideros diadema* (Diadem Roundleaf Bat) with 11 individuals. For non-volant mammals, five individuals were recorded through visual encountered which were *Ratufa affinis* (Cream-coloured Giant Squirrel), *Callosciurus notatus* (Plantain Squirrel), *Trachypithecus obscurus* (Dusky Langur), *Hylobates lar* (White-handed Gibbon) and *Sus scrofa* (Eurasian Wild Pig).

For volant-mammals survey, *Hipposideros bicolor* was the most abundant species recorded from the total of 54 individuals bats captured. This is due to the fact that *H. bicolor* is said to forage in understory of lowland forest (Francis 2008), and most of the mist nets and harp traps were set up at understory level of the forest. Within these 11 individuals of bats caught, only one species, *Rhinolophus robinsoni*

Table 3 Taxonomic checklist of birds at the foot of Gunung Tebu forest reserve

Order / Family	Species	Common name	Method	Total	Conservation status: WCA (2010)	Conservation status: IUCN (2016)
Accipitriformes						
Accipitridae		Eagle	LT	1	NP	NE
Columbiformes						
Columbidae	*Chalcophaps indica*	Emerald dove	MN	4	P	LC
	Geopelia striata	Zebra dove	PC, LT	8	NP	LC
Cuculiformes						
Cuculidae	*Centropus sinensis*	Greater coucal	PC, LT	2	TP	LC
	Centropus bengalensis	Lesser coucal	PC	1	TP	LC
	Phaenicophaeus chlorophaeus	Raffles's malkoha	LT	2	TP	LC
	Phaenicophaeus diardi	Black-bellied malkoha	PC, LT	2	TP	NT
	Eudynamys scolopaceus	Asian koel	PC	1	NP	LC
	Surniculus lugubris	Drongo cuckoo	PC	1	NP	LC
Strigiformes						
Tytonidae	*Phodilus badius*	Oriental bay owl	MN	3	TP	LC
Strigidae	*Strix leptogrammica*	Brown wood owl	MN	1	TP	LC
Caprimulgiformes						
Apodidae		Swift	PC	1	NP	NE
Trogoniformes						
Trogonidae	*Harpactes diardii*	Diard's trogon	MN	2	TP	NT
Bucerotiformes						
Bucerotidae		Hornbill	PC	1	NP	NE
	Berenicornis comatus	White-crowned hornbill	PC, LT	2	TP	NT
	Buceros bicornis	Great hornbill	PC, LT	3	TP	NT
	Anorrhinus galeritus	Bushy-crested hornbill	PC, LT	4	TP	LC

Coraciiformes						
Alcedinidae	*Ceyx erithaca*	Oriental-dwarf kingfisher	MN	3	NP	LC
	Actenoides concretus	Rufous collared kingfisher	MN	1	TP	NT
Piciformes						
Megalaimidae	*Psilopogon spp.*	Barbet	PC, LT	5	NP	NE
	Psilopogon mystacophanos	Red-throated barbet	LT	9	TP	NT
	Psilopogon henricii	Yellow-crowned barbet	LT	1	TP	NT
	Psilopogon oorti	Black-browed barbet	LT, PC	1	TP	LC
Picidae	*Picus spp.*	Woodpecker	LT	2	TP	NE
	Picus puniceus	Crimson-winged woodpecker	PC, LT	7	TP	LC
	Meiglyptes tukki	Buff-necked woodpecker	MN, PC, LT	6	TP	NT
Passeriformes						
Eurylaimidae	*Calyptomena viridis*	Green broadbill	MN	8	TP	NT
	Eurylaimus ochromalus	Black-and-yellow broadbill	PC, LT	4	TP	NT
Campephagidae	*Pericrocotus igneus*	Fiery minivet	PC	1	TP	NT
Dicruridae	*Dicrurus paradiseus*	Greater racket-tailed drongo	MN, PC, LT	16	TP	LC
Corvidae	*Platylophus galericulatus*	Crested jay	LT	1	TP	NT
Hirundinidae	*Hirundo rustica*	Barn swallow	PC, LT	9	TP	LC
		Swallow	LT	2	NP	NE
Pycnonotidae	*Pycnonotus spp.*	Bulbul	LT, PC	6	NP	LC
	Pycnonotus flavescens	Flavescent bulbul	PC, LT	2	TP	LC
	Pycnonotus goiavier	Yellow-vented bulbul	PC, LT	5	NP	LC
	Pycnonotus brunneus	Red-eyed bulbul	MN, LT	2	TP	LC
	Alophoixus bres	Grey-cheeked bulbul	MN, PC, LT	7	TP	LC
	Alophoixus phaeocephalus	Yellow-bellied bulbul	MN	5	TP	NE
Cisticolidae	*Orthotomus sutorius*	Common tailorbird	PC, LT	7	TP	LC
Timaliidae	*Cyanoderma erythropterum*	Chestnut-winged babbler	LT	1	TP	LC
	Stachyris nigricollis	Black-throated babbler	MN	1	TP	NE

(continued)

Table 3 (continued)

Order Family	Species	Common name	Method	Total	Conservation status WCA (2010)	 IUCN (2016)
Pellorneidae	*Pellorneum capistratum*	Black-capped babbler	MN	2	TP	LC
Muscicapidae		Philentoma	PC	1	NP	NE
	Enicurus ruficapillus	Chestnut-naped forktail	MN, PC	4	TP	NT
	Copsychus saularis	Oriental Magpie robin	PC, LT	15	P	LC
	Copsychus malabaricus	White rumped shama	MN	1	TP	LC
	Ficedula hyperythra	Snowy brow flycatcher	PC	1	TP	LC
	Philentoma pyrhoptera	Rufous-winged philentoma	MN, PC, LT	4	NP	LC
Chloropseidae	*Chloropsis spp.*	Leafbird	LT	1	NP	NE
	Chloropsis cyanopogon	Lesser green leafbird	PC, LT	3	TP	NT
	Chloropsis hardwickii	Oriental bellied leafbird	PC	1	TP	LC
Dicaeidae	*Prionochilus percussus*	Crimson-breasted flowerpecker	MN, PC, LT	4	TP	LC
	Dicaeum concolor	Plain flowerpecker	PC, LT	2	TP	LC
Nectariniidae	*Arachnothera sp.*	Spiderhunter	PC	1	NP	NE
	Anthreptes singalensis	Ruby-cheeked Sunbird	PC, LT	2	TP	LC
	Arachnothera crassirostris	Thick-billed Spiderhunter	PC, LT	2	TP	LC
	Arachnothera robusta	Long-billed spiderhunter	PC, LT	2	TP	LC
	Arachnothera longirostra	Little spiderhunter	LT	1	TP	LC
	Arachnothera affinis	Streaky-breasted spiderhunter	MN	1	TP	LC
	Hypogramma hypogrammicum	Purple-naped sunbird	MN, PC, LT	3	TP	LC
		Sunbird	LT	2	TP	LC
Passeridae	*Passer montanus*	Eurasian tree sparrow	PC, LT	12	NP	LC
Total order				10		
Total family				25		
Total species				64		
Total individuals				217		

WCA Wildlife Conservation Act
LT Line transect, *PC* Point count, *MN* Mist net, *NP* Not Protected, *P* Protected, *TP* Totally Protected, *NE* Not Evaluated, *LC* Least Concern, *NT* Near Threatened, *EN* Endangered

Table 4 Taxonomic checklist of mammals at the foot of Gunung Tebu forest reserve

Order					Conservation status	
Family	Species	Common name	Method	Total	WCA (2010)	IUCN (2016)
Chiroptera						
Pteropodidae	*Cynopterus brachyotis*	Short-nosed fruit bat	MN	9	NP	LC
	Cynopterus horsfieldii	Horsfield's fruit bat	MN	3	NP	LC
	Penthetor lucasi	Dusky fruit bat	MN	1	NP	LC
	Balionycteris maculata	Spotted-winged fruit bat	MN	3	NP	LC
Rhinolophidae	*Rhinolophus Lepidus*	Blyth's horseshoe bat	MN	1	NP	LC
	Rhinolophus affinis	Intermediate horseshoe bat	MN	4	NP	LC
	Rhinolophus robinsoni	Peninsular horseshoe bat	MN	1	NP	NT
Hipposideridae	*Hipposideros bicolour*	Bicoloured roundleaf bat	MN, HT	19	NP	LC
	Hipposideros larvatus	Intermediate roundleaf bat	MN	1	NP	LC
	Hipposideros diadema	Diadem roundleaf bat	MN	11	NP	LC
Vespertilionidae	*Scotophilus kuhlii*	Lesser Asian house bat	MN	1	NP	LC
Rodentia						
Sciuridae	*Ratufa affinis*	Cream-coloured giant squirrel	VE		TP	NT
	Callosciurus notatus	Plantain squirrel	VE		NP	LC
Primate						
Cercopithecidae	*Trachypithecus obscures*	Dusky langur	VE		P	NT
Hylobatidae	*Hylobates lar*	White-handed gibbon	VE		TP	EN
Artiodactyla						
Suidae	*Sus scrofa*	Eurasian wild pig	VE		P	LC
Total order				4		
Total family				8		
Total species				16		
Total individuals				54		

WCA Wildlife Conservation Act, *MN* Mist net, *HT* Harp trap, *VE* Visual encountered, *NP* Not Protected, *P* Protected, *TP* Totally Protected, *LC* Least Concern, *NT* Near Threatened, *EN* Endangered is listed as near threatened by IUCN (2016) red list. In Kingston et al. (2006) this species is considered becoming vulnerable due to habitat destruction as it is identified to dwell in forested habitats. From five species of observed non-volant mammals, *H. lar* is listed as Endangered while *R. affinis* and *T. obscurus* are listed as near threatened by the IUCN (2016) red list. Moreover, habitat destruction of the primates, similar with other mammals is also said to be one of the reason that these species are threatened (Adrus et al. 2013). Both *R. affinis* and *H. lar* are listed as totally protected by Wildlife Conservation Act (2010).

Comparison with Similar Study at Gunung Ledang National Park

This study was compared with a study by Adrus et al. (2013) at Gunung Ledang National Park (GLNP) (Table 5). Sampling periods at GLNP was only 5 days and 6 nights, while this study at GTFR was conducted in 14 days 12 nights. Due to the less sampling days and sampling effort, the species in GLNP documented was lower than GTFR although most samplings were covered at the same forest type. There could have been more species documented with at least 5–7 days of sampling at each elevation to increase the sampling effort at GLNP (Adrus et al. 2013). Despite of its national park status, it is a tourist attraction site allowing the public to access the site easily. The total number of tourists visited GLNP was 63,060 visitors from 2010 to 2012, showing an increasing pattern every year (Sanmargaraja and Seow 2013). Another factor that makes GLNP having lower species diversity and abundance is due to the overexploitation of land around the national park itself. According to Idris (2011), one type plantation which in this case is the oil palm plantation surrounding GLNP decreases the biodiversity level of the area. Although there is a massive oil palm plantation area nearby GTFR, the distance between the forest reserve and the plantation is favourable to maintain the species richness and abundance at GTFR.

Conservation Status of Vertebrates of Gunung Tebu Forest Reserve

Generally, the vertebrates of GTFR have the highest conservation status of least concern (77%) and 21% of the vertebrates are categorized under near threatened by the IUCN (2016) red list (Fig. 2). About 2% from GTFR species is listed as endangered (EN), which are *K. palmatissimus* (Lowland Grainy Frog) and *Hylobates lar* (White-handed Gibbon).

Table 5 Description of study conducted at Gunung Tebu Forest Reserve (GTFR) and Gunung Ledang National Park (GLNP)

	GTFR, Terengganu (this study)	GLNP, Johor (Adrus et al. 2013)
Methodology	T, O	T, O
Total traps	14MN, 2HT	20MN, 2HT, 35CT
Sampling site	Peladang Setiu Agro	Gunung Ledang
	Resort, Setiu	National Park
Habitat/type of vegetation	Oil palm plantation and lowland dipterocarp forest	Montane and hill forest and lowland dipterocarp
		Forest
Sampling period (nights)	12	6
Total species	99	50
Total individuals	470	98

T Traps, *O* Observation, *MN* Mist nets, *HT* Harp traps, *CT* Cage traps

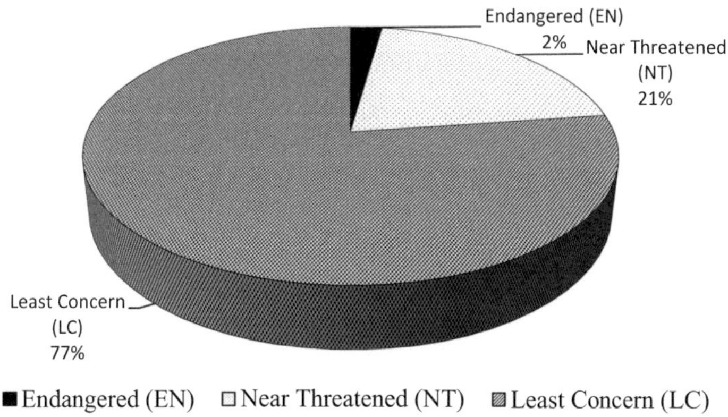

Fig. 2 Summary of vertebrates' conservation status in Gunung Tebu Forest Reserve according to IUCN (2016) red list

According to Wildlife Conservation Act (2010), there are three protected species of anuran which are *Limnonectes malesianus* (Malesian Frog), *Limnonectes pamacrodon* (Lesser Swamp Frog) and *K. palmatissimus* (Lowland Grainy Frog). Although there are no volant mammals listed in Wildlife Conservation Act (2010), two from non-volant mammals are protected which are *T. obscurus* (Dusky Langur) and *Sus scrofa* (Eurasian Wild Pig) while two species are totally protected which are *Ratufa affinis* (Cream-coloured Giant Squirrel) and *H. lar* (White-handed Gibbon). Forty-five species of bird are listed under totally protected and two bird species are protected.

Conclusion

This study established the first checklist of anurans, birds and mammals that is found in GTFR, within the foothill area. Further studies need to be conducted with an increase in the sampling effort such as sampling period or enlarge transect area to record occurrences of the species. Moreover, an increase of the number of certain traps (mist nets and harp traps) would probably increase the information on distributions of vertebrate species in particular habitat that could be documented in the future.

Previous study provided only little information about biodiversity of fauna in GTFR. Hence, this knowledge is important for future research aimed for selected species and those who are interested in knowing what species populate at the foothill of GTFR. This sampling shows that the foot of GTFR have a relatively high diversity of vertebrates. The wildlife management authorities should also maintain regular surveillance over the area as it can provide long term protection and management for these numerous kinds of vertebrates.

Appendix

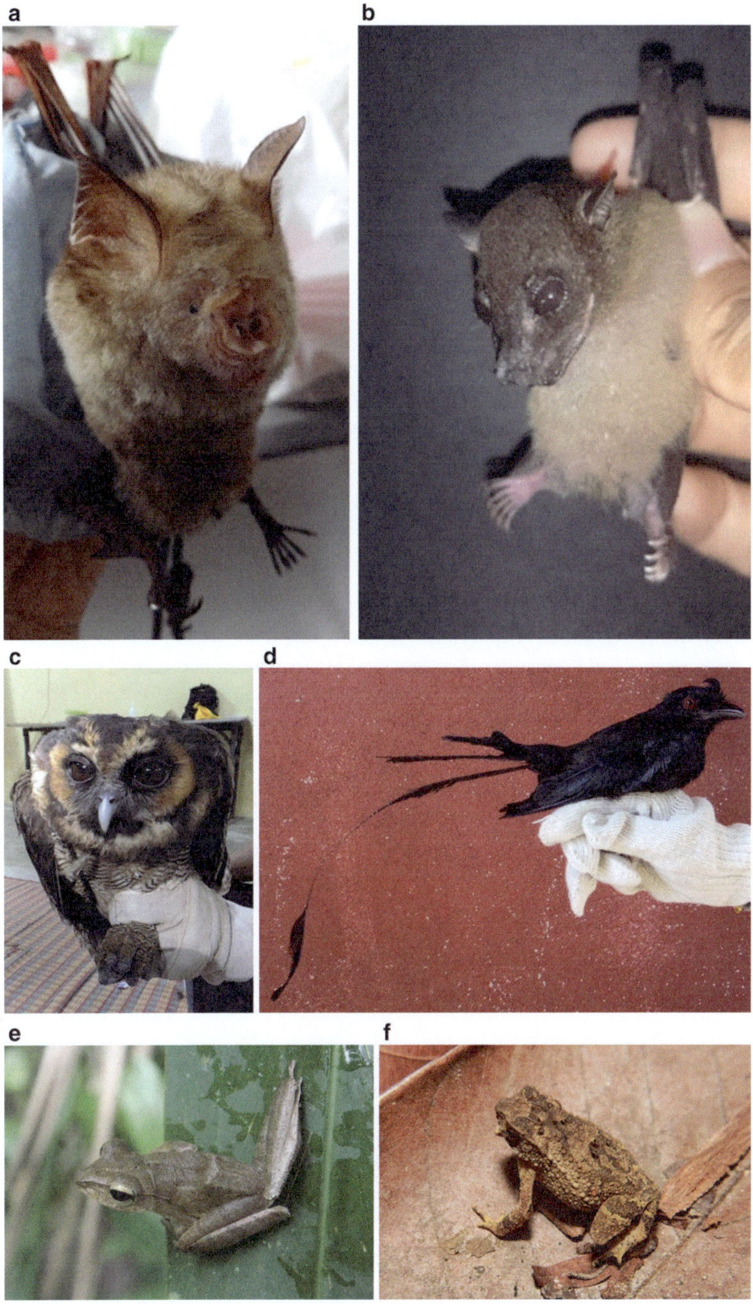

Some photos of vertebrate fauna recorded at Gunung Tebu Forest Reserve (GTFR). (**a**) *Hipposideros diadema* (Diadem Horseshoe-bat) (Photograph by Nur Iema Omar) and (**b**) *Balionycteris maculata* (Spotted-winged Fruit Bat) (Photograph by Nur Farhana Azmi) netted from understorey mist nets, (**c**) *Strix leptogrammica* (Brown Wood Owl) (Photograph by Habibah Anuar) and (**d**) *Dicrurus paradiseus* (Greater Racket-tailed Drongo) (Photo by Nursamiyah Shamsul) captured at canopy mist net deployed around the campsite, (**e**) *Chalcorana labialis* (White-lipped Frog) and (**f**) *Phrynoidis asper* (Asian Giant Toad) (Photograph by Amirul Asyraf Mazlan)

Acknowledgements We would like to express our appreciation to staffs of Kenyir Research Institute and Universiti Malaysia Terengganu (UMT), Mr. Azuan Roslan, Mr. Muhamad Aidil Zahidin, Ms. Gertrude David, Ms. Elizabeth Pesiu, Ms. Amirah Azizah, Ms Nur Amalina Adanan and Ms. Nur Aisyah Rahim for assisting us throughout this sampling period. We would like to thank the staff of PSAR for various assistances throughout our stay at the resort. Financial support from Trans Disciplinary Research Grant Scheme (TRGS), No Vot: 59373 led by Professor Dr. Mohd Tajuddin Abdullah is acknowledged.

References

Adrus, M., Mohd Azhar, M. I., Ramji, M. F. S., Mat Daud, M. H. R., Amran, M. F., Mat Yasin, S., & Abdullah, M. T. (2013). Rapid assessment of vertebrate fauna in Gunung Ledang National Park Johor. In A. A. Tuen, M. Jayasilan, & J. Grinang (Eds.), *Rimba 3 sustaining livelihood through prudent utilization and management of nature resources* (pp. 116–122). Kota Samarahan: Universiti Malaysia Sarawak.

Bhattacharya, H., Cirillo, J., Subba, B. R., & Todt, D. (2007). Song performance rules in the oriental magpie robin. *Our Nature, 5*, 1–13.

Burnham, K. P., Anderson, D. R., & Laake, J. L. (1980). Estimation of density from line transect sampling of biological population. *The Wildlife Society, 72*, 3–202.

Craig, R. (2000). *A field guide to the birds of Southeast Asia*. London: New Holland Publishers (UK) Ltd.

Davison, G. W. H., & Chew, Y. F. (1996). *A photographic guide to birds of Borneo, Sabah, Sarawak, Brunei and Kalimantan*. London: New Holland Publishers (UK) Ltd.

Davison, G. W. H., & Chew, Y. F. (2003). *A photographic guide to birds of Peninsular Malaysia and Singapore*. London: New Holland Publishers (UK) Ltd.

Francis, C. M. (2008). *A field guide to mammals of South-East Asia*. London: New Holland Publishers (UK) Ltd.

Fuller, R. J., & Langslow, D. R. (1984). Estimating number of birds by point counts. *Bird Study, 31*(3), 195–202.

Grismer, L. L., Anuar, S., Muin, M. A., Quah, E. S. H., & Wood, P. L., Jr. (2013). Phylogenetic relationship and description of a new upland species of Bent-toed Gecko (*Cyrtodactylus* Gray, 1827) of the *C. sworderi* complex from northeastern Peninsular Malaysia. *Zootaxa, 3613*(3), 239–252.

Idris, A. G. (2011). A comparison of the diversity and abundance of Braconidae in three highlands of Peninsular Malaysia. *Malaysian Journal of Science, 30*(2), 177–181.

Inger, R. F. (2005). *The systematic and zoogeography of the amphibia of Borneo*. Kota Kinabalu: Natural History Publication.

Ingram, J., Antis, M., & Corben, C. J. (1975). Observation of the Australian leptodactylid frog *Assa darlingtoni*. *Herpetologica, 31*, 425–429.

IUCN. (2016). The IUCN red list of threatened species. Version 2016-2. Retrieved from http://www.iucnredlist.org/. Retrieved on 04 Sept 2016.

Kingston, T., Lim, B. L., & Akbar, Z. (2006). *Bats of Krau Wildlife Reserve*. Bangi: Universiti Kebangsaan Malaysia.

Medway, L. (1978). *The mammals of Malaya (Peninsular Malaysia) and Singapore* (2nd ed.). Petaling Jaya: Oxford University Press.

Morten, S., & Jeyarajasingam, A. (1993). *Birds: A photographic guide to the birds of Peninsular Malaysia and Singapore*. Singapore: Sun Tree Publishing (Singapore) Pte Ltd.

Norhayati, A. (2016). Amphibia My: Amphibians and reptiles of Peninsular Malaysia. Retrieved from http://amphibia.my on 29 Jul 2016.

Payne, J., Francis, C. M., & Phillips, K. (2007). *A field guide to the mammals of Borneo*. Kota Kinabalu: The Sabah Society.

Sanmargaraja, S., & Seow, T. W. (2013). Constraints of alternative tourism in Malaysia. *International Journal of Advance in Management, Technology and Engineering Science, 2*(12), 9–12.

Sreekar, R. (2010). A study on the habits of Oriental Magpie Robin *Copsychus saularis*. *Indian Birds, 5*(5), 152–153.

Sulaiman, N., Mohd Bakri, M. A., Kahar, K. M., Yaacob, M. Z., & Boler, I. (2014). Moth fauna (Lepidoptera: Heterocera) of Gunung Tebu Forest Reserve, Terengganu, Malaysia. *Malayan Nature Journal, 66*(4), 376–389.

Sumarli, A. X., Grismer, L. L., Anuar, S., Muin, M. A., & Quah, E. S. H. (2015). First report on the amphibians and reptiles of a remote mountain, Gunung Tebu in northeastern Peninsular Malaysia. *Check List, 11*(4), 1679. https://doi.org/10.15560/11.4.1679.

Tingga, R. C. T., Anwarali, F. A., Mohd Ridhwan, A. R., Senawi, J., & Abdullah, M. T. (2012). Small mammals from Kuala Atok, Taman Negara Pahang, Malaysia. *Sains Malaysiana, 41*(6), 659–669.

Wildlife Conservation Act. (2010). Law of Malaysia. Act 716, Wildlife Conservation Act 2010. Retrieved from: https://www.unodc.org/res/cld/document/wildlife-conservation-act-2010_html/Wildlife_Conservation_Act_2010.pdf.

Wong, T. S. (2012). *A naturalist's guide to the birds of Borneo, Sabah, Sarawak, Brunei and Kalimantan*. Oxford: John Beaufoy Publishing Limited.

Diversity and Abundance Fish Assemblages in the Setiu Wetlands, Terengganu, Malaysia

Siti TafzilMeriam Sheikh Abdul Kadir, Meii Mohamad-Norizam, Nor Bakhiah Baharim, Takaomi Arai, Hiroyuki Motomura, Mohd-Lokman Husain, Mazlan Abd. Ghaffar, and Mohd Azmi Ambak

Abstract The sampling programme was conducted to determine the diversity and abundance in fish assemblages from June 2011 to July 2012 using gill and trawl nets and traps. A total of 13,132 individuals of fishes comprising 116 species and 50 families were recorded. Family Siganidae had the highest family abundance recorded with 43.0%. *Siganus javus* was determined as the most abundant species with a percentage value of 38.8%. Ninety species were caught and considered to be commercially valuable. Four species were recorded as vulnerable in the IUCN Red List, which were *Hippocampus spinosissimus, Hippocampus trimaculatus, Himantura uarnak*, and *Pegasus laternarius*; while *Himantura walga, Favonigobius melanobranchus, Favonigobius rechei. Epinephelus coioides,* and *Epinephelus diachanthus* were listed as near threatened. Overall values of diversity, richness and evenness indices were found to be 2.60, 12.12 and 0.55, respectively. Estimates

S. T. Sheikh Abdul Kadir (✉) · M.-L. Husain
South China Sea Repository & Reference Centre, Institute of Oceanography & Environment (INOS), Universiti Malaysia Terengganu, Kuala Nerus, Terengganu, Malaysia
e-mail: sititafzil@umt.edu.my

M. Mohamad-Norizam · N. B. Baharim
School of Marine and Environmental Sciences, Universiti Malaysia Terengganu, Kuala Nerus, Terengganu, Malaysia

T. Arai
Faculty of Science, Universiti Brunei Darussalam, Bandar Seri Begawan, Brunei Darussalam

H. Motomura
The Kagoshima University Museum, Kagoshima, Japan

M. A. Ghaffar
Office of Deputy Vice Chancellor (Research and Innovation) (PNC), Universiti Malaysia Terengganu, Kuala Nerus, Terengganu, Malaysia

M. A. Ambak
School of Fisheries and Aquaculture, Universiti Malaysia Terengganu, Kuala Nerus, Terengganu, Malaysia

M. T. Abdullah et al. (eds.), *Greater Kenyir Landscapes*,
https://doi.org/10.1007/978-3-319-92264-5_20

from these indices were an indication of high fish species composition, richness, and evenness in the population of fishes in Setiu wetlands. It is recommended that government should gazette the Setiu Wetlands as a state park in Terengganu for future planning.

Keywords Fish richness · Vulnerable · Shannon index · Margalef index

Introduction

Successful conservation management of mangrove ecosystem requires a clear understanding of factors structuring the fish assemblage. Apart from sustaining the biodiversity and ecosystem, the wetlands ecosystem contributes to socio-economic values mainly for the coastal communities (Woodroffe and Davies 2009; Mann 2000). The wetlands ecosystem is documented as a dynamic ecosystem that directly or indirectly contributes to aquatic and terrestrial biodiversity (Woodroffe and Davies 2009). They are among the richest and most productive ecosystems that provide numerous ecosystem services categorised as ecological, physical, chemical and socioeconomic (Ramsar Convention on Wetlands 1996).

This mangrove is part of wetland plays an important role for the fishery production of adjacent neritic waters by exporting organic and inorganic nutrients whereas root habitats provide abundant food for the fishes (Nagelkerken et al. 2000; Carr and Adams 1973). Study on the fish assemblage in the mangrove ecosystem had been reported in many parts of the world such as Ghana (Levy et al. 2015) and Brazil (Castellanos-Galindo and Krumme 2014). In Malaysia, limited study has been conducted on mangrove fishes from the east coast of Peninsular Malaysia such as Chong et al. (2010) who identified mangrove fishes in Bachok, Kelantan.

In comparison, most of the studies focused on the west coast of Peninsular Malaysia such as in Langkawi (Chong et al. 2005a) and Merbok in Kedah (Khoo 1990); Matang in Perak (Chong 2005; Kiso and Mahyam 2003; Sasekumar et al. 1994); Klang (Chong et al. 1990, 2005b) and Kuala Selangor (Sarpedonti and Chong 2008) in Selangor and Sungai Pulai and Sungai Johor in Johor (Chong and Sasekumar 2002). These studies concluded that there are two main factors controlling the mangrove fishes which are habitats and seasons (Hajisamae et al. 2006; Hajisamae and Chou 2003; Harris et al. 2001; Jenkins et al. 1997; Gray et al. 1996; Santos and Nash 1995; Ali and Hussain 1990; Nash 1988; Horn 1980).

Only limited studies were conducted on the abundance of fishes in Setiu Wetlands. The study of biodiversity indices-based research effort on fish was conducted by Siti Tafzil Meriam et al. (2008) and Ahmad Najdi (2005). However, Siti Tafzil Meriam et al. (2015) compiled the checklist of the fish community in the area. The present study was to determine the diversity and abundance in fish assemblages and the status of diversity based on biodiversity indices and ecological status of all fish species. The final data gained will support in the proper implementation and development of a comprehensive sustainable management plan for wetlands in the future.

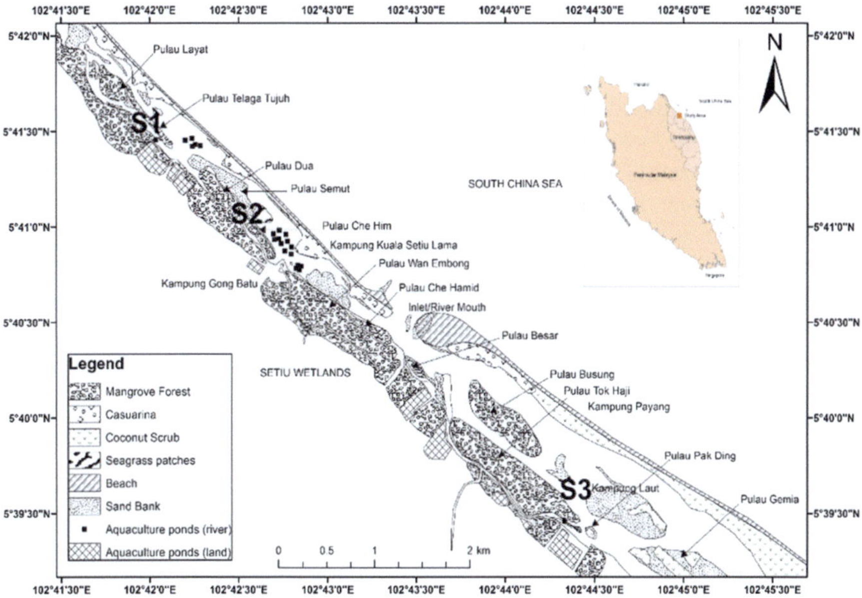

Fig. 1 Location of the sampling sites

The Setiu Wetlands is located in the northeast of Peninsular Malaysia, in the state of Terengganu. Setiu Wetlands is part of the Setiu River Basin and the larger Setiu-Chalok-Bari-Merang Basin wetland located along the northern state of Terengganu in Peninsular Malaysia facing the South China Sea. This wetland covers an area of about 23,000 ha and comprises riparian forests lining the riverbanks, freshwater *Melaleuca* and peat swamps, mangroves, brackish water lagoon with vegetated sand islands, seagrass beds, and sandy beaches (Nakisah and Fauziah 2003).

The Setiu Wetlands is important and well-known as one of the fish resources and provides a source of income to the communities around. Mangrove forests and sea grasses patches provide natural nursery grounds for the fish larvae and other animal species in Setiu Wetlands. The wetlands support diverse aquatic organisms which are ecologically and economically important (Salam 1997) and need to be conserved for future generation (WWF 2013). Fishes were collected at three different locations; Site 1 (Pulau Telaga Tujuh), Site 2 (Pulau Che Him), and Site 3 (Pulau Tok Haji) (Fig. 1).

Fish Sampling

Fishes were collected monthly from June 2011 to July 2012. The average water depth in study sites during the sampling period is between 0.1 and 4.0 m with the maximum of 7 m at the river mouth. The sampling methods used to catch all sizes of fishes were multifilament gill nets (1, 2, 3, 4, and 5-in. mesh size, 40 m long and 2–3 m depth), trawl, and fish trap. The nets was laid for one night. The fish samples were collected monthly for 14 months with covered both the monsoon and non-monsoon season. Based on measurements in 2011–2012 (Malaysia Meteorology Department, http://www.met.gov.my), the highest annual rainfall occurred in September 2011 until February 2012.

All fishes were identified to species level using available taxonomic keys (Matsunuma et al. 2011; Ambak et al. 2010; Mohsin and Ambak 1983) and any fishes whose species are still dubious were sent to Kagoshima University, Japan for identification. The fishes were then measured (total length, standard length, and fork length), and weighed (body weight). The length was measured to the nearest 1 mm and body weight to the nearest 0.01 g. The fishes were fixed in 10% formalin solution and preserved in 70% alcohol. The number of individuals were then counted in order to determine the species composition. Specimens of each species were deposited in the collections of the South China Sea Repository and Reference Centre (RRC), Institute of Oceanography and Environment (INOS), Universiti Malaysia Terengganu. The conservation status of all fish species was determined based on the International Union for Conservation of Nature Red List (IUCN Red List of Threatened Species 2017) and had the following threat categories: (1) extinct (EX), (2) extinct in the wild (EW), (3) critically endangered (CR), (4) endangered (EN), (5) vulnerable (VU), (6) near threatened (NT), (7) least concern (LC), (8) data deficient (DD), and (9) not evaluated (NE) (IUCN Red List of Threatened Species 2017).

Biodiversity Parameters

The fish analysis was done using Plymouth Routines in Multivariate Ecological Research (PRIMER) (PRIMER, UK) (Clarke and Warwick 2001). Three community indices were utilised to determine fish community structure according to the formula below:

1. Shanon (H') diversity index (Magurran 2004),

$$H' = -\sum_{s}^{i=1} pi.\ln pi$$

Where:

S = total number of species,
p_i = the ration of the total number of individual

2. Pielou evenness index (J') (Pielou 1977), and

$$J' = \mathrm{H}' / \ln(\mathrm{S})$$

Where:

H′ = diversity index
S = total number of species
ln = normal log

3. Species richness (Margalef 1958),

$$d = (S-1)/\log(N)$$

Where:

S = total number of species
N = total number of individuals of all species

Fish Assemblages

A total of 13,132 individuals of fishes, comprising 50 families and 116 species were collected from the three study sites (Table 1). The result showed that the fish species recorded 3% of over 3365 fish species recorded by Randall and Lim (2000) from the South China Sea, 5% of over 2243 fish species recorded by Ambak et al. (2010) in Malaysian waters and 26% of over 441 fish species recorded by Matsunuma et al. (2011) in Terengganu water.

In the present study, the number of species identified was higher compared to the previous studies, excluding the checklist of fish community reported by Siti Tafzil Meriam et al. (2015) about 150 fish species in the Setiu Wetlands. Other studies in the Setiu Wetlands by Siti Tafzil Meriam et al. (2008) and Ahmad Najdi (2005) reported 37 and 56 fish species, respectively. The number of fish species recorded in Setiu Wetlands are high compared with previous studies from certain areas in Malaysia. Chong et al. (2010) stated that in a recent review of the Malaysian fishes, the total number of mangrove fishes is 296 species including 47 species collected in Semerak estuary and adjacent inshore waters in Kelantan. In mangrove and coastal waters of Northeastern Langkawi, 91 fish species were reported (Chong et al. 2005a). Chong et al. (2005b) collected 114 species in Klang mangrove, Selangor while Chong and Sasekumar (2002) collected 130 species in Johor estuary. At other places, for example, the Matang mangrove, Perak, the number of species is higher with 138 species managed to be collected (Chong 2005).

A few studies in Southeast Asia collected a high number of species, for examples, Jaafar et al. (2004) collected 91 species in the coastal area at Pasir Ris, in the eastern part of Singapore; Hong and San (1993) collected 105 species in two mangrove areas, Quang Ninh, Hai Phong and Thanh Hoa, Nghe An, Ha Tinh; Haris et al. (2008) collected 133 species in estuaries of Mimika District, Papua Province, Indonesia and Tongnunui et al. (2002) as well as 135 species in Sikao Creek mangrove estuary. Blaber (2000) and Nagelkerken et al. (2008) stated that there were at

Table 1 Taxonomic list of species of fishes with the IUCN status and percentage abundance in Setiu Wetlands, Terengganu

No.	Family	No.	Species	English name	Local name	IUCN status	Percentage abundance
1	Ambassidae	1	*Ambassis interrupta*[a]	Long-spined glass perchlet	Seriding	LC	4.37
	Ambassidae	2	*Ambassis kopsii*[a]	Freckled hawkfish	Seriding	NE	2.15
	Ambassidae	3	*Ambassis urotaenia*	Banded-tail glassy perchlet	Seriding	LC	0.48
	Ambassidae	4	*Ambassis vachelii*	Vachelli's glass perchlet	Seriding	LC	0.18
2	Apogonidae	5	*Apogon amboinensis*	Amboina cardinalfish	Sebekah sejalur	DD	0.11
	Apogonidae	6	*Apogon hyalosoma*	Humpbacked cardinalfish	Sebekah	LC	2.63
	Apogonidae	7	*Ostorhinchus aureus*[a]	Ring-tailed cardinalfish	Sebekah	NE	0.01
3	Ariidae	8	*Arius maculatus*[a]	Spotted catfish	Duri	NE	1.11
	Ariidae	9	*Hexanematichtys sagor*[a]	Sagor catfish	Pedukang	NE	0.43
4	Bagridae	10	*Mystus gulio*[a]	Long whiskers catfish	Baung	LC	0.01
	Bagridae	11	*Mystus nigriceps*[a]	Twospot catfish	Baung	NE	0.01
5	Batrachoididae	12	*Allenbatrachus grunniens*	Grunting toadfish	Depu	NE	0.31
6	Belonidae	13	*Strongylura strongylura*[a]	Spottail needlefish	Todak	NE	0.95
	Belonidae	14	*Tylosurus acus melanotus*[a]	Keel-jawed needle fish	Todakbanang	NE	0.02
7	Callionymidae	15	*Callionymus belcheri*[a]	Flathead dragonet	Baji	NE	0.02
8	Carangidae	16	*Alepes djedaba*[a]	Shrimp scad	Pelata keledek	NE	0.04
	Carangidae	17	*Atule mate*[a]	Yellowtail scad	Pelata	NE	0.02
	Carangidae	18	*Carangoides praeustus*[a]	Brownback trevally	Demuduk abu-abu	NE	0.01
	Carangidae	19	*Caranx sexfasciatus*[a]	Bigeye trevally	Belokok putih	LC	0.02
	Carangidae	20	*Scomberoides tol*[a]	Needle scaled queenfish	Talang	NE	0.01

9	Clupeidae	21	*Herklotsichthys dispilonotus*[a]	Blacksaddle herring	Tamban	NE	0.19
	Clupeidae	22	*Herklotichthys quadrimaculatus*[a]	Bluestripe herring	Bilis	NE	0.03
	Clupeidae	23	*Sardinella gibbosa*[a]	Goldstripe sardinella	Tamban sisik jalur emas	NE	0.04
	Clupeidae	24	*Sardinella fimbriata*[a]	Fringe scale sardinella	Tamban	NE	0.01
10	Cynoglossidae	25	*Cynoglossus bilineatus*[a]	Fourlined tonguesole	Lidah pasir	NE	0.02
	Cynoglossidae	26	*Cynoglossus puncticeps*[a]	Speckled tonguesole	Lidah	NE	0.05
11	Dasyatidae	27	*Himantura uarnak*[a]	Honeycomb stingray	Pari rimau	Vu	0.02
	Dasyatidae	28	*Himantura walga*[a]	Dwarf whipray	Pari ketuka tanjung	NT	0.11
12	Drepaneidae	29	*Drepan punctata*[a]	Spotted sicklefish	Daun baru bintik	NE	0.01
13	Echeneidae	30	*Echeneis naucrates*[a]	Live sharksucker	Tapak kasut	NE	0.01
14	Eleotridae	31	*Butis butis*[a]	Duckbill sleeper	Ubi muncung itik	LC	0.06
	Eleotridae	32	*Ophiocara porocephala*[a]	Northern mud gudgeon	Ubi belantok	LC	0.47
	Eleotridae	33	*Oxyeleotris marmorata*[a]	Marble goby	Gobi	LC	0.01
	Eleotridae	34	*Qxyeleotris urophthalmus*	Goby	Gobi	NE	0.02
15	Engraulidae	35	*Encrasicholina punctifer*[a]	Buccaneer anchovy	Bilis	NE	0.02
	Engraulidae	36	*Stolephorus dubiosus*[a]	Thai anchovy	Bilis	NE	0.02
	Engraulidae	37	*Stolephorus indicus*[a]	Indian anchovy	Bilis	NE	0.17
	Engraulidae	38	*Thryssa hamiltonii*[a]	Hamilton's thryssa	Kasai minyak	NE	0.02
16	Ephippidae	39	*Oxyurichthys ophthalmonema*	Eyebrow goby	Gobi	NE	0.01
17	Gerreidae	40	*Gerres abbreviatus*[a]	Deep-bodied mojarra	Kapas	NE	0.30
	Gerreidae	41	*Gerres filamentosus*[a]	Whipfin silver-biddy	Kapas	LC	0.21
	Gerreidae	42	*Gerres limbatus*[a]	Saddleback silver-biddy	Kapas	NE	0.52
	Gerreidae	43	*Gerres oyena*[a]	Common silver-biddy	Kapas	NE	0.22

(continued)

Table 1 (continued)

No.	Family	No.	Species	English name	Local name	IUCN status	Percentage abundance
18	Gobiidae	44	*Acentrogobius viridipunctatus*	Spotted green goby	Gobi bintik hijau	NE	0.08
	Gobiidae	45	*Arcygobius baliurus*	Isthmus goby	Gobi	NE	0.01
	Gobiidae	46	*Aulopareia unicolor*	Goby	Gobi	NE	0.05
	Gobiidae	47	*Bathygobius* sp.[a]	Goby	Gobi	–	0.02
	Gobiidae	48	*Drombus key*	Key goby	Gobi	NE	0.02
	Gobiidae	49	*Favonigobius melanobranchus*	Blackthroat goby	Gobi	Lr-nt	0.01
	Gobiidae	50	*Favonigobius reichei*	Indo-Pacific tropical sand goby	Gobi	Lr-nt	0.14
	Gobiidae	51	*Glossogobius aureus*	Golden tank goby	Gobi	LC	0.48
	Gobiidae	52	*Glossogobius circumspectus*	Circumspect goby	Gobi	NE	0.16
	Gobiidae	53	*Glossogobius giuris*[a]	Tank goby	Gobi	LC	0.01
	Gobiidae	54	*Periothalmodon schlosseri*[a]	Giant mudskipper	Tembakul	NE	0.02
	Gobiidae	55	*Psammogobius biocellatus*	Sleepy goby	Gobi	LC	0.41
19	Haemulidae	56	*Pomadasys argenteus*[a]	Silver grunt	Gerut-gerut perak	LC	0.01
	Haemulidae	57	*Pamadasys kaakan*[a]	Javelin grunter	Gerut-gerut ompakan	NE	0.53
20	Hemiramphidae	58	*Hyporhamphus limbatus*[a]	Congaturi halfbeak	Jolong sungai	LC	0.09
	Hemiramphidae	59	*Zenarchopterus dunckeri*	Duncker's river garfish	Jolong	NE	0.58
21	Latidae	60	*Lates calcarifer*[a]	Barramundi	Siakap putih	NE	0.83
22	Leiognathidae	61	*Gazza minuta*[a]	Toothpony	Kekek	LC	0.75
	Leiognathidae	62	*Karalla daura*[a]	Goldstripeponyfish	Kekek	NE	0.02
	Leiognathidae	64	*Leiognathus equulus*[a]	Common ponyfish	Kekek	LC	7.70
	Leiognathidae	65	*Nuchequula longicornis*	Ponyfish	Kekek	NE	14.87
	Leiognathidae	63	*Photopectoralis bindus*[a]	Orangefinponyfish	Kekek	NE	0.05
	Leiognathidae	66	*Secutor ruconius*[a]	Deep pugnoseponyfish	Kekek	NE	0.41
23	Lethrinidae	68	*Lethrinus genivittatus*[a]	Longspine emperor	Landok	LC	0.71
	Lethrinidae	67	*Lethrinus lentjan*[a]	Pink ear emperor	Landok calit merah	NE	0.86

24	Lutjanidae	69	*Lutjanus argentimaculatus*[a]	Mangrove red snapper	Kakap merah	NE	0.05
	Lutjanidae	70	*Lutjanus fulviflamma*[a]	Dory snapper	Jenahak	NE	0.02
	Lutjanidae	71	*Lutjanus johnii*[a]	John's snapper	Jenahak tanda	NE	0.13
	Lutjanidae	72	*Lutjanus russelli*[a]	Russell's snapper	Tanda	NE	2.69
25	Megalopidae	73	*Megalops cyprinoides*[a]	Indo-Pacific tarpon	bulan-bulan	DD	0.03
26	Monodactylidae	74	*Monodactylus argenteus*[a]	Silver moony	Bawal perak	NE	0.05
27	Monacanthidae	75	*Paramonacanthus pussillus*	Faintstripe filefish	Tanduk	NE	0.02
28	Mugilidae	76	*Chelon melinopterus*[a]	Otomebora mullet	Belanak	LC	0.01
	Mugilidae	77	*Chelon subviridis*[a]	Greenback mullet	Belanak	NE	0.16
	Mugilidae	79	*Moolgarda perusii*[a]	Longfinned mullet	Belanak	NE	0.03
	Mugilidae	78	*Moolgarda seheli*[a]	Bluespot mullet	Belanak	NE	0.18
29	Muraenesocidae	80	*Muraenesox cinereus*[a]	Daggertooth pike conger	Malong	NE	0.02
30	Ophichthidae	81	*Pisodonophis cancrivorus*[a]	Longfin snake-eel	Malong	NE	0.02
31	Paralichthyidae	82	*Pseudorhombus arsius*[a]	Largetooth flounder	Sebelah bulat	NE	0.60
32	Pegasidae	83	*Pegasus laternarius*	Brick seamoth	Paddle fish	Vu	0.04
33	Platycephalidae	84	*Platycephalus cultellatus*	Flathead	Baji	NE	0.05
	Platycephalidae	85	*Platycephalus indicus*[a]	Bartail flathead	Baji ekor jalur	DD	0.23
34	Plotosidae	86	*Plotosus lineatus*[a]	Striped Eel catfish	Sembilang	NE	0.35
35	Pomacanthidae	87	*Pomacanthus arcuatus*[a]	Gray angelfish		LC	0.01
36	Pristigasteridae	88	*Ilisya elongata*[a]	Elongate ilisha	Puput	NE	0.01
	Pristigasteridae	89	*Ilisha megaloptera*	Bigeyeilisha	Puput	NE	0.11
37	Scatophagidae	90	*Scatophagus argus*[a]	Spotted scat	Kitang	LC	0.18
38	Scombridae	91	*Rastrelliger kanagurta*[a]	Indian mackerel	Kembung borek	DD	0.01
39	Serranidae	92	*Epinephelus coioides*[a]	Orange-spotted grouper	Kerapu bunga	NT	0.56
	Serranidae	93	*Epinephelus diacanthus*[a]	Spinycheek grouper	Kerapu	NT	0.01
	Serranidae	94	*Epinephelus sexfasciatus*[a]	Sixbar grouper	Kerapu bebeh	DD	0.01

(continued)

Table 1 (continued)

No.	Family	No.	Species	English name	Local name	IUCN status	Percentage abundance
40	Siganidae	95	*Siganus fuscescens*[a]	Mottled spinefoot	Dengkis	NE	1.21
	Siganidae	96	*Siganus guttatus*[a]	Goldlinedspinefoot	Dengkis	NE	3.44
	Siganidae	97	*Siganus javus*[a]	Streaked spinefoot	Dengkis jawa	NE	38.81
	Siganidae	98	*Siganus vermiculatus*[a]	Vermiculated spinefoot	Dengkis	LC	0.02
41	Sillaginidae	99	*Sillago aeolus*[a]	Oriental sillago	Bulus-bulus	NE	0.14
	Sillaginidae	100	*Sillago asiatica*[a]	Asian sillago	Bulus-bulus	NE	0.02
	Sillaginidae	101	*Sillago parvisquamis*[a]	Sillago	Bulus-bulus	NE	0.02
	Sillaginidae	102	*Sillago sihama*[a]	Silver sillago	Bulus-bulus	NE	0.34
42	Soleidae	103	*Brachirus orientalis*[a]	Oriental sole	Lidah – bronok timur	NE	0.36
43	Sphyraenidae	104	*Sphryaena forsteri*[a]	Bigeye barracuda	Kacang	NE	0.01
	Sphyraenidae	105	*Sphyraena putnamae*[a]	Sawtooth barracuda	Kacang	NE	0.05
44	Synodontidae	106	*Synodus tectus*	Tectus lizardfish	Mengkarong	NE	0.02
45	Synanceiidae	107	*Trachicephalus uranoscopus*[a]	Stargazing stonefish	Depu	NE	0.01
46	Syngnathidae	108	*Hippocampus spinosissimus*	Sea horse	Kuda laut	Vu	0.01
	Syngnathidae	109	*Hippocampus trimaculatus*	Sea horse	Kuda laut	Vu	0.01
47	Terapontidae	110	*Pelates quadrilineatus*[a]	Fourlinedterapon	Kerong empat jalur	NE	1.28
	Terapontidae	111	*Terapon jarbua*[a]	Jarbuaterapon	Kerong jalur	LC	0.01
48	Tetraodontidae	112	*Arothron immaculatus*[a]	Immaculate puffer	Buntal	NE	0.04
	Tetraodontidae	113	*Chelanodon patoca*[a]	Milkspotted puffer	Buntalsusu	NE	0.43
	Tetraodontidae	114	*Tetraodon nigroviridis*[a]	Spotted green pufferfish	Buntal	NE	2.48
49	Tetrarogidae	115	*Vespicula trachinoides*	Waspfish	Depu	NE	0.97
50	Toxotidae	116	*Toxotes jaculatrix*[a]	Banded archerfish	Sumpit belang	LC	0.33
51	Triacanthidae	117	*Tripodichthys blochi*	Triplespines	Lembu muncung	NE	0.28
	No. of individuals (13134)						
	No. of families (51)						
	No. of species (117)						

Vu Vulnerable, *NT* Near threatened, *Lr-nt* Lower risk near threatened, *LC* least concern, *DD* Data deficient, *NE* Not evaluated
[a]Commercially important species

least 600 species in mangrove areas in the Indo-West Pacific region, stretching from the east coast of Africa through the South and Southeast Asia to Australia and the Central Pacific. Nagelkerken et al. (2008) stated that the higher diversity still contains at least 100 species in the subtropical mangrove habitat and these supporting that the Setiu Wetlands are higher fish species with compared to other areas. However, those studies the researcher used different sizes of the area and a short period of times that might cause different results compared to this study. Furthermore, the function of the habitat as places in the food tropic might contribute to the diversity and abundance of the fish species in the areas. Chong et al. (1994) stated that the populations of juvenile penaeid prawns comprising the mangrove (70–90%) and mudflats (40–90%) were the major nursery areas of the fishes.

The present study shows that the families with the greatest number of species were Gobidae (14 species), followed by Leiognathidae (6 species), and Carangidae (5 species). This was different from the findings by Ahmad Najdi (2005) who determined that the dominant families with the greatest number of species were Gobiidae, Lutjanidae, and Tetraodontidae, which were three species respectively. Furthermore, Carangidae (five species) was the most diverse family followed by family Gobiidae (four species) and Tetraodontidae (three species) (Siti Tafzil Meriam et al. 2008).

The gobies were found in various habitats from torrential freshwater rivers shelf waters but were most common in brackish waters and shallow coastal waters (Matsunuma et al. 2011). Some fish species may remain in the mangrove at low tide. Although gobies may remain at low tide, they may leave their burrows and be caught during high tide. They have an adhesive organ such as modified ventral fins. The Gobiidae were widely distributed in Bogani Nani Wartabone National Park, North Sulawesi (Haryono et al. 2002) and Aceh Water, Northern-Sumatra, Indonesia (Muchlisin and Siti Azizah 2009). The pony and scad fishes whose were the second and third greatest number of species were widely distributed and commonly available in tropical to temperate marine, estuarine, and sometimes freshwater (Matsunuma et al. 2011).

Most of the fishes collected were juveniles of all families and adults of small-size fishes, especially in the family of Carangidae and Leiognathidae. This is the reason why the wetlands should be protected and managed as a nursery area to ensure the survival of juveniles commercially important fish species viewed from the sampling of 90 species caught and considered to be commercially valuable species as listed in Table 1. The highly economically important species including the families of Carangidae, Cluepidae, Latidae, Leiognathidae, Lethrinidae, Lutjanidae, Muraenesocidae, Pristigasteridae, Scombridae, and Serranidae.

Thirty most dominant species were accounted 93.7% of the total catch (Table 2). The result was similar to the finding by Siti Tafzil Meriam et al. (2008) who stated that *Siganus javus* was the most abundant species (66.0%) compared to the present study (38.8%) which was relatively low while Ahmad Najdi (2005) stated that *Ambassis commersoni* was the most abundant in the area. Carpenter and Niem (1998) and Mohsin and Ambak (1996) stated that siganids and rabbitfish were the species recorded from coastal water, brackish, and freshwater. This family can adapt in different types of water at different levels of water salinity, from fresh water to the

Table 2 Percentage abundance (%) and rank (in parentheses) of 30 most dominant species caught in the lagoon area of Setiu Wetlands, Terengganu between June 2011 and July 2012

Species	Percentage abundance (%)				Total length (min)	Total length (max)	Average	Habitat
	Overall	Site 1	Site 2	Site 3	(mm)	(mm)	Length (mean ± SE)	
Siganus javus	38.81 (1)	54.19 (1)	38.90 (1)	14.84 (2)	23	200	44.85 ± 17.10	M-F-B
Nuchequula longicornis	14.87 (2)	14.57 (2)	12.49 (2)	21.05 (1)	20	107	48.69 ± 13.83	M
Leiognathus equulus	7.7 (3)	7.85 (3)	5.52 (4)	12.69 (3)	22	179	53.22 ± 18.91	M-F-B
Ambassis interrupta	4.37 (4)	2.73 (5)	3.56 (6)	8.85 (4)	34	128	69.42 ± 11.37	M-F-B
Siganus guttatus	3.44 (5)	0.78 (10)	6.29 (3)	0.72 (22)	25	182	75.27 ± 41.14	M-B
Lutjanus russelli	2.69 (6)	0.71 (11)	4.14 (5)	2.26 (8)	19	238	78.69 ± 34.77	M-B
Apogon hyalosoma	2.63 (7)	1.61 (7)	3.55 (7)	2.00 (11)	55	811	114.84 ± 45.78	M-F-B
Tetraodon nigroviridis	2.48 (8)	2.75 (4)	1.87 (10)	3.54 (7)	47	190	109.41 ± 30.62	F-B
Ambassis kopsii	2.15 (9)	2.07 (6)	2.42 (8)	1.62 (12)	7	216	63.28 ± 18.71	M-F-B
Pelates quadrilineatus	1.28 (10)	0.34 (22)	2.42 (9)	–	4	95	50.01 ± 17.90	M-B
Siganus fuscescens	1.21 (11)	0.85 (9)	1.54 (13)	0.98 (18)	21	78	30.62 ± 9.62	M-B
Arius maculatus	1.11 (12)	0.68 (12)	0.30 (26)	3.73 (5)	48	535	159.43 ± 86.67	M-F-B
Vespicula trachinoides	0.97 (13)	0.51 (16)	1.63 (12)	–	17	64	40.35 ± 8.65	M-B
Strongylura strongylura	0.95 (14)	0.58 (13)	1.02 (15)	1.36 (13)	282	445	331 ± 24.81	M-B
Lethrinus lentjan	0.86 (15)	–	1.68 (11)	–	27	143	72.74 ± 25.87	M-B
Lates calcarifer	0.83 (16)	–	1.37 (14)	0.53 (25)	60	488	182.70 ± 93.86	M-F-B
Gazza minuta	0.75 (17)	–	–	3.73 (6)	30	66	48.59 ± 6.83	M-B
Lethrinus genivittatus	0.71 (18)	0.56 (14)	1.07 (15)	–	25	101	51.92 ± 15.66	M-B
Pseudorhombus arsius	0.6 (19)	0.34 (23)	0.49 (21)	1.28 (14)	39	200	107.47 ± 38.88	M-B
Zenarchopterus dunckeri	0.58 (20)	0.90 (8)	–	1.02 (17)	79	242	195.20 ± 21.91	B
Epinephelus coioides	0.56 (21)	0.39 (19)	0.61 (20)	0.68 (23)	25	305	150.56 ± 54.36	M-B
Pamadasys kaakan	0.53 (22)	–	–	2.11 (9)	34	154	84.74 ± 27.06	M-B
Gerres limbatus	0.52 (23)	0.41 (18)	0.28 (29)	1.24 (15)	25	133	41.02 ± 18.08	M-F-B

Ambassis urotaenia	0.48 (24)	–	0.83 (16)	–	34	88	49.82 ± 13.87	M-F-B
Glossogobius aureus	0.48 (25)	–	0.72 (17)	–	24	223	106.40 ± 47.03	F-B
Ophiocara porocephala	0.47 (26)	0.54 (15)	0.63 (18)	–	51	146	103.92 ± 17.99	M-F-B
Chelanodon patoca	0.43 (27)	0.46 (17)	0.47 (22)	–	25	144	75.40 ± 20.86	M-F-B
Hexanematichtys sagor	0.43 (28)	0.39 (20)	–	0.98 (19)	147	535	336.55 ± 116.02	M-B
Psammogobius biocellatus	0.41 (29)	–	0.63 (19)	0.49 (28)	28	174	69.31 ± 26.14	M-F-B
Secutor ruconius	0.41 (30)	–	–	2.03 (10)	28	60	45.81 ± 8.37	M-F-B
Brachirus orientalis	–	0.27 (29)	0.30 (27)	0.64 (22)	32	218	94.52 ± 28.51	M-F-B
Plotosus lineatus	–	0.32 (25)	–	1.17 (16)	128	560	193.10 ± 65.44	M-B
Sillago sihama	–	–	–	0.87 (21)	42	248	111.62 ± 65.73	M-B
Toxotes jaculatrix	–	0.34 (24)	0.38 (23)	–	70	360	157.70 ± 54.48	F-B
Allenbatrachus grunniens	–	0.29 (28)	–	0.94 (20)	111	257	185.36 ± 33.17	M-B
Gerres abbreviatus	–	0.22 (30)	0.36 (24)	–	21	133	66.5 ± 25.42	M-B
Triacanthus blochii	–	0.32 (26)	0.36 (25)	–	22	93	57.85 ± 20.35	M
Platycephalus indicus	–	–	0.30 (28)	–	49	350	200.83 ± 88.64	M-B
Gerres oyena	–	–	–	0.60 (24)	25	128	47.96 ± 20.30	M-B
Gerres filamentosus	–	–	–	0.53 (26)	28	126	83.05 ± 26.66	M-F-B
Herklotsichthys dispilonotus	–	0.32 (27)	–	–	76	143	112.76 ± 20.48	M-B
Scatophagus argus	–	–	–	0.49 (29)	102	217	153.49 ± 32.71	M-F-B
Ambassis vachelii	–	–	–	0.53 (27)	32	85	62.58 ± 12.69	M-F-B
Stolephorus indicus	–	–	–	0.45 (30)	24	72	53.00 ± 9.35	M-B
Favonigobius reichei	–	–	0.26 (30)	–	33	189	73.83 ± 58.17	M-F-B
Himantura walga	–	0.37 (21)	–	–	150	334	254.10 ± 39.55	M

M-B marine-brackish, *M* marine, *M-F-B* marine-freshwater-brackish, *F-B* freshwater-brackish, *B* brackish marine environment. The extensive sea grass beds function as nursery and feeding grounds for various species of siganids (United Nations Environment Programme [UNEP] 2004). *Siganus javus* are Oceanodromous fish bred near nursery ground and then drift onto the ocean currents as larvae before settling as juveniles to grow into adults before migrating back to nursery ground (Fish Base 2014).

The second most abundant species was *Nuchequula longicornis* (14.9%) followed by *Leiognathus equulus* (7.7%), *Ambassis interrupta* (4.4%), *Siganus guttatus* (3.4%), *Lutjanus russelli* (2.7%), *Apogon hyalosoma* (2.6%), *Tetraodon nigroviridis* (2.4%), *Ambassis kopsii* (2.2%), *Pelates quadrilineatus* (1.3%), *Siganus fuscescens* (1.2%) and *Arius maculatus* (1.1%) while others contributed less than 1%. The second and third most abundant fishes originated from the family Leiognathidae. The leiognathids ponyfish is a demersal fish and commonly available in shallow coastal waters, estuaries, and mangrove areas.

Occasionally, the leiognathids can adapt to freshwater and usually forms big feeding schools on the shallow sea floor (Mazlan and Seah 2006). *Leiognathus equulus* is an amphidromous fish spawned in freshwater/estuaries then drifted into the ocean as planktonic larvae before migrating back into freshwater/estuaries to grow into adults and spawn (Fish Base 2014).

The fourth most abundant species dominating the fish assemblages were *Ambassis interrupta.* In this study, the most juveniles of small size fish were resident and pelagic species. The high abundance of *Siganus javus* and *Nuchequula longicornis* from the present study showed a marked difference in species domination compared to other studies. *Ambassis kopsii* and *Sillago sihama* dominated the sublittoral zones of the eastern Johor Strait (Hajisamae and Chou 2003) while *Ambassis gymnocephalus*, *Thryssa hammalensis* and *Thryssa hamiltonii* were found in the mangrove inlets and creeks on the Selangor coast (Sasekumar et al. 1992). Saint-Paul and Schneider (2010) stated that the resident fish community is functionally important as an intermediate trophic level for many consumers. Siti TafzilMeriam et al. (2008, 2015) were showed in juvenile stages mainly from the families Engraulidae and Carangidae, suggesting that the area is a suitable nursery ground and habitat for fishery resources. The small fishes were suitable occupies the intertidal forest as the predatory protection (Ronnback et al. 1999).

Site 1 was dominated by *Siganus javus* with the percentage abundance at 54.2%. *Nuchequula longicornis, Leiognathus equulus, Tetraodon nigroviridis, Ambassis interrupta, Ambassis kopsii* and *Apogon hyalosoma* made up 14.6%, 7.9%, 2.8%, 2.7%, 2.1%, 1.6% respectively, while other species had less than 1.0% abundance. Site 2 consisted of *Siganus javus* (38.9%), *Nuchequula longicornis* (12.5%), *Siganus guttatus* (6.3%), *Leiognathus equulus* (5.5%), *Lutjanus russelli* (4.1%), *Ambassis interrupta* and *Apogon hyalosoma* (3.6%), and others (less than 3.0%) (Table 2). Site 3 consisted of *Nuchequula longicornis, Siganus javus, Leiognathus daura, Ambassis interrupta, Arius maculatus,* and *Gazza minuta, Tetraodon nigroviridis* with 21.1%, 14.8%, 12.7%, 8.9%, 3.7%, 3.7%, and 3.5%, respectively and other species (less than 3.0%).

Family Siganidae had the highest family abundant recorded with 43.0% followed by Family Leiognathidae (24.0%), Family Ambassidae (7.0%), Family

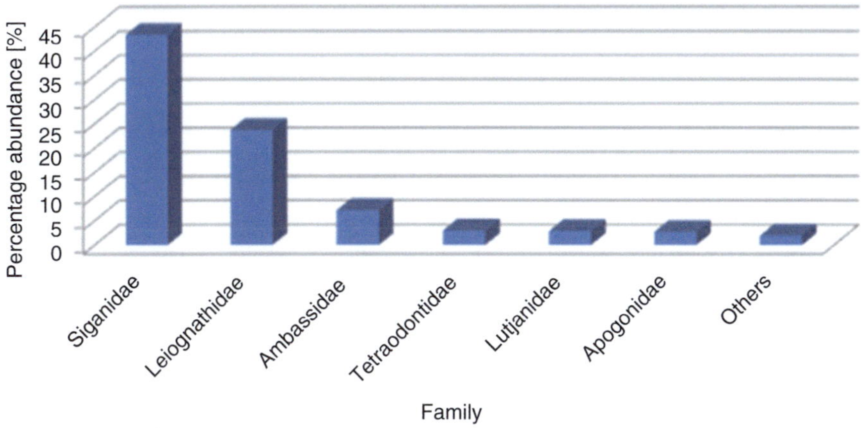

Fig. 2 Family abundance of fishes in Setiu Wetlands

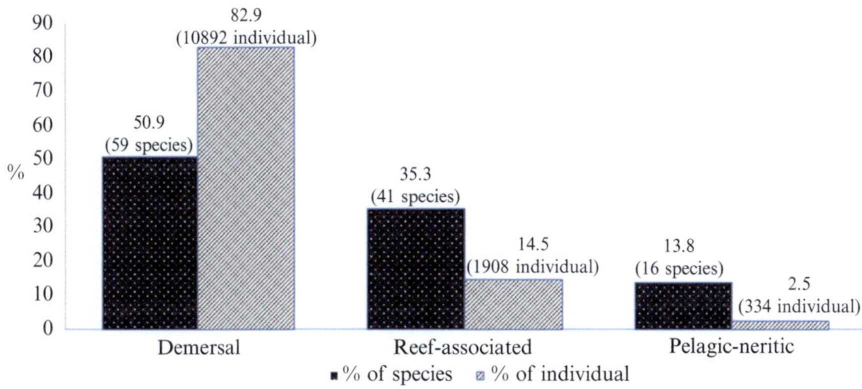

Fig. 3 Percentages of total number of fish species and individual fishes based on habitat of three ecological groups; demersal, reef-associated and pelagic-neritic species of fishes collected from Setiu Wetlands during June 2011 and July 2012

Tetraodontidae, Family Apogonidae and Family Lutjanidae (3.0%) and others (less than 2.0%) (Fig. 2). The highest species was collected at Site 2 (83 species) followed by Site 1 (78 species) and Site 3 (74 species).

This study showed the dominant groups of 59 demersal fish species (50.9%) were caught between two others group, 41 fish species of reef-associated (35.3%) and 16 fish species of pelagic-neritic (13.8%) (Fig. 3). The higher commercial demersal fish species caught including those from the families Carangidae, Serranidae, Dasyatidae, Haemulidae Leiognathidae and Mugilidae. It may be noted that the Setiu Wetlands were found to be an important habitat for the groups of demersal type species. In addition, there is a great need for conservation action for many fish species which provide benefits such as food, livelihoods and recreation.

Diversity, Richness and Evenness

The value of diversity, richness and evenness indices was calculated according to sites (Table 3). The values of diversity, richness and evenness indices for all the fishes studied were found to be 2.60, 12.10 and 0.55, respectively. The species richness (*d*) values were revealed the highest at Site 3 (13.91), and the lowest at Site 2(13.40). The diversity (*H′*) indices of fish samples from Site 3 (2.965) were higher than Site 2 (2.559) and Site 1 (1.969). The evenness (*E*) values were recorded high at Site 3 (0.689) followed by Site 2 (0.579) and Site 1 (0.452).

Overall, Site 2 showed the highest number of individual fishes (6373 fishes) and species (83 species), different from Site 3 which showed the lowest number of individuals and species but the highest values in all diversity indices between the Site 1 and Site 2. Site 3 showed that the individuals in the community were distributed more equitably among these species. There were other reasons why certain sites were found to be dominated by a particular species of fish. This might be due to the location of these both two sites which were located farthest from the estuarine inlet/river mouth, which is the seawater movement into the wetland.

The abundance and distribution of species were usually influenced by the local extinction previously documented in the marine environment, although their natural range and capability to move can help against extinction (Go et al. 2015). Several spatial factors may play a role in adding variation to the local fish assemblage compositions; e.g. creek size and order, topographical height, distance to low-tide resting sites, estuarine salinity zone (Saint-Paul and Schneider 2010). The geographical analysis showed that the movement of the sand spit was about 7.0 m/year for the long beach and 0.8 m/year for small islands in Setiu Wetlands during the year 1980 until 2004, and the critical areas were found around the river mouth. The islands had slowly eroded due to the wave, current, and dominant wind actions during the monsoon loaded (Kasawani et al. 2010). All the spatial factors may lead to the loss of coastal resources and damage of the fragile marine habitats.

Another reason for this is that the seagrass beds in the Sites 2 and 3 may promote more diverse in fish assemblages. They are also one of the important contributors to the organic matter in coastal environments (Canuel et al. 1997). In addition, seagrass detritus supports consumer fish production through bacterial and fungal

Table 3 Diversity indices of fish species in the Setiu Wetlands

Diversity index	Site 1	Site 2	Site 3
Number of species	78	83	74
Number of individuals	4104	6373	2657
Mean number of species (species ± SD)*	0.21 ± 0.4	0.27 ± 0.4	0.21 ± 0.4
Mean number of individuals (individual ± SD)*	297.4 ± 667.2	457.3 ± 706.5	190.4 ± 345.5
Margalef Richness, *d* = (s-1)/Log(N)	13.55	13.4	13.91
Shannon, *H′* = -SUM (Pi*Log(Pi))	1.969	2.559	2.965
Pielou Eveness, *J′* = H′/Log(S)	0.452	0.579	0.689

Significant different (*P < 0.001)

Table 4 Diversity index of fish species by month in the Setiu Wetlands

Month	Number of species	Number of individuals	Mean number of species (species ± SD)	Mean number of individu al (individual ± SD)	Margalef richness, *d* = (s-1)/ Log(N)	Shannon, *H'* = -SUM (Pi*Log(Pi))	Pielou eveness, *J'* = H'/ Log(S)
11-Jun	52	718	0.26 ± 0.4	239.3 ± 225.7	9.312	2.969	0.7514
11-Jul	45	427	0.24 ± 0.4	142.6 ± 164.5	8.873	3.276	0.8605
11-Aug	39	306	0.20 ± 0.4	102.3 ± 109.4	8.216	2.896	0.7905
11-Sep	49	565	0.24 ± 0.4	188.3 ± 221.6	9.166	2.987	0.7676
11-Oct	51	769	0.27 ± 0.4	257.0 ± 290.6	9.017	2.628	0.6683
11-Nov	46	764	0.23 ± 0.4	254.6 ± 350.2	8.124	2.73	0.7131
11-Dec	55	741	0.27 ± 0.4	247.3 ± 226.7	9.62	2.883	0.7228
12-Jan	40	426	0.20 ± 0.40	142.0 ± 167.2	7.872	2.84	0.7698
12-Feb	43	606	0.20 ± 0.41	206.7 ± 228.4	7.914	2.11	0.5611
12-Mar	48	801	0.22 ± 0.42	267.0 ± 283.0	8.414	2.605	0.6729
12-Apr	52	2193	0.26 ± 0.4	731.3 ± 671.8	7.734	2.067	0.5231
12-May	38	3848	0.20 ± 0.4	1286.7 ± 1041.9	5.17	0.8214	0.2258
12-Jun	48	638	0.24 ± 0.4	212.6 ± 166.2	8.773	2.34	0.6044
12-Jul	42	330	0.24 ± 0.5	132.7 ± 124.3	8.726	2.896	0.7749

intermediates or remain unutilised in a trophic dead end as particulate organic carbon in marine sediments (Jones et al. 2003).

The diversity indices of fish species by months indicated the highest fish species diversity in Setiu Wetlands, Terengganu (Table 4). The diversity indexes as estimated were highest species richness (d) in December 2012 (early rainy season), followed the diversity (H) and evenness, E were highest in July 2011 (late dry season). The effect was clearly showed of this study whereby by lowest species richness, diversity and evenness index values were recorded in May 2012. Only 38 species of fish were captured and may be that the other species are required to leave their habitat in this month due to the low water level condition. This may be due to the temporal fluctuations throughout the year whereas that are consideration of the run off tidal range, and sea water movement in the site. Similarly, study by Barletta-Bergan et al. (2002) in the Caete River in north Brazil, noted that the highest species richness, diversity and evenness in the rainy season when salinity started to increase.

The majority of the fish species still dominate the rivers and the areas near the estuarine inlet/river mouth where nutrients are found in abundance (Department of Fisheries Malaysia [DOF] 1994). Sedimentation and siltation, food, nutrients, and a high concentration of dissolved oxygen were fluxed together with the currents of the seawater movements into the wetland. The net organic matter production may accumulate in local sediments or can be transported to adjacent coastal waters through the export of the detrital material depending on the hydrodynamic conditions (Madsen et al. 2001).

These factors are related to the fish movements into the habitats. According to Chong (2007), the fish movements between mangrove and offshore waters concern the type of movement, purpose of the movement, and time and developmental stage, which achieve this movement. Marine species move to coastal shallow waters before reaching maturity, where they contribute to the coastal fisheries delay (Saint-Paul and Schneider 2010; Barletta-Bergan et al. 2002). According to Flannery et al. (2002), peripheral fishes move to the low salinity and freshwater habitats for feeding and shelter from marine predators.

Furthermore, Blaber (1997) reported that most estuarine fishes could tolerate salinity fluctuations but their adaptability and distribution varied among species depending on physiological tolerances, which may influence their distributions. Chong (2007) stated that their distributions were related to the requirements of maximum survival, growth, body condition, and conditions prevailing in the nursery and spawning areas.

IUCN Status

Among all species mentioned in the study, four of them were recorded as vulnerable (VU) in the IUCN Red List of Threatened Species (2017), which are *Hippocampus spinosissimus, Hippocampus trimaculatus, Himantura uarnak,* and *Pegasus laternarius, Himantura walga, Favonigobius melanobranchus, Favonigobius rechei. Epinephelus coioides,* and *Epinephelus diachanthus* were listed as NT (Table 1). It is interesting to note that the fish species must be conserved and protected because some of them might be beneficial for some the purposes of indicator of the health and biodiversity of habitats.

In the Southeast Asia, *Himantura* species are caught and landed in the inshore fisheries and commercially. This species prefers inshore waters, shallow estuaries and lagoons, which means that it is also threatened by extensive habitat degradation and loss of mangrove forests. In Indonesia and Malaysia, the mangrove forest areas have been lost through conversion of land for shrimp farming, excessive logging, and urban development and to a lesser extent, for agriculture of salt pans (Manjaji and White 2009).

The wetlands have been reported to supply 50.0–60.0% juveniles of wild grouper (*Epinephelus* spp.) for Peninsular Malaysia (Harban Singh 2010). The local people who live around the wetlands are greatly dependent on the wetland resources for their survival and to generate income, particularly through fisheries. Based on preliminary findings from a recent study on the economic valuation of the Setiu Wetlands ecosystem, 77% of the 200 locals from the Setiu district interviewed agreed that Setiu Wetlands has economic value and needs to be conserved for future generation (WWF 2013).

In Malaysia, 12.0% of *Epinephelus coioides* was lost from 1980 to 1990 in population and stock assessment (Spalding et al. 1997). Overfishing and habitat destructions were the two most possible factors (Cornish and Harmelin-Vivien 2004) of the loss. Overfishing of certain species near coral reefs can easily damage the reef ecologies. Habitat destructions due to human activities (population growth, urbanisation, industrialisation and tourism) can damage the coral reefs and mangrove habitats. It has been identified as the main threat to 85% of all species described in the IUCN Red List (those species officially classified as threatened and endangered) (IUCN Red List of Threatened Species 2017). In Indian waters, the greatest threat to *Epinephelus diachanthus* is overfishing and exploitation of juveniles and possibly by trawling (Sadovy et al. 2008).

Malaysia plays an important role in the international trade of syngnathids. The syngnathids targeted by fishers supplies a substantial trade in seahorses for traditional medicine, curios, and aquarium uses (Martin Smith and Vincent 2006; Martin Smith et al. 2003). Both *Hippocampus spinosissimus* and *Hippocampus trimaculatus* are listed to have declined in number at least 30% due to high levels of exploitation, accidental capture and habitat degradation (Wiswedel 2012, 2015). Lim et al. (2011) stated that most syngnathids were listed as vulnerable in the IUCN Red List of Threatened Species while all seahorses were listed the Appendix II (Convention on the International Trade in Endangered Species of Wild Flora and Fauna [CITES] 2002). It is recommended that the state government should gazette the wetlands to protect and preserve the area as a state park in Terengganu.

Conclusion

This study showed that the Setiu Wetlands is an important habitat for freshwater and marine fishes as a nursery and breeding ground. The sea grass and mangrove habitats provide the best environment for maintaining biodiversity in the Setiu Wetlands. The biggest threats faced by the Setiu Wetlands are palm oil plantations, aquaculture, and from urban development that can destroy the vegetation and the habitat for a wide variety of flora and fauna. The wetlands must be conserved for the sustainability of fisheries resources. This study provided important data and information that can be used for proper implementation and development of a comprehensive sustainable management plan for wetland estuarine.

Acknowledgements The authors would like to thank the management of the Institute of Oceanography and Environment (INOS), UMT for funding this study, and allowing them to use the facility and staffs (Mrs. Nur Fatiha Che Wail, Mr. Ahmad Nazila Ali and Mrs. Noratikah Ab. Manaf), assisting them during the field trips, and collecting specimens. Great appreciation is also extended to Mr. Mohd Nasir b. Mohamad for providing the map to the researchers. The authors would also like to thank Prof. Madya Dr. Zainudin Bachok, and Dr. Amirrudin Ahmad from UMT for their useful comments and advises.

References

Ahmad Najdi, A. K. (2005). *Kepelbagaian komuniti ikan pada musim tengkujuh di Setiu Wetland, Terengganu.* Unpublished undergraduate thesis, KolejUniversitiSainsdanTeknologi Malaysia, Kuala Terengganu, Terengganu, Malaysia.

Ali, T. S., & Hussain, N. A. (1990). Composition and seasonal fluctuations of intertidal fish assemblage in Kohr al-Zubair, Northwestern Arabian Gulf. *Journal of Applied Ichthyology, 6*(1), 24–36.

Ambak, A. A., Mansor, M. I., Mohd Zaidi, Z., & Mazlan, A. G. (2010). *Fishes of Malaysia*. Kuala Terengganu: Universiti Malaysia Terengganu Press.

Barletta-Bergan, A., Barletta, M., & Saint-Paul, U. (2002). Structure and seasonal dynamics of larval fish in the Caete River in north Brazil. *Estuarine, Coastal and Shelf Science, 54*, 193–206.

Blaber, S. J. M. (1997). *Fish and fisheries of tropical estuaries fish and fisheries (Series 22).* London: Chapman and Hall.

Blaber, S. J. M. (2000). *Tropical estuarine fishes: Ecology, exploitation and conservation* (1st ed.). Malden: Blackwell Science.

Canuel, E. A., Freeman, K. H., & Wakeham, S. G. (1997). Isotopic compositions of lipid biomarker compounds in estuarine plants and surface sediments. *Limnology and Oceanography, 42*(7), 1570–1583.

Carpenter, K. E., & Niem, V. H. (1998). *FAO species identification guide for fishery purposes. The living marine resources of the Western Central Pacific*. Rome: Food and Agriculture Organization of the United Nations.

Carr, W. E., & Adams, C. A. (1973). Food habits of juvenile marine fishes occupying seagrass beds in the estuarine zone near Crystal River, Florida. *Transactions of the American Fisheries Society, 102*(3), 511–540.

Castellanos-Galindo, G. A., & Krumme, U. (2014). Long-term stability of tidal and diel-related patterns in mangrove creek fish assemblages in North Brazil. *Estuarine, Coastal and Shelf Science, 149*, 264–272.

Chong, V. C. (2005). Fifteen years of mangrove fisheries research in Matang: What have we learnt? *Sustainable Management of Matang Mangroves, 100*, 1–20.

Chong, V. C. (2007). Mangroves-fisheries linkages-the Malaysian perspective. *Bulletin of Marine Science, 80*, 3,755–3,772.

Chong, V. C., & Sasekumar, A. (2002). Fish communities and fisheries of Sungai Johor and Sungai Pulai Estuaries (Johor, Malaysia). *Malayan Nature Journal, 56*(3), 279–302.

Chong, V. C., Sasekumar, A., Leh, M. U. C., & D'Cruz, R. (1990). The fish and prawn communities of a Malaysian coastal mangrove system, with comparisons to adjacent mud flats and inshore waters. *Estuarine Coastal and Shelf Science, 31*(5), 703–722.

Chong, V. C., Sasekumar, A., Atmadja, W. S., & Low, J. K. Y. (1994). *Status of mangrove prawn fisheries in the ASEAN countries*. Bangkok: Chulalongkorn University Press.

Chong, V. C., Affendy, N., Ooi, A. L., & Chew, L. L. (2005a). Mangrove fishes of northeastern Langkawi Island. *Malayan Nature Journal, 57*(2), 193–208.

Chong, V. C., Sasekumar, A., & Zgozi, S. W. (2005b). Ecology of fish and shrimp communities. In A. Sasekumar & V. C. Chong (Eds.), *Ecology of Klang Strait* (pp. 174–200). Kuala Lumpur: University of Malaya Press.

Chong, V. C., Jamizan, A. R., Mohamad Yazid, Z., Rizman-Idid, Muhammad Ali, S. H., & Natin, P. (2010). Diversity and abundance of fish and invertebrates of Semerak Estuary and adjacent inshore waters, Kelantan. *Malaysian Journal of Science, 29*, 95–110.

Clarke, K. R., & Warwick, R. M. (2001). *Change in marine communities: An approach to statistical analysis and interpretation* (2nd ed.). Plymouth: PRIMER-E Ltd.

Convention on the International Trade in Endangered Species of Wild Flora and Fauna. (2002). *Amendments to appendices I and II of the convention adopted by the conference of the parties at its 12th meeting (CITES)*. Santiago: Convention on the International Trade in Endangered Species of Wild Flora and Fauna.

Cornish, A., & Harmelin-Vivien, M. (2004). Grouper & wrasse specialist group *Epinephelus coioides*. Retrieved from The IUCN Red List of Threatened Species Version 2014.2. website http://www.iucnredlist.org.

Department of Fisheries Malaysia. (1994). *Fishes of Kenyir*. Terengganu: Department of Fisheries Malaysia.

Fish Base. (2014). *Siganus javus* (Linnaeus, 1766). Retrieved from http://www.fishbase.org/summary/4618.

Flannery, M. S., Peebles, E. B., & Montgomery, R. T. (2002). A percent-of-flow approach for managing reductions of freshwater inflows from unimpounded rivers to southwest Florida estuaries. *Estuaries, 25*(6), 1318–1332.

Go, K. T. B., Anticamara, J. A., de Ramos, J. A. J., Gabona, S. F., Agao, D. F., Hererra, E. C., & Bitara, A. U. (2015). Species richness and abundance of non-cryptic fish species in the Philippines: A global center of reef fish diversity. *Biodiversity and Conservation, 24*(10), 2475–2495.

Gray, C. A., McElligott, D. J., & Chick, R. C. (1996). Intra-and inter-estuary differences in assemblages of fishes associated with shallow seagrass and bare sand. *Marine and Freshwater Research, 47*(5), 723–735.

Hajisamae, S., & Chou, L. M. (2003). Do shallow water habitats of an impacted coastal strait serve as nursery grounds for fish? *Estuarine, Coastal and Shelf Science, 56*(2), 281–290.

Hajisamae, S., Yeesin, P., & Chaimongkol, S. (2006). Habitat utilization by fishes in a shallow, semi-enclosed estuarine bay in southern Gulf of Thailand. *Estuarine, Coastal and Shelf Science, 68*(3), 647–655.

Haris, A., Lala, D., & Setyadi, G. (2008). Fishes of the estuaries of Mimika District, Papua Province. *Marine Research in Indonesia, 33*(2), 155–166.

Harris, S. A., Cyrus, D. P., & Beckley, L. E. (2001). Horizontal trends in larval fish diversity and abundance along an ocean-estuarine gradient on the northern KwaZulu-Natal coast, South Africa. *Estuarine, Coastal and Shelf Science, 53*(2), 221–235.

Haryono, A. H., Tjakrawidjaja, D., & Riyanto, A. (2002). Iktiofauna di perairansekitar Gunung Kabela Taman Nasional Bogani Nani Wartabone Sulawesi utara. *Jurnal Iktiologi Indonesia, 2*(2), 31–40.

Hong, P. N., & San, H. T. (1993). *Mangroves of Vietnam* (Vol. 7). Bangkok: IUCN.

Horn, M. H. (1980). Diel and seasonal variation in abundance and diversity of shallow-water fish populations in Morro Bay, California. *Fishery Bulletin, 78*(3), 759–769.

IUCN. (2017). Red list of threatened species. Retrieved from www.iucnredlist.org.

Jaafar, Z., Hajisamae, S., Chou, L. M., & Yatiman, Y. (2004). Community structure of coastal fishes in relation to heavily impacted human modified habitats. *Hydrobiologia, 511*(1), 113–123.

Jenkins, G. P., May, H. M. A., Wheatley, M. J., & Holloway, M. G. (1997). Comparison of fish assemblages associated with seagrass and adjacent unvegetated habitats of Port Phillip Bay and Corner Inlet, Victoria, Australia, with emphasis on commercial species. *Estuarine Coastal and Shelf Science, 44*(5), 569–588.

Jones, A. B., O'Donohue, M. J. O., Udy, J., & Dennison, W. C. (2003). Assessing ecological impacts of shrimp and sewage effluent: Biological indicators with standard water quality analyses. *Estuarine, Coastal and Shelf Science, 52*, 91–109.

Kasawani, I., Kim Asbiyallah Shah, A., Sulong, I., MohdHasmadi, I., & Kamaruzaman, J. (2010). Coastal change detection using GIS in Setiu lagoon, Terengganu, Malaysia. *Journal of GIS Trends, 1*(1), 20–26.

Khoo, K. H. (1990). The mangrove fisheries in Matang, Perak and Merbok, Kedah. In B. H. R. Othman (Ed.)., (Compiler) *Report of ASEAN Australia. Cooperative program on marine science. Coastal living resources project (phase I)* (pp. 115–144). Selangor: Faculty of Sciences, University Kebangsaan Malaysia.

Kiso, K., & Mahyam, M. I. (2003). Distribution and feeding habits of juvenile and young John's snapper *Lutjanus johnii* in the Matang mangrove estuary, west coast of Peninsular Malaysia. *Fisheries Science, 69*(3), 563–568.

Levy, E. S., Asare, N. K., Yankson, K., & Wubah, D. A. (2015). Effects of hydrographic conditions of ponds on juvenile fish assemblages in the Kakum mangrove system, Ghana. *Regional Studies in Marine Science, 2*, 19–27.

Lim, A. C. O., Chong, V. C., Wong, C. S., & Choo, C. K. (2011). Diversity, habitats and conservation threats of syngnathid (Syngnathidae) fishes in Malaysia. *Tropical Zoology, 24*(2), 193–222.

Madsen, S. D., Nielsen, T. G., & Hansen, B. W. (2001). Annual population development and production by *Calanus finmarchicus*, *C. glacialis* and *C. hyperboreus* in Disko Bay, western Greenland. *Marine Biology, 139*(1), 75–93.

Magurran, A. E. (2004). *Measuring biological diversity*. Malden: Blackwell Publishing.

Manjaji, B. M., & White, W. T. (2009). *Himantura uarnak*. Retrieved from The IUCN Red List of Threatened Species. Version 2014.2. website: www.iucnredlist.org.

Mann, K. H. (2000). Ecology of coastal waters with implications for management. *Marine Biology and Ecology, 96*, 199–212.

Margalef, R. (1958). Information theory in ecology. *General Systems, 3*, 36–71.

Martin-Smith, K. M., & Vincent, A. C. (2006). Exploitation and trade of Australian seahorses, pipehorses, sea dragons and pipefishes (family Syngnathidae). *Oryx, 40*(2), 141–151.

Martin-Smith, K. M., Lam, T. F., & Lee, S. K. (2003). Trade in pipehorses *Solegnathus* spp. for traditional medicine in Hong Kong. *Traffic Bulletin, 19*(3), 139–148.

Matsunuma, M., Motomura, H., Matsuura, K., Shazili, N. A. M., & Ambak, M. A. (2011). *Fishes of Terengganu: East coast of Malay Peninsula, Malaysia*. Terengganu: Universiti Malaysia Terengganu Press.

Mazlan, A. G., & Seah, Y. G. (2006). Meristic and length-weight relationship of ponyfishes (Leiognathidae) in the coastal water of Pulau Sibu-Tinggi, Johor, Malaysia. *Malaysian Applied Biology, 35*(1), 27–35.

Mohsin, A. K. M., & Ambak, M. A. (1983). *Freshwater fishes of Peninsular Malaysia*. Serdang: Universiti Pertanian Malaysia Press.

Mohsin, A. K. M., & Ambak, M. A. (1996). *Marine fishes & fisheries of Malaysia and neighbouring countries*. Serdang: Universiti Pertanian Malaysia Press.

Muchlisin, Z. A., & SitiAzizah, M. N. (2009). Diversity and distribution of freshwater fishes in Aceh waters, northern Sumatra Indonesia. *International Journal of Zoological Research, 5*(2), 62–79.

Nagelkerken, I., Van der Velde, G., Gorissen, M. W., Meijer, G. J., Van't Hof, T., & Den Hartog, C. (2000). Importance of mangroves, seagrass beds and the shallow coral reef as a nursery for important coral reef fishes, using a visual census technique. *Estuarine, Coastal and Shelf Science, 51*(1), 31–44.

Nagelkerken, I., Blaber, S. J. M., Bouillon, S., Green, P., Haywood, M., Kirton, L. G., & Somerfield, P. J. (2008). The habitat function of mangroves for terrestrial and marine fauna: A review. *Aquatic Botany, 89*(2), 155–185.

Nakisah, M. A., & Fauziah, A. H. (2003). *Setiu wetlands: Tranquility amidst plenty*. Kuala Terengganu: Kolej Universiti Sains dan Teknologi Malaysia.

Nash, R. D. (1988). The effects of disturbance and severe seasonal fluctuations in environmental conditions on north temperate shallow-water fish assemblages. *Estuarine, Coastal and Shelf Science, 26*(2), 123–135.

Pielou, E. C. (1977). *Mathematical ecology*. New York: Wiley.

Ramsar Convention on Wetlands. (1996). *Proceedings of the 6th meeting of the conference of the contracting parties*. Brisbane: Ramsar Convention on Wetlands.

Randall, J. E. & Lim, K. K. P. (Eds.) (2000). A checklist of the fishes of the South China Sea. *Raffles Bulletin Zooogy Supply, 8*, 569–667.

Ronnback, P., Troell, M., Kautsky, N., & Primavera, J. H. (1999). Distribution pattern of shrimp and fish among Avicennia and Rhizophora microhabitats in the Pagbilao mangroves, Philippines. *Estuarine Coastal and Shelf Science, 48*, 223–234.

Sadovy, Y., Pollard, D., Russell, B., & Heemstra, P. C. (2008). *Epinephelus diacanthus*. The IUCN Red List of Threatened Species. Version 2014.2 website: www.iucnredlist.org.

Saint-Paul, U., & Schneider, H. (Eds.). (2010). *Mangrove dynamics and management in North Brazil* (Vol. 211). London: Springer Science & Business Media.
Salam, M. N. A. (1997). *Coastal and estuarine land use management around some major river systems in Peninsular Malaysia. Case study III: Sg. Setiu, Terengganu. Project MYS 304/94.* Petaling Jaya: WWF Malaysia.
Santos, R. S., & Nash, R. D. (1995). Seasonal changes in a sandy beach fish assemblage at Porto Pim, Faial, Azores. *Estuarine Coastal and Shelf Science, 41*(5), 579–591.
Sarpedonti, V., & Chong, V. C. (2008). Abundance and distribution of *Stolephorus baganensis* (Hardenberg 1933) and *Thryssa kammalensis* (Bleeker 1849) larvae in relation to ontogeny and environmental factors in a Malaysian estuary. *Tropical Zoology, 21*(2), 195–208.
Sasekumar, A., Chong, V. C., Leh, M. U., & D'Cruz, R. (1992). Mangroves as a habitat for fish and prawns. *Hydrobiologia, 247*, 195–207.
Sasekumar, A., Chong, V. C., Lim, K. H., & Singh, H. R. (1994). *The fish community of Matang mangrove waters, Malaysia*. Bangkok: Chulalongkorn University Press.
Singh, H. (2010, February 16). Wetlands not wastelands: Setiu Wetlands in Terengganu. *New Straits Times*. p.
SitiTafzil Meriam, S. A. K., Mohd Helmi, A., Yusuf, Y., & Ahmad, A. (2008). Preliminary surveys of fishes at Setiu Wetland. In A. T. Ahmad, A. M. A. Bolong, A. A. Nuruddin, F. A. Latiff, C. T. Thye, Y. A. Hua, H. Y. Siang, K. K. K. Yaacob, & W. M. Arshaad (Eds.), *Proceedings of the 5th national fisheries symposium 2008* (p. 102). Kuala Lumpur: Perpustakaan Negara Malaysia.
SitiTafzil Meriam, S. A. K., Yusuf, Y., Ahmad, A., Mohamad-Norizam, M., Mohd Lokman, H., Arai, T., Motomura, H., Mazlan, A. G., & Ambak, M. A. (2015). *The checklist of fish fauna of Setiu Wetlands, East Coast, Terengganu, Malaysia.* Poster presented at International Seminar on the Straits of Malacca and the South China Sea, Universiti Malaysia Terengganu, Terengganu, Malaysia.
Spalding, M. D., Blasco, F., & Field, C. D. (Eds.). (1997). *World mangrove atlas*. Okinawa: The International Society for Mangrove Ecosystems (ISME).
Tongnunui, P., Ikejima, K., Yamane, T., Horinouchi, M., Medej, T., Sano, M., & Taniuchi, T. (2002). Fish fauna of the Sikao Creek mangrove estuary, Trang, Thailand. *Fisheries Science, 68*(1), 10–17.
United Nations Environment Programme. (2004). *Seagrass in the South China Sea* (UNEP/GEF/SCS Technical Publication No. 3–14). Bangkok: UNEP/GEF: United Nations Environment Programme.
Wiswedel, S. (2012). *Hippocampus spinosissimus*. Retrieved from The IUCN Red List of Threatened Species 2012 website: https://doi.org/10.2305/IUCN.UK.2012.RLTS.T10084A17251369.
Wiswedel, S. (2015). *Hippocampus trimaculatus*. Retrieved from The IUCN Red List of Threatened Species 2015 website: https://doi.org/10.2305/IUCN.UK.2015.RLTS.T10087A17252219.
Woodroffe, C. D., & Davies, G. (2009). The morphology and development of tropical coastal wetlands. In G. M. E. Perillo, E. Wolanski, D. R. Cahoon, & M. M. Brinson (Eds.), *Coastal wetlands: An integrated ecosystem approach* (pp. 65–88). Amsterdam: Elsevier.
WWF. (2013). *Setiu wetlands: A perfect choice to be Terengganu's First State Park*. Retrieved from http://www.wwf.org.my

Part III
Management

Trade-Off Analysis for Eco-Tourism of the Tasik Kenyir Protected Area

Muhamad Safiih Lola, Mohd Noor Afiq Ramlee, Mohd Fadli Hussin, Muhamad Na'eim Abdul Rahman, Mohd Tajuddin Abdullah, Anton Abdulbasah Kamil, Izham Mohamad Yusof, Yahaya Ibrahim, and Nur Zafirah A. Khadar

Abstract Continuous pandemic of sustainable development rise numerous concern, hence resulting towards integration of multidimensional principle as an underlay in order to form sound decision-making process especially in ecological-sensitive area such as Tasik Kenyir. This study develops the structural framework for decision-making inclusive of all variables in order to strive for sustainable development of Tasik Kenyir in order to promote responsible tourism practices. Several criteria are

M. S. Lola (✉)
School of Informatics and Applied Mathematics, Universiti Malaysia Terengganu, Kuala Nerus, Terengganu, Malaysia

Institute of Tropical Biodiversity and Sustainable Development, Universiti Malaysia Terengganu, Kuala Nerus, Terengganu, Malaysia
e-mail: safiihmd@umt.edu.my

M. N. A. Ramlee · M. F. Hussin
Institute of Tropical Biodiversity and Sustainable Development, Universiti Malaysia Terengganu, Kuala Nerus, Terengganu, Malaysia

M. N. Abdul Rahman
School of Informatics and Applied Mathematics, Universiti Malaysia Terengganu, Kuala Nerus, Terengganu, Malaysia

M. T. Abdullah
Institute of Tropical Biodiversity and Sustainable Development, Universiti Malaysia Terengganu, Kuala Nerus, Terengganu, Malaysia

School of Marine and Environmental Sciences, Universiti Malaysia Terengganu, Kuala Nerus, Terengganu, Malaysia

A. A. Kamil
School of Industrial Engineering, Telkom University, Bandung, Bandung, Indonesia

I. Mohamad Yusof · N. Z. A. Khadar
School of Distance Education, Universiti Sains Malaysia, Penang, Malaysia

Y. Ibrahim
Faculty of Applied Social Sciences, Universiti Sultan Zainal Abidin, Kuala Terengganu, Terengganu, Malaysia

Darul Iman Institute, Universiti Sultan Zainal Abidin, Kuala Nerus, Terengganu, Malaysia

M. T. Abdullah et al. (eds.), *Greater Kenyir Landscapes*,
https://doi.org/10.1007/978-3-319-92264-5_21

selected and analyzed using Multi-criteria Analysis (to show the corresponding trade-off); ranged from economic, ecological and social variables such as economic revenue, employment, conservation of flora and fauna and environmental quality. The results show that under different scenarios, the score of different type of variables will change accordingly.

Keywords Multi-criteria analysis · Trade-off analysis · Tasik Kenyir · Economic · Social · Environment and eco-tourism

Introduction

Degradation in the quality of Tasik Kenyir resources and amenities in recent years has called for enhanced management in order to improve resource quality and sustainability (Kamarudin et al. 2011). However, environments depreciation in many of the designated Protected Area, has not been arrested by designation and management of the protected area (Hodgson 1997; Dixon et al. 1993). Hence, delicate approaches have to be taken to ensure the direction of developments is in line with the interests of all stake-holders involved in the Tasik Kenyir ecosystem.

Capturing deliberation and evaluation for decision making to take places in real time are the fundamental cores of trade-off analysis (Yoe et al. 2002). The decision will be built based on consideration of specific attributes or criteria. To be precise, the consideration will debate certain attributes of what will exists (made or increase) and what attributes will cease to exist (gone or reduce). There are value of trade-offs that cannot be avoided; choosing one thing simultaneously means not choosing the other.

The human reality is multidimensional, which means it consists of branches of choices (Srinivasan 1988). In developing Tasik Kenyir as a sub-urban and vibrant Eco-tourism place with appeal to a certain type of tourist, we need to consider certain aspects before plans are made; economic, services, environment and socio-cultural developments aspects and their interactions effects on the ecosystem should be highlighted. In brief, this study explores the cost of relaxing previous aspects in order to increase eco-tourism-based goals.

Tasik Kenyir History, Flashback and Decision

Prior to the formation of Tasik Kenyir, this area was a center of early civilization (Mustafa et al. 2013; Chia 2003; Gin 2009). According to Taha (1991), caves around the Tasik Kenyir area, namely recognized as Batu Tok Bidan and Gua Bewah, were proved to have produced significant archaeological discoveries; stone tool artifacts, axes and weapons dating back to the Neolithic era (estimated roughly around 10,000 years ago).

When the area was inundated with water between 1978 and 1985, most of the hilltops remained above the water level and fortuitously creating 340 man–made islands (Shahrom 2012). Many archaeological artifacts and unexplored caves, along with Batu Tok Bidan were believed had submerged during the creation of Tasik Kenyir (Mustafa et al. 2013; Chia 2003). Nowadays, there were Gua Bewah and Gua Taat that are remain accessible to the public.

Prior to its submerge during the flooding of the reservoir, Batu Tok Bidan cave was excavated in 1959 by R. Noone, and later by the Malaysian Historical Society in 1976. Mollusk shells indicating signs of human consumption (tips broken off) discovered at the site suggest that this site were frequently used as a shelter in the prehistoric era (Chia 2003). A Neolithic burial place was also found at the site along with broken pottery laid at the foot of the deceased.

In 2010, human remains believed dated from the Mesolithic Age were found in Gua Bewah. The female skeletal remains were confirmed to be dated back from 13,400 years old. In 2012, the media reported the discovery of a second prehistoric skeleton, also retrieved from Gua Bewah, not far from where the first skeleton was found.

Decision

The finding of archeological artifacts are proof that many intellectual treasure was buried inside when the Kenyir dam was built. And many items or natural resources were sacrificed for the greater good; to provide sources of electricity and fresh water for the human population in Terengganu (Shahrom 2012; Kamaruddin et al. 2011; Zakaria et al. 2000). This is the first example of trade-off that Tasik Kenyir undergoes in the early years. Hence, now, the scenarios are different, but the weight of the decisions are actually pretty much the same.

Based on previous research, the Tasik Kenyir trade-off analysis should include several variables. Some variables could be further separated into independent and dependent attributes such as the following:

Independent Variables

Economic Benefits

This group of five 5-scale Likert-type items was utilized to ask potential respondents about their perception of general economic benefits brought by nature-based eco-tourism such as marketing, stabilized revenues, local tax revenues, and development of related businesses (Zambrano et al. 2010; Stronza and Pegas 2008; Wunder 2000).

Preservation and Conservational Benefits

The assessments of conservational benefits of nature-based eco-tourism according to the past literature (Hill and Gale 2009; Orams 1995).

Socio-cultural Benefits

Incorporated evaluation of socio-cultural benefits for enrichment (promoting cultural for eco-tourism attraction) that are heavily discussed by the previous researchers (Stronza and Pegas 2008; Jamal et al. 2006).

Ecotourism Involvement

The estimation of tour-related revenues derived from ecotourism, spill-over benefits of developing tourism (Berkemer et al. 1993).

Dependent Variables

Conservation Behaviors of Planning and Management Approaches

The value-driven project (Economic boosts, environment exploitation and socio-cultural enrichment) versus value-driven preservation or conservation.

Trade-Off Value Interactions Between Variables

The comparison evaluation between all the variable related towards the research; Ecotourism involvement, socio-cultural benefits, preservation and conservational benefits, economic benefits.

However, available data is limited and certain barriers exist in the research, this study allocate the evaluation into three major variables; Stakeholder (socio-economic), Economic (Tourism and development) and Ecology (Environment).

Multi-criteria Analysis (MCA)

Trade-off analysis is a method which measures the weighing of corresponding respondents' utilities for various product features (Agrell 1995). Hence, respondents are asked to choose or consider alternatives and state a likelihood of purchase or preference for each alternative. Trade off study was performed by using multi-criteria analysis (MCA). In MCA, a set of scenarios and criteria are required in order to construct framework for the MCA within our trade-off analysis. Both the criteria and the scenarios are developed by consultation with the relevant stakeholders and involved discussions, interviews and public meetings (Agrell 1995; Katrina et al. 2001).

Data collection methods that rely on written descriptions or verbal consists of all product attributes by assuming that the behavior being modeled is cognitive, this attributes are because of the process in understanding a verbal or written description is itself a cognitive behavior (McCullough 1998; François et al. 1991; Luce 1959). Corresponding steps for MCA development and data collection could be seen in Fig. 1.

From Fig. 1, a set of scenario and criteria are required to construct MCA framework within trade-off analysis. Both the criteria and the scenarios are developed in consultation with the relevant stakeholders and involved public meetings, interviews and discussions. MCA has been widely applied in planning (Buchanan and Daellenbach 1987; Macmillan et al. 1998; Malczewski et al. 1997; Joubert et al. 1997) and vigorously attempted to incorporate all stakeholders in the process (Tiwari et al. 1999).

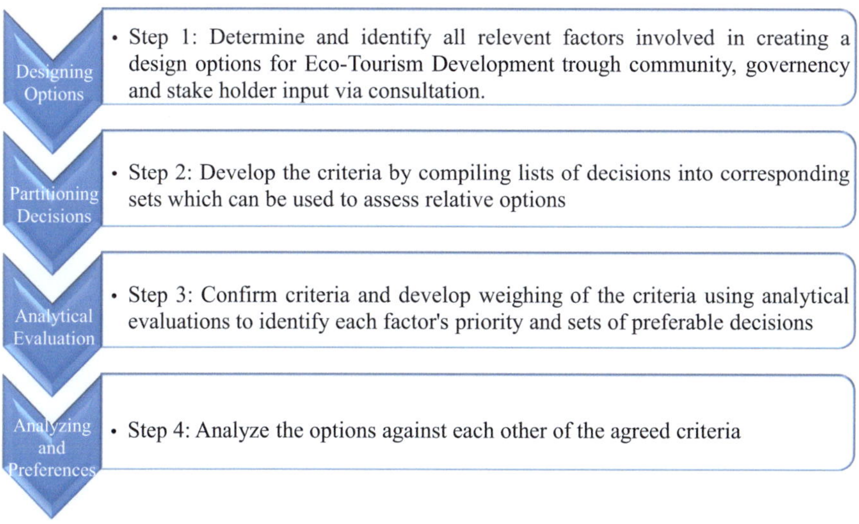

Fig. 1 Steps for trade off – multicrateria analysis (Yoe et al. 2002; Katrina et al. 2001; Brown et al. 2000; Malczewski et al. 1997; Joubert et al. 1997)

Stakeholder

Stakeholders are a group who has direct interest or concern in an organization (Business Dictionary 2016). Other than the corrosponding governing bodies that rule and upper management that are involved in constructing the rules, the masses could be interprated as one of the stakeholders in this matter (Randall and Richard 1982). Delli Priscoli (1989) indicated that a new trend in public involvement was occurring for the development of project that potentially could affect the masses. Hence, obeying the thumb rules of involvement, the stake holder will consist of government bodies, the population and policy makers. Therefore, the study will obey the needs to analyze all stakeholders in order to complete the MCA for determining trade-off of Tasik Kenyir Protected Area (TKPA). By referring to Fig. 2 by Brown et al. (2000), we can conclude the steps necessary to conducting MCA for Tasik Kenyir.

However, to harvest the essential data from the stake holder, we need certain approach to make it work. Therefore, by referring to Table 1, Brown et al. (2001) proposed certain techniques to be used in order to collect sample data from the respective stakeholders.

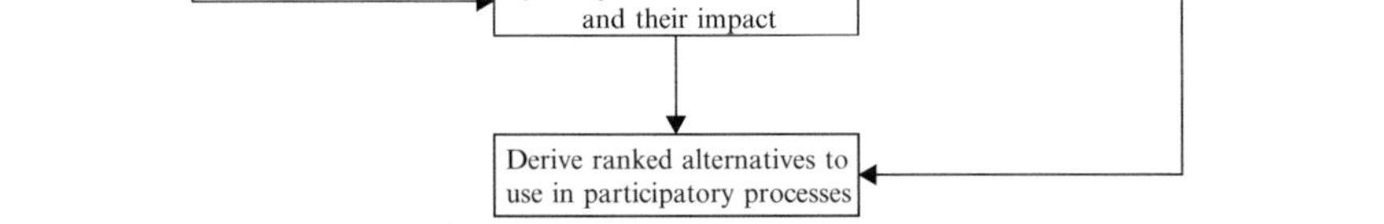

Fig. 2 Stakeholder analysis used for MCA and TDA suggested by Brown et al. (2000)

Table 1 Suggested methods of engaging different types of stakeholder groups (Brown et al. 2001)

Type of stakeholder group	Example of group	Method of engagement
Cohesive organisation with formal structure	Village council	Focus group
Cohesive organisation without formal structure	Informal trade group	Focus group
Mobile individuals, time-limited	Tourists	Questionnaire
Mobile individuals, frequent users	Informal sector worker	Individual interviews
Leaders of hierarchical organisations	Policy makers	Individual interviews
Workers within hierarchical organisations	Government departments	Structured group interviews

Economic

This criterion focuses on recreational benefits and macro-economic benefits. The first concerns are to aim for the maximization of revenue towards all participating sectors (Brown et al. 2001). The maximization of revenue is very essential objective for business owners and stakeholders to continue and expending their operations. Revenues are one of the indirect indicators for the sustainability of tourism.

To evaluate the total revenue for tourism activity in TKPA, we use the formula to determine the volume of tourist spending per day during their stay in TKPA (Sun and Styres 2006; Agarwal and Yochum 1999).

$$Per\ Day\ Spending\left(P_{MYR}\right)=\sum_{n}^{i=1}\frac{Trip\ Spent\left(MYR\right)}{Length\ of\ Stay\left(n\right)} \tag{1}$$

Sun and Styres (2006) propose *Per Day Spending* formula in order to determine the average spending per day/night based on tourist's total spending and the length of stay. By understanding Eq. (1), we applied the data gathered from the tourists against duration of stays to determine what is the average of spending in TKPA. We then applied the numerical data to Eq. (2) to determine the estimated total revenue generated.

$$Total\,Estimate\,Revenue=\left(No.Tourist\right).\left(P_{MYR}\right) \tag{2}$$

However, the total estimated revenue are applied to the total spending of accommodation or basic provided tourism package that was applied in TKPA. To understand further the value of cents and dollars that circulate in tourism activity in Tasik Kenyir, we have to further analyze the willingness to pay (WTP), as an indicator of tourist satisfaction or visitation enjoyment (Breidert et al. 2006; Schiffner et al. 2002; Rodgers 2001). Hence, we refer to Eq. (3);

$$WTP_i=\alpha+\beta X_i+v_i \tag{3}$$

Where, α is the spending per night, β is the total tourist or respondent per group of X, X is the matrices value of spending for tourist and v represent the value of. While $i = 1, \ldots, n$.

Ecological and Environmental

Ecology and environment are the cog and gear that grind all aspects into place. Without the attraction of nature, Tasik Kenyir will lose its brilliance. Hence, it is important to manage and preserve the natural environment of Tasik Kenyir while producing development plans to align with economical aspects that correspond to

Table 2 Possible impacts of scenario drivers on income based onto ecological value (Brown et al. 2001)

Scenario driven	Enhanced environment management	Impact	Without enhanced environment management	Impact
Extensive tourism development	1. Larger hotels, houseboats, more amenities produce a larger range of tourists	+	1. Larger hotels, houseboats, more amenities produce a larger range of tourists	+
	2. Increase in management of environment leads to strict regulations and developments of research and conservation bodies	+	2. No change in Park or Environmental Management	–
Restricted tourist development	1. Smaller hotels, houseboats, some amenities restricted for adventurous tourists	–	1. Smaller hotels, houseboats, some amenities restricted for adventurous tourists	–
	2. Increase in management of environment leads to strict regulations and developments of research and conservation bodies	+	2. No change in Park or Environmental Management	–

Table 3 Summary report on Tasik Kenyir (Ketengah 2015)

Year Item	2010	2011	2012	2013	2014
Total tourists arriving	221,302	265,241	309,908	467,678	649,94
Total spending for TKPA (MYR '000)	1605	2507	4500	3735	3640

Sources: Annual Report for 2012–2014, Lembaga Kemajuan Terengganu Tengah (Ketengah 2015)

the expectations of stakeholders. Based on Table 2, there are some scenarios that are used to determine the direction of MCA's in Tasik Kenyir.

The scenario in Table 2 was divided into two categories; Development with Enhanced Environmental Management or Development without Enhanced Environmental Management. This two scenarios were introduced by Brown (2002) as indicators of development direction. Hence, by using these indicators as a framework of evaluation, the following variable was developed; accommodation (Hotels and houseboat), amenities, management and restrictions. The impact was noted using positive (+) and negative (−) based on its effects towards the economy and environment. Hence, the overall picture of evaluation could be seen.

Discussion

Based on data that was provided by the Terengganu Tengah Authorities (KETENGAH), the basic situation of TKPA was as shown in Table 3.

Table 4 Estimate impact of four possible scenarios and one existing scenario on TKPA

	Scenario				
Criteria	A	B	C	D	E
Economic					
1. Economic revenue to Tasik Kenyir (MYR'000)	25,000	35,000	40,000	120,000	90,000
2. Visitor enjoyment of TKPA (MYR '000)	–	–	–	–	–
3. Rate of revisits (%)	−1.31	0.921	−4.032	3.0	1.0
Social					
4. Local employment (jobs)	30	50	84	84	42
5. Informal benefits (score)	2	4	4	6	3
6. Local excess (score)	6	5	6	7	6
Ecological					
7. Water quality (score)	3	1.2	2.2	1.9	2.2
8. Management of protected species (score)	2	7	4	7	5
9. Management of protected land (score)	3	6	1	5	4

This research part describes an entry point into stakeholders led negotiations on priorities for management. The set of standardized order information for the MCA's is adopted in order to engage with all stakeholders for evaluating their priorities in terms of decision-making criteria based on development scenarios and outcomes that have been introduced (Randall and Richard 1982). Evaluation of repercussion for all scenarios and outcome on the criteria as shown in the effects table (Table 2) is the initial step in MCA and help to generate order ranking for the advancement of the scenarios, the highest scoring scenario can be treated as the most preferred scenario. Table 4 shows the ranking order of characters for a range based on preferences. These ranking were compared with a base case of equal weighting of economic, social and ecological criteria.

Based on Table 4, the different scenario schemes contribute to different outcomes. By using scenario E as the controller, the classification of results could be organized from poor to excellent based on their performance. Scenario A proves to be the poorest approach as it produces a total of MYR 25mill of revenue per year based on the same number of tourists in 2014. This, however, worsens as the amount of revisits plummeted to −1.31% from 1% total revisits. Due to the limited development, the amount of spill-over are especially low as scenario A provides 30, 2 and 6 for employment, benefits and local accessibility. However, scenario B produces more pleasant results as the total revenues are MYR 35mill from the same amount of tourist arrival. However, as the development strategies are very limited, the amount of revisit was only 0.921% due to the limitation of amenities and 50, 4 and 5 for employment, benefits and local accessibility, respectively.

For scenario C performance, the total revenues are MYR 40mill and the rate of revisiting are −4.032%. The negative rate of revisiting are due to the overall performance of the environment and amenities in TKPA that received minimal or no maintenance. However, the employment, benefits and local accessibility score are relatively high at 84, 4 and 6, respectively, due to the demand. However, for scenario

D, the total performance are excellent compare to scenario A, B, C and E. The revenues are MYR 120mill yearly with the same amount of tourists arriving. The relative revisit rate of 3% represent the satisfaction of the tourists from the services and nature evaluation. Scenario D also provide a satisfactory score for both social spillover and ecological variable. Hence, scenario D; Extensive tourism development with complementary environmental management are the best direction of development of TKPA.

Type of Scenarios Involved in estimation:

A: Restricted tourism development without complementary environmental management.
B: Restricted tourism development with complementary environmental management.
C: Extensive tourism development without complementary environmental management.
D: Extensive tourism development with complementary environmental management.
E: Stagnant tourism development and environmental management.

Conclusion

By embarking this concept, it means that not only can stakeholders be definitive about their arrangements of preferences for decision-making, but they could also see the potential outcomes and impacts in terms of the ranking of development strategies based on these priorities. In a nutshell, they can be notified about the trade-offs inherent on management decisions for resource use. The trade-off approach that was used to understand the different introduced schemes that can be used to determine the direction of development in Tasik Kenyir could be enhance with further assessment using other method in Multi-Criteria Analysis. Findings from this study and the approaches on Multi-Criteria Analysis could use to sizing up the ripple effects onto biological, ecological and economical aspect of Tasik Kenyir and this could be used by other researchers as a baseline comparison or even guideline for their study in this area. However, this study approach are more onto relational approach on limited aspect from the study and trade-off analysis was used to bring together diverse quantitative and qualitative information for decision-making to rank development scenarios on the basis of stakeholder values. Hence, a deeper study focusing on wider reciprocal relationship between economic, heritage, ecology, society, tourism and their values should be applied. The state activities, projection governance and policy intervention that effluence onto all the variable mention seldom shows the effects on short-term period of time. Hence, a continuation of the study and comprehensive research are needed to develop further understanding of what happened and what will happen in the future.

Acknowledgements This research study was financially supported by both the Trans Disciplinary Research Grant Scheme (TRGS) under Vote No.: 59373, Kenyir Research Institute (IPK), Universiti Malaysia Terengganu and Universiti Sains Malaysia (USM) Research Grant (RUI-2015 under Vote No: 1001/PJJAUH1/811299). Appreciation and acknowledgement for those who participate in the study, especially the staff and post-graduate students of the Kenyir Research Institute (IPK), Terengganu Tengah Authorities (KETENGAH), JAKOA, Terengganu Economic Planning and Development Unit (UPEN), Terengganu State Authorities, all respondents and those who involved in this study.

References

Agarwal, V. B., & Yochum, G. R. (1999). Tourist spending and race of visitors. *Journal of Travel Research, 38*(2), 173–176.

Agrell, P. J. (1995). *Interactive multi-criteria decision making in production economics*. Linkoeping: Production Economic Research.

Berkemer, R., Makowski, M., & Watkins, D. (1993). *A prototype of a decision support system for river basin water quality management in Central and Eastern Europe*. Laxenburg: WP-93-049.

Breidert, C., Hahsler, M., & Reutterer, T. (2006). A review of methods for measuring willingness-to-pay. *Innovative Marketing, 2*(4), 8–32.

Brown, S. (2002). Measuring carbon in forests: current status and future challenges. *Environmental Pollution, 116*(3), 363–372.

Brown, K., Tompkins E., Peter, B., Shim, D., Young K., & Adger, W. N. (2000). *Trade-off analysis for marine protected area management* (CSERGE Working Paper GEC (2)). London: University Collage London.

Brown, K., Tompkins, E., & Adger, W. N. (2001). *Trade-off analysis for participatory coastal zone decision-making*. Norwich: University of East Anglia.

Buchanan, J. T., & Daellenbach, H. G. (1987). A comparative evaluation of interactive solution methods for multiple objective decision models. *European Journal of Operation Research, 3*(29), 353–359.

Business Dictionary. (2016). Retrieved from http://www.businessdictionary.com/definition/stakeholder.html

Chia, D. (2003*). Preservation and conservation of archaeological sites and artefacts in Malaysia*, presented at the Third International Convention of Asia Scholars, Singapore, 2003. Singapore: National University of Singapore.

Delli Priscoli, J. (1989). Public involvement in risk assessment, keynote speech at Workshop, University of Southern California, Los Angeles, December 10, 1989.

Dixon, J. A., Scura, L. F., & van't Hof, T. (1993). Meeting ecological and economic goals: marine parks in the Caribbean. *Ambio, 22*(2–3), 117–125.

François, P., MacLachlan, D. L., & Jacobs, A. (1991). *Bridging designs for conjoint analysis: The issue of attribute importance*. Leuven: Katholieke Universiteit Leuven.

Gin, O. K. (2009). *Historical dictionary of Malaysia* (Vol. 71). Maryland: Scarecrow Press.

Hill, J., & Gale, T. (2009). *Ecotourism and environmental sustainability*. Burlington: Ashgate.

Hodgson, G. (1997). Resource use: conflicts and management solutions. In C. Birkeland (Ed.), *Life and death of coral reefs* (pp. 386–410). New York: Chapman and Hall.

Jamal, T., Marcos, B., & Stronza, A. (2006). The Institutionalization of ecotourism: Certification, cultural equity and praxis. *Journal of Ecotourism, 5*(3), 145–175.

Joubert, A. R., Leiman, A., de Klerk, H. M., Katau, S., & Aggenbach, J. C. (1997). Fynbos vegetation and the supply of water: a comparison of multicriteria decision analysis and cost benefit analysis. *Ecology Economic, 22*(1), 123–140.

Kamaruddin, I. S., Mustafa Kamal, A. S., Christianus, A., Daud, S. K., & Yu Abit, L. (2011). Fish community in Pengkalan Gawi – Pulau Dula section of Kenyir Lake, Terengganu, Malaysia. *Journal of Sustainability Science and Management, 6*(1), 89–97.

Katrina, B., Adger, W. N., Tompkins, E., Bacon, P., Shim, D., & Young, K. (2001). Trade-off analysis for marine protected area management. *Ecological Economics, 37*, 417–434.

Ketengah. (2015). Laporan Tahunan 2012–2014. Retrieved from http://www.ketengah.gov.my/laporan-tahunan.

Luce, D. (1959). *Individual choice behavior: A theoretical analysis*. Mineola: Dover Publication.

Macmillan, D. C., Harley, D., & Morrison, R. (1998). Cost-effectiveness analysis of woodland and ecosystem restoration. *Ecology Economic, 27*, 313–324.

Malczewski, J., Moreno-Sanchez, R., Bojorquez, L. A., & OngayDelhumeau, E. (1997). Multicriteria group decision-making model for environmental conflict analysis in the Cape region, Mexico. *Journal of Environmental Planning and Management, 40*, 349–374.

McCullough, R. D.. (1998). Data use: Trade-off analysis: A survey of commercially available techniques. *Marketing Research Review*. ID: 19980203. Retrieved from http://www.quirks.com/articles/a1998/19980203.aspx.

Mustafa, N. A., Abdullah, N. C., Mohamed, B., & Bahauddin, A.. (2013). Preservation of cultural heritage in Malaysia: An insight of the National Heritage Act 2005. International Conference on Tourism Development, Penang, Malaysia, 2013. Penang: Sustainable Research Cluster, Universiti Sains Malaysia.

Orams, M. (1995). Towards a more desirable form of ecotourism. *Tourism Management, 16*(1), 3–8.

Randall, G. C., & Richard, S. (1982). Exploiting rank ordered choice set data within the stochastic utility model. *Journal of Marketing Research, 1*, 288–301.

Rodger, L. (2001). Willingness to pay; an examination of nonuse value. *Economic 6818, 2*(24), 1–28.

Schiffner, R., Brunnberg, S., Hohenleutner, U., Stolz, W., & Landthaler, M. (2002). Willingness to pay and time trade-off: Useful utility indicators for the assessment of quality of life and patient satisfaction in patients with Port Wine Stains. *British Journal of Dermatology, 146*(3), 440–447.

Shaharom, F. (2012). *Fish parasites of Lake Kenyir, Peninsular Malaysia*. Kuala Terengganu: Penerbit Universiti Malaysia Terengganu.

Srinivasan, V. (1988). A conjunctive-compensatory approach to the self-explication of multi-attributed preferences. *Decision Sciences, 19*(2), 295–305.

Stronza, A., & Pegas, F. (2008). Ecotourism and conservation: Two cases from Brazil and Peru. *Human Dimensions of Wildlife, 1*(3), 263–279.

Sun, Y.-Y., & Styres, D. J. (2006). A note on estimating visitor spending on a per-day/night basis. *Tourism Management, 27*(4), 721–725.

Taha, A. H. (1991). Archaeological discoveries in Peninsular Malaysia (1987–1990). *Journal of the Malaysian Branch of the Royal Asiatic Society, 64*(1/260), 75–96.

Tiwari, D. N., Loof, R., & Paudyal, G. N. (1999). Environmental economic decision-making in lowland irrigated agriculture using multi-criteria analysis techniques. *Agricultural Systems, 60*, 99–112.

Wunder, S. (2000). Ecotourism and economic incentives; an empirical approach. *Ecological Economics, 32*, 465–479.

Yoe, C., Orth, K. D., & Pietrowsky, R. A. (2002). *Trade-off analysis planning and procedures guidebook – final*. Alexandria: Planning and Management Consultants, Ltd..

Zakaria, M. Z., Jalal, K. C. A., & Ambak, M. A. (2000). Length weight relationship and relative condition factor of Sebarau, *Hampala macrolepidota* (Van Hasselt) in Kenyir Lake, Malaysia. *Pakistan Journal of Biological Science, 3*(5), 721–724.

Zambrano, A. M. A., Broadbent, E. N., Schmink, M., Perz, G. Z., & Asner, G. P. (2010). Deforestation drivers in Southwest Amazonia: Comparing smallholder farmers in Inapari, Peru, and Assis Brasil, Brazil. *Conservation and Society, 8*(3), 157–170.

Sustainable Development of Tasik Kenyir Eco-Tourism Using System Dynamic

Muhamad Safiih Lola, Mohd Noor Afiq Ramlee, Mohd Fadli Hussin, Mohd Tajuddin Abdullah, Anton Abdulbasah Kamil, Izham Mohamad Yusof, Yahaya Ibrahim, and Nur Zafirah A. Khadar

Abstract The sustainable development of Tasik Kenyir eco-tourism depends on the availibility of spatially explicit information on the state and trends of ecosystems and their services. Thus, we develop an interactive dynamic model of Tasik Kenyir Eco-Tourism that links ecological and economic systems, and generated the effects of the short and long terms besides to determine the direction of development and conservation in Tasik Kenyir development policies. In this study, Tasik Kenyir eco-tourism coins Macro Management Framework for Tasik Kenyir Tourism Dynamics consist of four different subsystem; Toursim Activity, Development of Tasik Kenyir (infrastucture and tourism activities), Waste Generated and

M. S. Lola (✉)
School of Informatics and Applied Mathematics, Universiti Malaysia Terengganu, Kuala Nerus, Terengganu, Malaysia

Institute of Tropical Biodiversity and Sustainable Development, Universiti Malaysia Terengganu, Kuala Nerus, Terengganu, Malaysia
e-mail: safiihmd@umt.edu.my

M. N. A. Ramlee · M. F. Hussin
Institute of Tropical Biodiversity and Sustainable Development, Universiti Malaysia Terengganu, Kuala Nerus, Terengganu, Malaysia

M. T. Abdullah
Institute of Tropical Biodiversity and Sustainable Development, Universiti Malaysia Terengganu, Kuala Nerus, Terengganu, Malaysia

School of Marine and Environmental Sciences, Universiti Malaysia Terengganu, Kuala Nerus, Terengganu, Malaysia

A. A. Kamil
School of Industrial Engineering, Telkom University, Bandung, Indonesia

I. Mohamad Yusof · N. Z. A. Khadar
School of Distance Education, Universiti Sains Malaysia, Penang, Malaysia

Y. Ibrahim
Faculty of Applied Social Sciences, Universiti Sultan Zainal Abidin, Kuala Terengganu, Terengganu, Malaysia

Darul Iman Institute, Universiti Sultan Zainal Abidin, Kuala Terengganu, Terengganu, Malaysia

M. T. Abdullah et al. (eds.), *Greater Kenyir Landscapes*,
https://doi.org/10.1007/978-3-319-92264-5_22

Environment Sustainability (flora and fauna). The Model developed reveals that sustainable development of Tasik Kenyir eco-tourism requires continuous prolonged effort and insight towards creating a balance between social, economic and environmental developments.

Keywords Tasik Kenyir · Eco-tourism · Dynamic model · Development and conservation

Introduction

Eco-tourism is a word that comes from two terms, "eco" and "tourism". "Eco" brings two logical meanings, "ecology and economy" which shows that it is possible to make "good tourism" and reconsile the imperatives of ecology and economy (Villepontoux 2013). Originally, eco-tourism is a term invented by biologists in 1980s, who were concerned by environmental degradation caused by the development of tourism practices (Villepontoux 2013). Since the last decades, the eco-tourism industry has been growing rapidly and is a major source of foreign exchange earnings for many developing contries (UNEP 2002). Nowadays, the total values of international tourism market in the world in 2014 is estimated around USD1.25 trillion in revenue (Statista 2015). In Malaysia, this industry will generate an estimated MYR72 billion in revenue in 2014 (Tourism Malaysia 2015), which is 1.47% from the total world revenue. With that figure, eco-tourism is a real industry that seeks to take advantage of market trends and lead to many developing countries are now embracing it and including it in their economic development and conservation strategies (Kiper 2013). Besides that, eco-tourism is preceived as having potential as an effective tool for suitable development.

However, the global growth of tourism poses a significant threat to cultural and bilogical diversity (Butler 1980). Hence the need for a sustainable tourism that seek for conservation niches are in great demand for parks and nature tourism. Therefore, sustainable tourism has become an issue in the debate on environmentally-integrated tourism development and the environmental consequences of this rapidly growing industry can no longer be ignored (Gossling et al. 2005; Hunter and Green 1995).

In this paper, we shall discuss sustainable tourism in an exotic and rich tourism spot which is in Terengganu, Malaysia. Terengganu has the charms and splendor compared to other states in Malaysia. Terengganu has stood still as the state retains all the rustic and idyllic charms despite rapid development and modernization (Tourism Terengganu 2009). The abundance of coastal areas with pearly white beaches and breathtaking islands promote Terengganu as one of the most popular island-related tourism destinations. Table 1 shows the number of tourists in Terengganu from 1990 to 2013. In monsoon seasons, which start from early November to mid-February, all activities related to the sea cease to function for a while, producing an income gap towards tourism activity in Terengganu.

Table 1 Terengganu tourism arrival from 1991 to 2013 (UPEN 2014)

Year	Domestic tourists	Foreign tourists	Total
1990	196,957	79,371	276,328
1991	228,153	121,948	350,101
1992	201,073	90,898	291,971
1993	277,724	97,572	375,296
1994	410,327	107,598	517,925
1995	730,020	127,124	857,144
1996	1,065,327	166,951	1,232,278
1997	1,626,392	227,646	1,854,038
1998	981,263	112,714	1,093,977
1999	1,002,324	146,713	1,149,037
2000	1,170,552	159,993	1,330,545
2001	1,248,607	142,041	1,390,648
2002	1,276,671	125,148	1,401,819
2003	1,284,761	97,743	1,382,504
2004	1,418,141	162,826	1,580,967
2005	1,624,726	197,952	1,822,678
2006	2,061,486	238,893	2,300,379
2007	2,572,299	295,084	2,867,383
2008	3,147,873	380,281	3,528,154
2009	2,963,250	341,526	3,304,776
2010	2,528,000	287,149	2,815,149
2011	2,587,735	578,476	3,166,211
2012	2,768,876	607,399	3,376,275
2013	3,229,704	807,426	4,037,130

Nevertheless, many local or international tourists do not know that Terengganu still offers unique gateways such as Tasik Kenyir and other attractive places to be explored.

Research Area

Tasik Kenyir is one of the artificial water catchment areas in Malaysia, created in 1985 by the damming of the surrounding rivers. Tasik Kenyir as shown in Fig. 1, is the biggest man-made lake in Southeast Asia. Covered by more than 340 islands spread out in a water catchment area of 260,000 ha (KETENGAH 2012). The biggest island is estimated to be as big as Singapore. Another interesting fact is that Tasik Kenyir is name originated from Sungai Kenyir that was sunken deep when Tasik Kenyir was almost filled by surrounding water sources (KETENGAH 2012; Nakisah and Fauziah 2003).

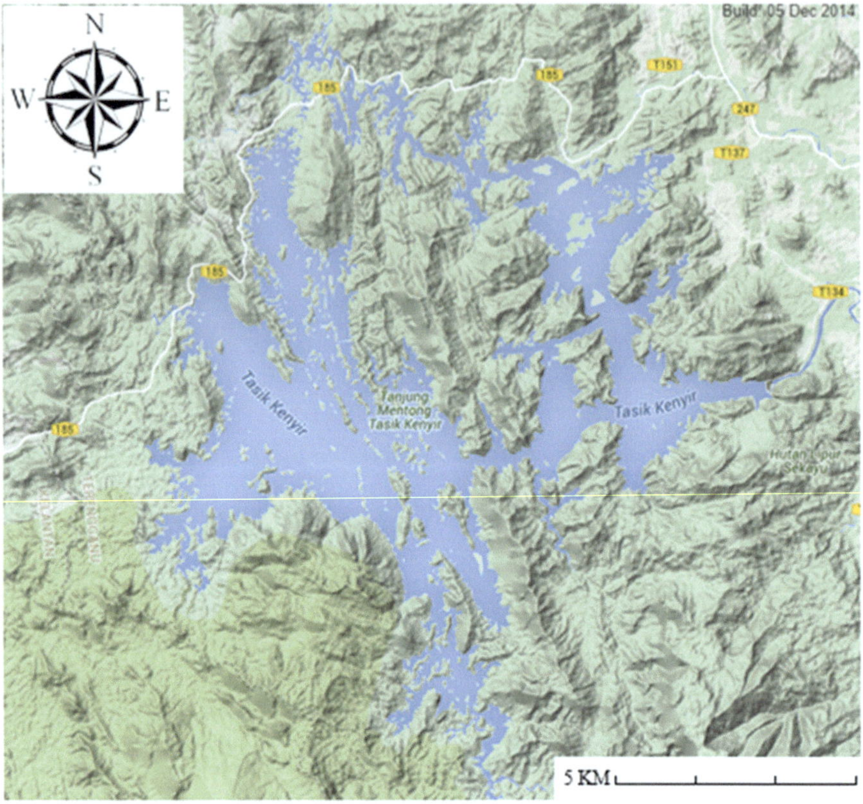

Fig. 1 Tasik Kenyir area (Google Maps 2014)

Massive development projects at Tasik Kenyir are in progress recently. Some are good, but others seem to be out of the supposedly Kenyir's "bigger picture". According to Tan (2014), the main idea for Tasik Kenyir development are focuses on conservation and development of geopark, prior to what some scientists believe that Kenyir has what it takes to warrant consideration as a geological site of global importance. Tan added that the idea of making it a geopark was announced by the state government. The proposal document is currently being prepared by the Kenyir Research Institute, Universiti Malaysia Terengganu (UMT). However, the development of a free-tax plaza on Pulau Bayas, one of the many Islands in Tasik Kenyir, seems to be out of place in terms of enhancing and highlighting 130 million years of the rich history and diversity of Kenyir; where sustainable tourism development (e.g.: eco-tourism and green-tourism) should take place. This is due to the logistic problems that could arise from visitor access to the plaza; which is to be built nearby to ease the access of locals and tourists to the plaza. However, the idea of a tax-free zone could be promoted for the advancement of socio-economic development of locals (Tan 2014).

All forms of tourism in all types of destinations are applicable to the concept of sustainable tourism including mass tourism and various niche-tourism segments

(Swarbrooke and Horner 2001; Orams 1995;). Sustainability principles refer to the environmental, economic, and socio-culturak ascpects of tourism development (Stefan 2008; Swarbrooke and Horner 2001; Orams 1995). The balance between these three dimensions are the fundamental to ensure long-term sustainable tourism development (Stefan 2008). Hence, by utilizing the sustainable concept in tourism development in Tasik Kenyir, the Eco-Tourism concept comes through.

The main objective of this study is to encourage the implementation of actions that promote sustainability in tourism by developing tools that support the various stakeholders. The analytical research that includes all aspects and policy recommendations that will address key stakeholders – government including local authorities, Orang Asli and local communities, the private sector, NGOs and consumers. There recommendations build upon these projects and lessons learnt during the research will be used in promoting sustainable tourism.

With eco-tourism, several elements should be highlighted and focussed; to create and increase demand within the tourism sector for the Tasik Kenyir natural ecosystem and cultural heritage. To better articulate the value of linking tourism supply with demand (services and etc.) through better policies that are driven by empirical studies. The governement must engage at all levels international organizations, bussinesses NGOs and consumers to intergrate open-ended processes engaging the government at all levels. The recommendations are structured around a simplified life-cycle of the tourism value chain, including tourism planning, tourism operations and management, tourism investment, tourism promotion and marketing, capacity building, consumption of products and services and monitoring and evaluation of tourism development (Dhanoa 2013; Butler 1980).

Prehistoric Resources and Historical Artifacts of Tasik Kenyir

Before particular area was turned into a reservoir, Kenyir had been listed as a center of civilization in the Neolithic era. Some artifacts had been uncovered by a group of archeologists in 1956 and 1970s (Nik et al. 1990). Such artifacts include kitchen utensils, axes and tools dating back to the Neolithic era. It is believed that during this era, the place was known as one of the bussiness activity centers. Among the relics discovered in previous research are:

Stone Tool

The stone tool was found in both of the caves at Tasik Kenyir. It was used by pre historic men and the purpose of the utensil was for cutting up animals (prey), preparing foods and other uses.

Stone Knife

The stone knife was used by pre historic men during Neolithic age in Gunung Bewah, Kenyir which was found in Kuala Sungai Bewah.

Seashells

Known as Siput Cengkerang, these cengkerang shells were source of food to the prehistoric man during the Neolithic age. This seashell was found in Gua Gunung Bewah, Kenyir. These samples were sent for date determination and were dated between 8920 + 120 B.P to 2630 + 80 B.P (Beta Analytic Inc.) (Nik et al. 1990).

Weapons

Some weapons which in Gua Gunung Bewah were made from iron casting and there are some of the artifacts that show that the civilization was blooming there (Nik et al. 1990). The existance of historic artifacts indicate that Tasik Kenyir was capable of becoming a renowned archeological site and should be preserved for the future (Tan 2014; Nik et al. 1990). Combining the tranquil and flora and fauna, and historical and biodiversity resources, the development of Tasik Kenyir should be done with precaution and focused on expressing its natural beauty rather that uncoordinated development for a singular objective.

Niche Tourism Activity

Despite abundance of natural resources and many places to be explored, Tasik Kenyir seems lifeless due to its limited segment tourism market. This particular area is vast and the movement from one place to another are limited. Hence, several groups such as the elderly, toddlers and certain members of the family are incapable of visiting and enjoying the benefits of vast natural resources that Tasik Kenyir has to offer. Hence, the objectives of this paper are:

1. To construct an Eco-Tourism System Dynamic Models (Eco-TSD) framework which integrated the economic, environment and ecological sysrems for understanding risks, potential and impacts of eco-tourism practices and development.
2. To evaluate the current policy on sustainable development of eco-tourism in Tasik Kenyir and its impact on current economic, environmental and community sectors.

System Dynamics

System Dynamic developed by Professor Forrester of MIT in the mid-1950s was applied in several research fields. This particular research method was chosen because system dynamics was widely used for economic, social and environmental studies became this particular method has proven useful to reveal the dynamic changes, feedback and other variable cause and effect onto its particular area (Sterman 2000; Stearns and Montag 1974). By utilizing this particular research method, we could provide insights for positive and negative effects that can be

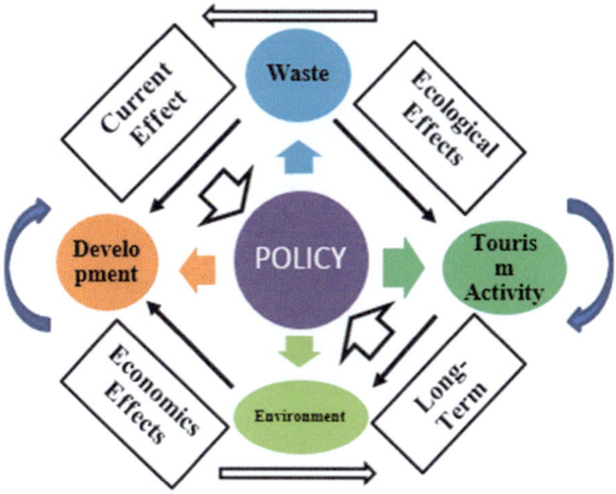

Fig. 2 Macro management framework for Kenyir tourism dynamics

derived from an artificial model based on an actual surrounding environment. Hence, this research paper discusses the utilization of System Dynamic Model for various fields such as tourism planning, agricultural and aquaculture practices, sustainable development and even survival of certain species based on its surroundings. In this study, the system dynamics application are mainly used in tourism development for sensitive areas.

Causal loop: The blue arrow indicates generally positive part of tourism ecosystem such as conservation, cultural representation, and attraction that prove fundamental to generate an economical balance for eco-tourism practice. The red arrow represents the negative part of tourism such as pollution, over development and non-strategic development. The yellow arrow represents implement of policy that used to formulate growth and promoting implementation of policy towards tourism practices. Stock Flow Diagram: Stock-flow diagram used to perform more detailed quantitative analysis, a casual loop diagram is transformed into a stock and flow diagram. A stock and flow model helps in studying and analyzing the system in a quantitative way; such models are usually built and simulated using computer software (Teegavarapu and Tangirala 2005).

The model: The dynamism and elasticity, together with mobility are characteristics specific to the tourism market. Hence, the tourism model should consist those aspects for the tourism indutry to bloom. However, given the Tasik Kenyir special traits, sustainability is the most important aspect that should be focused on. Therefore, a suitable framework should be made as a guideline that considers of macro management of overall Tasik Kenyir Tourism Dynamics. The macro management framework is shown in Fig. 2.

The macro management framework consists of four (4) different subsystems; Tourism Activity, Development (infrastructure and tourism contain), Waste Generated and Environment Sustainability (flora and fauna). These subsytems are developed from macro management frameworks based on the observation of the overall system in Tasik Kenyir (Safiih et al. 2016; Wang and Zhang 2001). The subsystem interaction with each change in one or more subsystems will affect others (Safiih et al. 2016). Hence, these changes in the subsystem are what is called the Tasik Kenyir Tourism dynamics. This dynamic will generate effects both in short and long terms in both economics and biological factors that will somehow determine the direction of development and conservation in Tasik Kenyir Development policies.

By taking into account of all the subsystems that exist in macro management framework, we will create a causal loop which represent almost overall system in Tasik Kenyir Tourism sector (Fig. 3a). Using the causal loop and macro management framework, a new stock-flow model will be generated and the current effects can be analyzed. The stock flow model is the tourism model (Fig. 3b), sustainability model (Fig. 3c) and development model (Fig. 3d).

Based on Macro Management Framework for Kenyir Tourism Dynamics, the probability of trend evolution for Tasik Kenyir eco-tourism for the next 42 years (2004–2047) and simulated and the natural condition forecasting are produced under the name Person. Meanwhile, the production simulated under preferred condition is named Simulation.

Tourism Spending

The appearance of a global consumer was identified in relationship with its drivers, such as; global effeciency (global organizations trying to benefit from the mutual features of consumers in order to enhance production, marketing and distribution effeciency) and clients' convergence (worldwide consumers more and mostly likely about consumer habits and patterns) (Butler 1980; Swarbrooke and Horner 2001). Hence, for the Tasik Kenyir consumer (Tourist) that comes from both local and international markets shows the strength of Tasik Kenyir tourism product. However, the local attraction that are segmented should be upgraded and changed to more user-friendly tourism activities (Schubert and Brida 2009). Hence, by these changes, the demand trend of visits from tourists will subsequently increase and produce more working opportunities for locals. Following the demands of the users in Fig. 4, the market rate will be high and the total spending of local and international tourists will be higher. This will positively increase the net worth of Tasik Kenyir tourism industries in the Future.

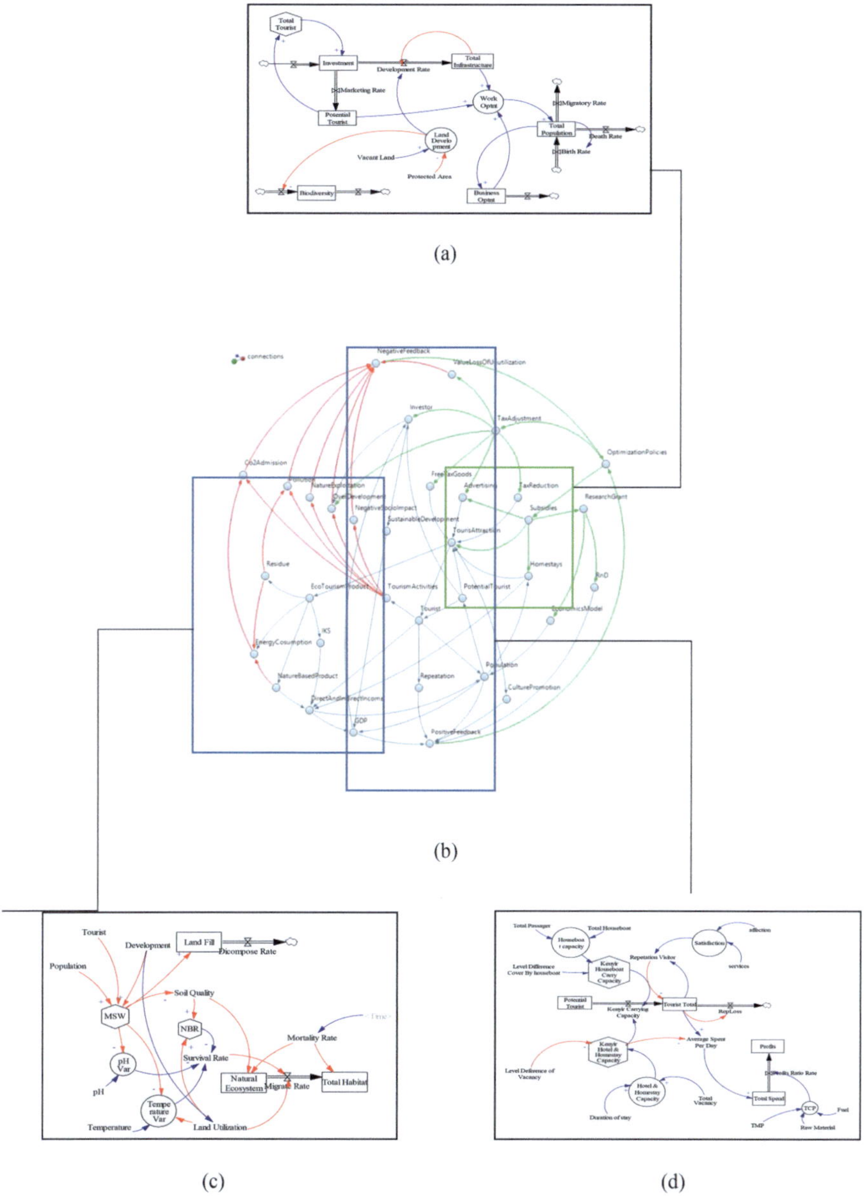

Fig. 3 Macro management framework; (**a**) Developed model of Tasik Kenyir tourism, (**b**) Causal loop of Tasik Kenyir tourism dynamics, (**c**) Sustainability model of Tasik Kenyir tourism dynamics, (**d**) Tourism model of Tasik Kenyir dynamics

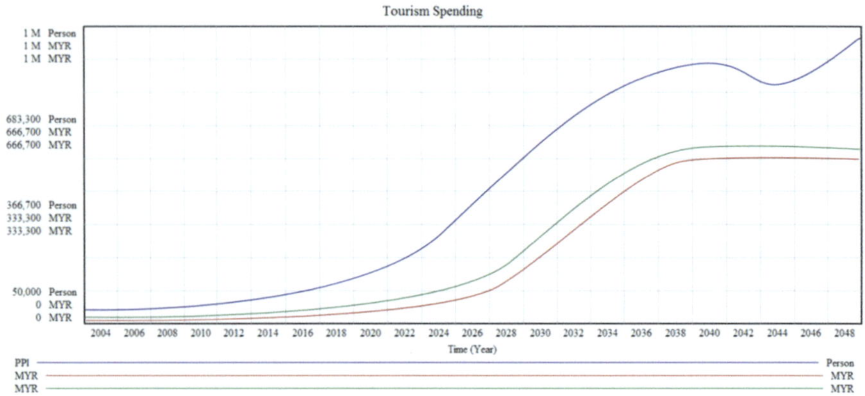

Fig. 4 Tourism spending (Tourism population growth, total local tourist spending and total spending for international tourist)

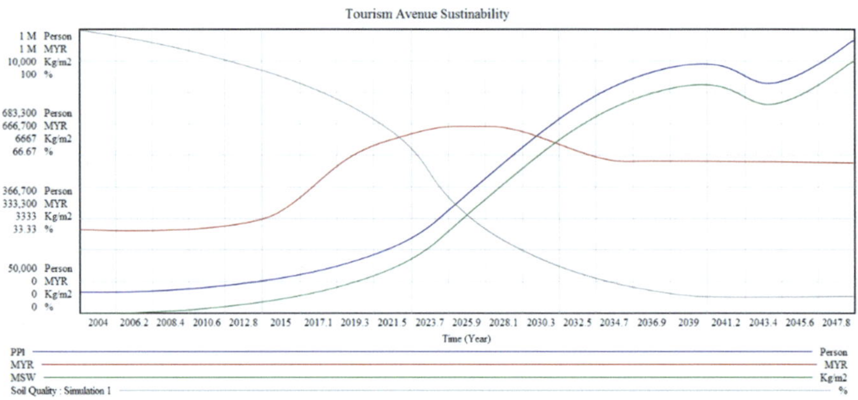

Fig. 5 Tourism spot sustainability (Total tourist visit, total management costs, municipal waste generated, and soil quality)

Tourism Spot Sustainability

A concern regarding the environmental sustainability, safety and protection represent another factor which are not at all less important (Bădulescu and Rusu 2009). The modelling generated for Tasik Kenyir Tourism Sustainability in Fig. 5 indicate that by 2019, despite of the tourist rate increase, there will be an increase in total suspended solid waste generated by the tourism activities, hence resulting in the increase of total management costs and deterioration of soil quality. The noticeable changes that will occur in the Tasik Kenyir environment are very alarming and precautions should be taken to avoid unnecessary damage in the future.

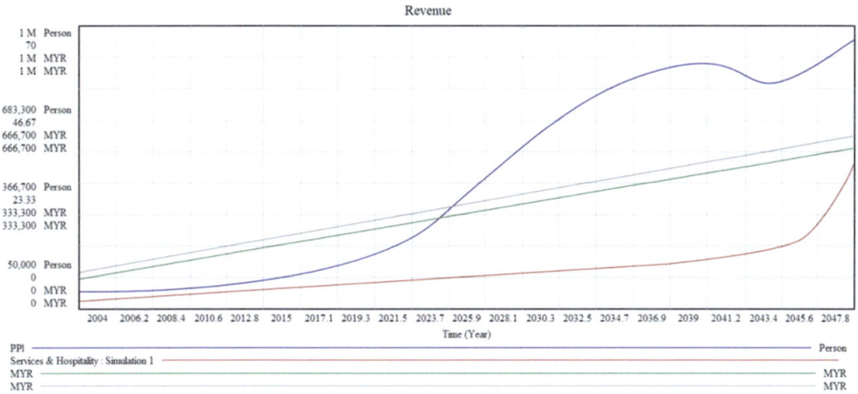

Fig. 6 Revenue (total tourist visit, number of services and hosptality, wholesale revenue, services and hospitality revenue)

Revenues

The business environment in tourism is in continuously changing due to several causes and factors (Schubert and Brida 2009; MIT 1997; Sterman 1992). The ageing of the tourism population (visiting tourist) affect very much the services provided and revenue that they generate towards locals and authorities that play key a role in maintaining and developing the area (Bădulescu and Rusu 2009). Increase of tourists and verification of their activities, the number of services providers such as hotels, boathouses and other related services would increase (Fig. 6).The trend will continue to satisfy the demand of the markets. Therefore, this will surely increase the revenue accumulated by the service providers and most importantly, to develop the Tasik Kenyir Tourism activity to the highest level.

Sustainable Development and Policies Recommendation

Innovation of technology is the key to improve the effeciency of resources use in tourism. In addition, public institutions that in charge the tourism should protect the biodiversity by tackling the challenges of climate change, minimize emmissions of greenhouse gases (GHG) as well as the waste production.

The legislature should be respected by tourism operations and management. The objectives related to tourism development and management should be planned as set out by local and national authorities and ensure that conditions related to the three dimensions of economy, environemnet and social concerns are included. Internationally-recognized standards must be used by the operations for sustainable tourism. Tourism actors should participate actively in the initiatives and processes put in place or supported by the Intenational Task Force on Sustainable Tourism

Development like the ‘Davos Process’ on climate change and tourism, the Sustainable Investment and Finance in Tourism (SIFT) Network, the Sustainabke Tourism Stewardship Council (STSC) etc.; and usage of tools developed by the International Task Force projects and other relevant voluntary initiatives.

Sustainable choices can be made by tourist if the efforts are made to communicate clear information on labels claiming sustainability. Therefore, better consistency between such labels and a clear validaton of content on labels. The public has the best possible information if a wide variety of communication techniques and being delivered in the most appropriate way taht lead the public to make the most sustainable choices in their tourism selection. The tourism industry need to inform, educate and work collaboratively to integrate sustainability into their policies and management practices, and secure their active participation in developing sustainable tourism.

Conclusion

Sustainability development requires continuous and prologed efforts and insight towards creating a balance between social, economic and environmental developments. While this preliminary study provides linkage for some of the factors in biodiversity and resiliency in Tasik Kenyir, it could be used as a baseline for future dynamical study of Tasik Kenyir cause and effects in future tourism study. While extensive developments frequently leads to a reversal of the development trend, the natural environment generally will not reveal its limits of exhaustible tolerance, not until it is too late. Since the intrinsic nature of mutual conflicts among sectoral or departmental development always exists in the path towards sustainability policies becomes a key duty of government. The current policy on planning tourism development, maintaining physical assets and promoting fisical growth seems inadequate and could be upgraded. Some enviromental policies are good but the capability of enforcing the policy need to be improved to cope with future challanges such as poaching. Hence, by accommodating the gaps in policy and its implementation, new changes can be seen and the quality of introduced policy could also be assessed based on its forecasting capabilities and the feasibility of implementation.

Acknowledgements This research study was financially supported by the University Sains Malaysia (USM) Research Grant (RUI-2015 under Vote No: 1001/PJJAUH1/811299) and the Kenyir Research Institute (IPK), University Malaysia Terengganu (UMT) by Trans Disciplinary Research Grant Scheme (TRGS) under Vote No.: 59373. Acknowledgement are given to those who participate in the study especially the Terengganu Tengah Authories (KETENGAH), Terengganu Economic Planning and Development Unit (UPEN), the Terengganu State Authorities, and those who are involved in this study.

References

Bădulescu, A., & Rusu, S. (2009). The dynamics of the international tourism market. Recent developments and challenges. *GeoJournal of Tourism and Geosites, 4*(2), 145–152.

Butler, R. W. (1980). The concept of the tourist area life-cycle of evolution: Implications for management of resources. *Canadian Geographer, 24*(1), 5–12. Retrieved from. https://doi.org/10.1111/j.1541-0064.1980.tb00970.x/abstract.

Dhanoa, R. (2013). Eco-tourism: Conservation or exploitation. *International Journal in Multidisciplinary and Academic Research (SSIJMAR), 2*(4), 22–37.

Google Maps. (2014). Tasik Kenyir, Terengganu, Malaysia. Retrieve from https://www.google.com/maps/place/Kenyir+Lake/@4.9975358,102.5101097,11z/data=!3m1!4b1!4m5!3m4!1s0x31b62bfa9e9475cd:0x250d30b733b4ae9e!8m2!3d5.0039622!4d102.6388451.

Gossling, S., Peeters, P., Ceron, J. P., Dubois, G., Patterson, T., & Richardson, R. B. (2005). The eco-efficiency of tourism. *Ecological Economics, 54*, 417–434.

Hunter, C., & Green, H. D. (1995). *Tourism and the environment: A sustainable relationship?* London: Routledge.

KETENGAH. (2012). Latar Belakang KETENGAH. Web Rasmi, Lembaga Kemajuan Terengganu Tengah (KETENGAH). Retrieved from http://www.ketengah.gov.my/latar-belakang.

Kiper, T. (2013). Role of ecotourism in sustainable development. Advances in Landscape Architecture, licensee InTech. Retrieved from https://doi.org/10.5772/55749

MIT. (1997). System dynamics in education project (SDEP). Massachusetts Institute of Technology. Retrieve from http://web.mit.edu/sysdyn/sd-intro/

Nakisah, M. A., & Fauziah, A. H. (2003). *Setiu wetlands: Tranquility admist plenty*. Kuala Nerus: Kolej Universiti Sains dan Teknologi Malaysia.

Orams, M. B. (1995). Towards a more desirable form of ecotourism. *Tourism Managment, 16*(1), 3–8.

Safiih, M. L., Mohd, N. A., Naeim, A. R., Ikhwanuddin, A. M., Madzli, H., Syerrina Z., & Marzuki I. (2016). System dynamics model for analyzing the eco-aquaculture system with policy recommendations: Case Study on Integrated Aquaculture Park (/-Sharp), Setiu, Terengganu, Malaysia. *Simposium Matematik Kebangsaan, SKSM-23*. Proceeding Paper.

Schubert, S. F., & Brida, J. G. (2009). In Á. Matias, P. Nijkamp, & M. Sarmento (Eds.), *A dynamic model of economic growth in a small tourism driven economy*. Heilderberg: Physica-Verlag.

Statista. (2015). Global international tourism revenue from 2000 to 2015 (in billion U.S. dollars). The Statistics Portal. Retrieve from https://www.statista.com/statistics/273123/total-international-tourism-receipts/.

Stearns, F., & Montag, T. (1974). *The urban ecosystem: A holistic approach*. New York: Dowden, Hutchinson & Ross.

Sterman, J. D. (1992). *System dynamics modeling for project managment*. Cambridge, MA: System Dynamics Group Sloan, School of Managment, Massachusetts Institute of Technology.

Sterman, J. D. (2000). *Business dynamics: Systems thinking and modeling for a complex world*. Boston: McGraw Hill Inc. Publication.

Swarbrooke, J., & Horner, S. (2001). *Business travel and tourism*. Routledge, Taylor & Francis Group, Oxfordshire: New York.

Tan, C. L. (2014). Kenyir Geopark: Protecting a geological wonder. The Star. Retrieved from http://www.thestar.com.my/lifestyle/features/2014/10/27/kenyir-geopark-protecting-a-geological-wonder/.

Teegavarapu, R. S. V., & Tangirala, A. K. (2005). Modeling water quality management alternatives for a nutrient impaired stream using system dynamics simulation. *Journal of Environmental Informatics, 5*(2),73–81.

Tourism Malaysia. (2015). Malaysia tourism statistics in brief. Tourism Malaysia Portal. Retrieved from http://mytourismdata.tourism.gov.my/.

Tourism Terengganu. (2009). Introduction of Terengganu. Terengganu tourism authorities. Retrieved from http://tourism.terengganu.gov.my/index.php/2013-03-25-09-25-49/islands-beaches.

Unit Perancangan Ekonomi Negeri (UPEN) Terengganu. (2014). Sektor Pelancongan Negeri Terengganu. (N. M., & A. R., Interviewers).

United Nations Environment Programme (UNEP). (2002). United Nations Environment Programme (UNEP) 2002 annual report. United Nations Environment Programme. Retrieved from http://www.unep.org/gc/gc22/Media/UNEP_Annual_Report_2002.pdf.

Villepontoux, S. (2013). Ecotourism, concept and reality. What are perspectives?. HAL Archives-Ouvertes, France. Retrieved from https://hal.archives-ouvertes.fr/hal-00834646v1/document.

Wang, Y., & Zhang, X. (2001). A dynamic modeling approach to simulating socio-economic effects on landscape changes. *Ecological Modelling, 140*(1–2), 141–162.

Livelihood Resilience: The Case of Sungai Terengganu Communities

Suriyani Muhamad, Mohd Nasir Nawawi, Noorhaslinda Kulub Abd Rashid, Suhal Kusairi, Nik Fuad Nik Mohd Kamil, and Hazman Samsudin

Abstract Livelihood in river settings may appear simple, but upon closer examination, it is multifaceted and dynamic. The elements of uncertainty in sustaining a satisfactory livelihood are always present. The concept of resilience is about individuals, households or groups making a living and attempting to satisfy their consumption and economic necessities, surviving with uncertainties and responding to new opportunities. The earlier approaches in livelihood studies regarded poor people as inactive or passive victims. However, the trend has changed with greater interest to study on the survival strategies of the poor. In particular, consideration is given to the living experiences of households and the community. The objective of this study is to apply the concept of resilience as a diagnostic approach to the understanding of rural communities' livelihoods. The analysis highlights the resilience aspects of the rural community and their livelihood strategies for comfortable living.

Keywords River community · Sustainable · Livelihood · Resilience

Introduction

In recent years 'rural resilience' has attracted the attention of researchers due to the notion that the welfare of the rural communities has become less favourable. Resilience is strongly related to the ability of an individual to face challenges in life. In general, rural communities face more challenges in their daily lives compared to those living elsewhere. The strength to meet these challenges varies among the community members. The resiliency of the rural communities to manage unanticipated or unpredictable social or environmental changes have been discussed by many researchers (McManus et al. 2012; Wilson 2012; Bristow 2010; Pendall et al. 2010).

S. Muhamad (✉) · M. N. Nawawi · N. Kulub Abd Rashid · S. Kusairi · N. F. Nik Mohd Kamil · H. Samsudin
School of Social and Economic Development, Universiti Malaysia Terengganu, Kuala Nerus, Terengganu, Malaysia
e-mail: suriyani_md@umt.edu.my

M. T. Abdullah et al. (eds.), *Greater Kenyir Landscapes*,
https://doi.org/10.1007/978-3-319-92264-5_23

Among researchers, the concept of community resilience is commonly related to disasters. It has become associated with enhancing the well-being of the community through adaptive behaviour that permits some level of influence over the future direction (McManus et al. 2012). Rural communities have also been urged to deal with the dynamic of the changes rather than passively watching uncontrollable external forces (Deveson 2003).

Community resilience concept as highlighted in this study is related to the ability and strength of rural communities for survival and livelihood. In studying community resilience, we begin with community-level resources that are embedded within or accessible to communities. These resources can offer communities the capacity to adapt in positive ways to risk (Mowbray et al. 2007). This study relies on the model of community resilience as proposed by Mowbray et al. (2007). The simplified model of community resilience as suggested by Maybery et al. (2009) is then used as a conceptual framework (Fig. 1).

Three types of resources considered as factors that contribute to the resiliency of communities (Fig. 1). The first component is social assets. Social assets are resources such as a relationship built within the communities. This relationship is based on shared values and trust which normally promote teamwork among community members. A relationship includes regularity and nature of interactions between neighbours, the social connections between community members and other forms of interactions. The second component is institutional resources or service agency assets which comprise of community organisations that support the needs of the community. It includes schools, medical services, politically based service centres, libraries and recreational programs. The third component of community resilience resources is economic resources which include employment opportunities, family income and assets, land use and property values and infrastructure. The model also outlines the community risk factors such as economic disadvantage and poor physical conditions and support. This model adopted for investigating the rural community resilience

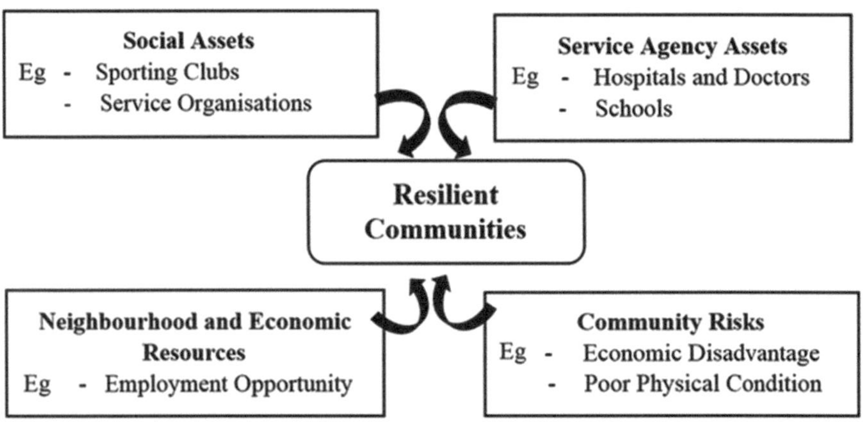

Fig. 1 Model of community resilience – social, service agency and neighbourhood and economic resources – and community risks. (Adapted from Maybery et al. 2009)

along the Sungai Terengganu with the focus on Hulu Terengganu communities. The relevance and suitability of the variables will be suited based on this case study.

Study Area

Hulu Terengganu district located south-west of Kuala Terengganu where it consists of eight subdivisions of a district, namely Kuala Berang, Hulu Berang, Kuala Telemong, Hulu Telemong, Tersat, Jenagur, Tanggol and Penghulu Diman. It is here that the long stretch of Terengganu river meets its gate into Tasik Kenyir from the mouth of South China Sea. Figure 2 shows the location of the Hulu Terengganu District.

Most of the river communities in the subdivision of the district centred along the Sungai Kuala Terengganu. Being near to Tasik Kenyir the land is fertile and therefore

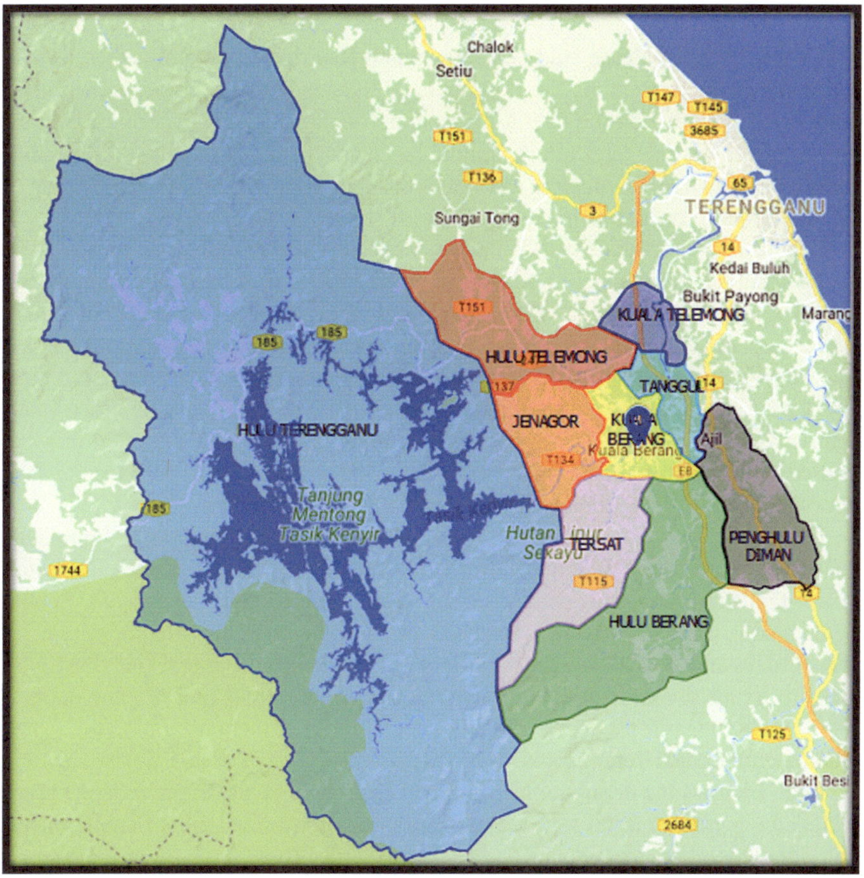

Fig. 2 The Hulu Terengganu District map. (Source: District and Land Office of Hulu Terengganu 2016)

suitable for agricultural and other land and water based activities. The products of these activities are traded in other areas, especially in Kuala Terengganu. Historically, the Sungai Terengganu is used as a pathway for transportation. Due to its strategic location which allows easy movement of goods via rivers, Kuala Berang was the place where the earliest river communities evolved in Terengganu (Moore 1998).

The evidence from Terengganu Inscription Stone discovered in Kuala Berang in 1887 (Ngah 2009) shows that Kuala Berang has been inhabited somewhere between March 1303 and February 1387 (Fatimi 1963). The exact date is somewhat arbitrary due to some damages on the original inscription stone which inflicted the date area written in the stone. In fact, Kuala Berang was the first capital city of the modern Sultanate of Terengganu. Later Kuala Terengganu was declared as the capital city somewhere in between 1708 and 1733 (Istana Terengganu 2016).

Terengganu Riverine Communities

This paper outlines a case study of Terengganu Riverine Communities focuses on the resilience factors among the communities. Respondents were selected based on households who rely on the river as a resource for carrying out economic activities. The activities include agriculture, animal husbandry and cage culture. All respondents reside in Hulu Terengganu District. A conceptual framework outlined in Fig. 1 has adopted as a resilience model for the case study.

The data for this study was obtained through a structured questionnaire. The questionnaire was designed to quantify the perceptions of the respondents towards the value of community assets and risks that affect their well-being. Four main aspects are social assets, service assets, economic resources and community risks. The items based on the research team's knowledge of the local perspective. Based on the pilot study carried out earlier, the demographic profile of the potential respondents was found to be relatively inferior in many aspects. Old age, low income, limited skills and low education level dominate the profile. Therefore, a face-to-face interview method was used to obtain more accurate information.

The survey instrument consists of semi-structured questionnaires and guidance questions to get more accurate information about the subject under discussion. Since the sampling unit comprised of communities that depend on Sungai Terengganu, interviews were conducted at morning markets which operate on Wednesdays and Thursdays. The majority of the traders at this market are those who are directly involved in producing agricultural commodities and in cage culture. These two groups of producers directly depend on the functions and services provided by the Sungai Terengganu.

Table 1 shows the demographic profile of the respondents. Eight variables were included – gender, marital status, age distribution, education, area of employment, household member, dependency and settlement area. There are 23% of the respondents are men while 77% of them are women; 3.3% single, 80% married and 16.7% widowed. The majority of the respondents are between 41 and 60 years old. Regarding employment, 80% of the respondents are farmers, 16.7% are involved in

Table 1 Demographic profile of respondents

	Frequency	Percent (%)
Gender:		
Male	7	23
Female	23	77
Marital status:		
Single	1	3.3
Married	24	80.0
Widow	5	16.7
Age distribution:		
30–40	2	6.7
41–50	2	6.7
51–60	12	40.0
61–70	13	43.3
71 or older	1	3.3
Education:		
No formal education	5	16.7
Primary school	15	50.0
Secondary school	6	20.0
High secondary school	4	13.3
Employment:		
Agriculture	24	80.0
Cage culture	5	16.7
Wholesale	1	3.3
Household member:		
Minimum	0	
Average	4.5	
Maximum	11	
Dependency:		
Average	2.73	
Maximum	8	
Subdivision of district:		
Kuala Berang	21	70.0
Jenagur	2	6.7
Tersat	1	3.3
Tanggol	1	3.3
Hulu Telemong	5	16.7

cage culture while the remaining 3.3% are involved in wholesaling. About 17% had no formal education, 50% obtained elementary school education, 20% attended secondary school, and 13.3% went to high schools. The majority (70%) of the respondents lived in Kuala Berang and 16.7% in the Hulu Telemong subdivisions. The others reside in Jenagur, Tersat, and Tanggul subdivisions. The average household size is 4.5 with a maximum of 11. The average size of dependents is 2.73.

The majority of the household heads is female and married. Regarding age, 54% of the respondents are below than 60 years old. Half of the respondents have at least pri-

Table 2 Dependence/children's profile

	Living with parents		Living elsewhere	
	Male	Female	Male	Female
Marital status (%)				
Single	90.52	76.92	11.54	14.81
Married	9.52	23.08	88.46	81.48
Widowed	–	–	–	3.70
Age average (year)	21.67	19.54	32.00	32.48
Education (%)				
No formal education	9.52	15.38	0.00	0.00
Primary school	19.05	7.69	7.69	0.00
Secondary school	47.62	46.15	38.46	51.85
High secondary school	14.52	23.08	34.62	18.52
Diploma	9.52	7.69	19.23	29.63
Type of employment (%)				
Own business	19.05	0.00	30.77	18.52
Private sector	23.81	7.69	30.77	3.70
Public sector	–	–	34.62	33.33
Unemployed	57.14	92.31	3.85	44.44
Income per month (RM)	1411	2600	2284	2707
No of children/dependence (person)	21 (62%)	13 (38%)	26 (49%)	27 (51%)

mary school education. The main economic activity is agriculture. The analysis of data collected implies that the younger segment of the communities is not interested in agricultural activities. Hence the sustainability of agriculture is uncertain. However, the situation may change with the current economic transformation program of the government with social innovation being pushed as an agenda for change. The transformation program covers all aspects of economic activities including agriculture, land-based fishery activities, tourism, and others (Economic Transformation Programme 2016).

Table 2 gives the profile of dependents or children of the respondents. The dependents (children) are classified into two – those who live with their parents and those who reside elsewhere. About 40% live with their parents in the study area, and 60% live elsewhere.

For the children/dependence who lives with their parents, 62% are male and 38% female. About 90% of the male are singles with an average age of 22 years. Regarding the type of employment, 19% of the male conduct their own business, 24% work in the private sector while the majority of them (57%) are unemployed. As for the education background, 19% goes to primary school, 47.62% are a secondary school, and 14.52% are in high secondary school. About 9.52% are diploma holders with an average monthly income of RM1411. As for the female, about 77% are single, and the rest are married. The average age of the female dependents is 20 years old. Only 8% of the female are employed. The high percentage of unemployed females (92%) are of concern but can be viewed positively as a potential labour force.

The profile of the children who do not live with their parents is as follows: male and female are about equally divided, 49% male and 51% female. For the male, about 11% are singles and 89% married with an average age of 32 years. About 38% of the male conduct their own business, 31% work in the private sector, 35% are government servants, and 3.85% are unemployed. Education wise about 7.69% are in primary schools, 38.46% in secondary school, and 34.62% in high secondary school. About 19% are diploma holders with an average monthly income of RM2284. About 15% of the female children are singles, 81% married, and 3.70% are divorced. The average age is 32 years old. About 18% of the female children have their own business, 3.70% work in private sectors, 33% are government servants, and 44% are unemployed. Female diploma holder earns an average monthly income of RM2707. In short, academic achievement is a good measure of the progress of communities. Among the respondents, the females are more educated than the male. Consequently, the average income of women is higher.

Economic Activities of the River Communities

Economic Activities

As discussed earlier, the river community in Hulu Terengganu is actively involved in agricultural (Fig. 3a, b) and aquaculture (Fig. 4a, b) related activities. Most of the activities conducted on the small to medium scale basis. The farmers are divided into three groups. The first group of farmers is involved in primary production only. The second group cultivates the farms and trade the products themselves. The third group comprises of traders of primary products (Table 3 and Fig. 5).

As shown in Table 3, about 20% of the respondents are farmers who produce the products and sell them to traders. They are not involved in trading due mainly to age factor. The second group of farmers (67%) produce and sell the products by themselves. The high percentage of this group shows that the majority of the farmers prefers to cultivate the crops and sell the products by themselves. They may double

Fig. 3 Farming activities. (Source: Museum Terengganu, 2016)

Fig. 4 Cage culture. (Source: Museum Terengganu, 2016)

Table 3 Three groups of river communities

Economic activities	Percentage	Own land (acre)	Lease land (acre)	Estimated net income (RM/month)
Farming/husbandry only	20	4	0.4	975
Farming/animal husbandry and sell directly to the market	67	2.8	1.8	1967
Traders of primary products	13	–	–	1075

Fig. 5 Direct selling at farmers' markets. (Source: Kamarudin 2016)

Table 4 Expected continuity of current economic activities

Economic activities	Inherited to descendant?		Estimated idle land (acre)
	Yes	No	
Farming/husbandry only	33%	67%	17.6
Farming/husbandry and direct selling of products	20%	80%	73.6
Middleman	0	100%	–

their income by trading the products themselves. About 13% of the community members are traders who purchase products from the first group. The majority of those in the first and second groups conducts activities on a small scale only. Most of the farmers work on their land.

The average monthly income of those involved in farming activities ranges from RM1000 to RM2000. It supports the evidence that the community here is small-scale producers. Not all smallholders are bona fide farmers; some are temporary agriculture wage earners while others derive income from non-agricultural, and some are in the form of remittances.

The Continuity of the Economic Activities

As explained in the profiling section, 47% of the Sungai Hulu Terengganu community who are involved in agriculture are above 60 years old. The questions are, in the future will the descendants of the river community continue with the agricultural activities?

Based on Table 4, the respondents indicated that 67% of the current economic activities in Hulu Terengganu would not be continued by their descendants. During the interview session with Kuala Berang District Headman, Alias (2016), the rather unstable and low income from traditional agricultural activities are the factors that drive the younger generation out of these sectors. Most the younger generations reside outside the area and are involved in modern economic activities. The estimated idle lands show that 91% of the agricultural land will be deserted in the future. How the traditional Hulu Terengganu River community looks like in 10–20 years' time could be one of the interest if the recent trend continues.

Community Resilience

The concept of resilience covers various dimensions such as ecological, social and economic. As such what do communities perceive as the most important characteristics to be resilient? To achieve the success of community resilience, it is important to know how to measure the sustainability of a community that responds to

Table 5 Inventory reliability test

Community asset and risk	Cronbach's alpha (N = 30)	No. of items
Social asset	.773	4
Service agency	.821	4
Economic source	.794	4
Community risk	.633	5

unexpected changes (Thulstrup 2015; Tran 2015; Nik Mohd Kamil et al. 2011). Over time how do these responses reflect shifts in community attitudes and behaviours towards risk? This section discusses in-depth the definition of a community asset and risk. The validity and reliability of the research instrument are important to measure the value and level of internal consistency by using Cronbach's Alpha (Chua 2006). Reliability analysis using the Cronbach's Alpha test revealed a value greater than 0.60. It means that the item measurements are valid and acceptable. Based on Table 5, the value of Cronbach alpha shows the internal consistency reliability is acceptable. Referring to Sekaran (1992) and Mujis (2004), the value of alpha often used as a benchmark is between 0.60 and 0.70. Therefore, the research instruments are qualified and acceptable based on the result provided.

Community Asset and Risk

The concept of resilience has become popular since it can measure how sustainable society is (Tran 2015; Wilson 2012; Walsh 2007; Walker and Salt 2006). Resilience and sustainable livelihood are two concepts that are related to each other. Both are important in promoting sustainable development with community participation as important criteria. Therefore, resilience can be described as the ability to apparently recover from the extremes of trauma, deprivation, threat or stress is known as resilience (Atkinson et al. 2009). While sustainable livelihoods (SL) consist of asset, natural processes and entitlements that enable people to make a living, SL also deals with risks and how people cope with the pressures and adapt to changes that affect livelihoods (Nik Mohd Kamil et al. 2011).

The discussion here emphasises on the challenges typically faced by the community and how they adapt to the current economic situation. With the definition above, this paper focuses on community asset and risk factors towards the community resilience. Table 6 shows the mean and standard deviation value of the community asset and risk in river communities. These mean values explain on how the respondent ratings the impact of a community asset and risk factors on them with using 5 Likert's scale.

Across these four components of a community asset and risks, service agencies are the highest scoring component, with a mean 2.14, compared to other categories. Vice versa, the lowest scoring component was the economic source, with 1.91. These two asset types scored as providing 'some effect'. There are four components

Table 6 Community asset and risk

No.	Community asset and risk	Mean (S.D)
1	**Social asset**	**2.10** (0.67)
	Committee Of The Village Development And Security (JKKK)	2.03 (0.96)
		2.43 (0.77)
	Committee Mosque	1.73 (0.87)
	Community Service Organisations or Youth Association	2.20 (0.85)
2	**Service agencies**	**2.14** (0.81)
	Hospital/clinic	2.90 (0.61)
	School	2.40 (1.16)
	Public transport	1.87 (1.07)
	Library	1.40 (1.07)
3	**Economic sources**	**1.91** (0.80)
	Employment opportunities	1.70 (1.09)
	Income	1.97 (0.96)
	Business ownership/land/wealth	2.13 (0.86)
	Tourism attraction	1.83 (1.15)
4	**Community risk**	**1.96** (0.62)
	Lack of employment opportunities	1.80 (1.10)
	Physical condition and severe infrastructure	1.80 (0.85)
	High rate of crime,	2.57 (0.97)
	Environmental pollution	1.73 (0.94)
	Natural disaster (e.g. floods and droughts)	1.90 (1.03)

'a lot of effects' (a score of 3 and 4), 'some effect' (a score of 2) and 'some difficulty' (a score of 0 and 1)

that explain community asset and risks, such as social asset; service agency; economic source and community risk. For the first component, social asset, in the context of resilience in the river community, refers to the Committee of The Village Development and Security (JKKK), Mosque Committee, Sports and Recreation Clubs and Community Service Organizations or Youth Association. The mean score for a social asset component is 2.10, indicated that the respondent evaluates the component with moderate value, with 'some effect' they obtain rather than nothing. Furthermore, the highest score of 2.43 in mosque committee indicated strong effect with religious activities and community wellbeing. It shows that religiosity is an important aspect of their life.

On the other hand, service agencies (second component) provide good facilities such as a clinic or hospital, school, public transport and library to support the community assets. The highest mean score of this component shows the value was 2.14 for hospital or clinic while the lowest score was the library, with the mean value, 1.40, due to lack of the number of libraries provided and how they appreciate the library as sources of knowledge.

Economic sources of community asset (the third component) refer to employment opportunities, sources for generating income, business or wealth ownership

Table 7 Aspects of resilience

Community asset and risk	Level of resilience (frequency) Low	Moderate	High	Total
Social asset	12	6	12	30
	(40%)	(20%)	(40%)	(100%)
Service agency	8	2	20	30
	(26.7%)	(6.7%)	(66.7%)	(100%)
Economic source	4	17	9	30
	(13.3%)	(56.7%)	(30%)	(100%)
Community risk	7	10	13	30
	(23.3%)	(33.3%)	(43.3%)	(100%)

and tourism attractions. The mean score for this category was 1.91 (some effect), slightly lower compared to the first and the second components. The item for business ownership, land or wealth had the highest mean score of 2.13 (some effect). In Hulu Terengganu, a lot of lands is being left idle because their children have migrated to the cities to look for better jobs based on communication with Kuala Berang District Headman, Alias (2016). According to Alias, the river community also faces many risks, obstacles and challenges in their daily life. These include lack of employment opportunities, physical condition and poor infrastructure, high crime rate, environmental pollution, natural disaster and others (the fourth component). These scenarios explained the negative environment exist in the community of Hulu Terengganu. The risk is a threat to the locals because they are unable to enjoy some of the life's pleasures, shows the mean value of 2.57 (a lot of effects). Many do not enjoy a good quality of life due to the high crime incidences such as theft cases.

There are four components of a community asset and risk, explaining the local aspects of resilience such as social asset; service agency; economic source and community risk, as mentioned in Table 7. Regarding the social asset, the majority of respondent scores 'low' and 'high' for a social asset with the same value, 40% respectively. On the other hand, service agency such as clinic or hospital, school, public transport and library with the highest score of the respondent is 66.7%, moderate with 6.7% and low with 26.7%. Hence, most respondents unanimously agreed that service agency provided positive benefits and advantages to them.

Regarding economic sources of a community asset, the level of resilience shows that the highest score of respondents is moderate with 56.7% compared to low (13.3%) and high (30%). It indicated that the majority the respondents are satisfied with their standard of living although the level of resilience is moderate.

The results of this study also revealed that the community of Hulu Terengganu faces many risks and challenges to sustain their livelihood. Table 7 indicates that seven respondents score low level of resilience while another 10 and 13 respondents score the moderate and high level of resilience respectively. The involvement in drugs among the youth is a cause for concern. This situation will be exacerbated if the youth does not have permanent jobs and are unemployed. Under the category of

Table 8 Overall level of resilience

Level of resilience	Frequency	Percentage (%)
Low	17	56.7
Moderate	5	16.7
High	8	26.7

'low' level of resilience, the social asset has been found to provide the most benefits compared to other components of the asset. However, under the category of moderate level of resilience, the economic source is the main factor that affects to community asset. Lastly, under the category of 'high' level of resilience, the service agencies scored the highest, meaning that service agencies gave 'a lot of effects' to the community.

Table 8 shows the level of resilience among the river community in Kuala Berang. Overall, the level of resilience is low as indicated by the percentage of 56.7%.

The percentages of the moderate and high level of resiliency are 16.7 and 26.7 respectively. Despite the low level of resilience, the respondents feel that their life is quite good, with the availability of facilities and a cleaner environment. In short, satisfaction, and happiness are not only the materialistic measurement but more importantly, how they appreciate happiness with their surroundings and community. These respondents rate higher in the sense of perceptions of personal connection and being part of the community. This finding suggests that 'social asset' is the dominant factor that promotes interpersonal interaction and *ukhuwwah* (social ties) among the communities in Hulu Terengganu. Materialism, status and wealth are secondary.

Conclusion

The findings of this study recommend that the model of community resilience is well sustained by the broader community development and constitutes a valuable, applied framework on the community resilience. The model of community resilience provides a basis that is useful for future policy development, planning, interventions and research. Policymakers should recognise the social and economic assets that contribute to the resilience and well-being of river communities. The resilience model developed in this study can be applied to other river communities in Malaysia and elsewhere.

The main question is; what holds the future of the Sungai Terengganu communities? Preserving the existing communities and promoting growth requires a belief in the capacity of the communities in Hulu Terengganu to participate fully in profitable economic activities. The grassroots communities need to develop a vision for their future. They need to weigh the options for economic development and identify how to mobilise internal and external resources to meet the challenges. They also need to understand the adjacent river ecosystem and how stewardship can lead to sustainability. Therefore the river communities of Hulu Terengganu must develop strategies to achieve community resilience.

The need for an intervention to end the current situation is pressing. Today, the elder community members of Hulu Terengganu will fear for their livelihood. Majority lamented that Hulu Terengganu was once a thriving agricultural area, but now the community has become somewhat marginalised.

In a larger context, it is recommended that a comprehensive study for long-term sustainability of the Sungai Terengganu communities should be conducted. The study would include all elements of sustainable livelihood and would provide answers to the following issues: (1) How can River communities be sustainable? What changes are needed? What are the impediments and challenges to effecting change?; and (2) Is agriculture a viable option for the future? How can we have a sustainable agriculture? To remain in agriculture may be a traumatic experience for some. How could one explain to a corn producer in Hulu Terengganu who receives a price as low as 20 cents per cob when he sees a price tag of RM2.50 for a nicely steamed corn in Kuala Terengganu? Something is excessively wrong with the distribution system; (3) Is aquaculture an option? Is it economically feasible and how should start-up be funded?; (4) What are other feasible industries along the Sungai Terengganu? Do these other options encounter with or complement the agriculture and aquaculture sectors? What are the encounters and how do we overcome them? These are the issues that should be addressed in the future planning of river communities.

Thus far, under the eleven Malaysia plan (RMK-11) abundance of incentives was introduced to develop the rural communities which include the river community. It is parallel with the theme which emphasises on community centred economic development. Among the strategies planned is to structure the community better at the macro level. With greater macroeconomic resilience, the economy at the micro level can be better managed hence promoting rural livelihood resilience. The main focus is about amplifying the private consumption and investment, promoting the economy through productivity and knowledge-based economy and better public finance management. In achieving these objectives, the development of the economy should not segregate to a certain area but inclusive. Therefore, the development between rural and urban are aiming to be a balance. In this sense, it is hoped that the agricultural land of the Hulu Terengganu River community may find its way forward in the economy with innovative used and not be abandoned. The community is also hoped to better benefit from the knowledge-based economy strategy which may improve their capabilities hence widened their skills. If the objectives are achieved, the economic activities of Hulu Terengganu River community will be wider and hence increase their livelihood resilience.

Nevertheless, the strategy to empower the local communities is not something new in Malaysia. Numerous of incentives and policy have been regulating to this end. Despite the effort, the rural sees less economic improvement thus far. Therefore, the main issue in developing rural communities depends very much on the effectiveness of implementation rather than policy regulation. A policy without efficient enforcements is meaningless.

Acknowledgements The authors acknowledge the assistance and support provided by YB Dato' Rozi Mamat, ADUN Telemong, an officer of the District and Land Office, Hulu Terengganu and En. Mohd Azrin bin Alias, Kuala Berang District Headman for this study.

References

Alias, M. A. (2016, June 21).*Personal interview*.

Atkinson, P. A., Martin, C. R., & Rankin, J. (2009). Resilience revisited. *Journal of Psychiatric and Mental Health Nursing, 16*(2), 137–145.

Bristow, G. (2010). Resilient regions: Re-placing regional competitiveness. *Cambridge Journal of Regions, Economy and Society, 3*, 153–167.

Chua, Y. P. (2006). *Kaedah dan statistik penyelidikan: Kaedah penyelidikan buku 1*. Kuala Lumpur: McGraw-Hill (Malaysia) Sdn. Bhd.

Deveson, A. (2003). *Resilience*. Sydney: Allen & Unwin.

District and Land Office of Hulu Terengganu. (2016). Laman rasmi Pejabat Tanah dan Daerah Hulu Terengganu. Retrieved from http://pdtht.terengganu.gov.my/index.php?option=com_content&view=article&id=80&Itemid=98.

Economic Transformation Programme. (2016). Supporting policies: Overview. Retrieved from http://etp.pemandu.gov.my/Supporting_Policies-@-Overview.aspx.

Fatimi, S. Q. (1963). *Islam comes to Malaysia*. Singapore: Malaysian Sociological Research Institute.

Istana Terengganu. (2016). Laman rasmi istana Terengganu. Retrieved from https://web.archive.org/web/20150407123820/http://www.istana.terengganu.gov.my/kesultanan_terengganu.php.

Kamarudin, S. (Photographer)(2016). *Direct selling at farmers' markets*. Kuala Berang: Personal Photo.

Maritime: Terengganu River in Kuala Berang. (2016). [Photo] Museum Terengganu, (13–506). Kuala Terengganu.

Maybery, D., Pope, R., Hodgins, G., Hitchenor, Y., & Shepherd, A. (2009). Resilience and well-being of small inland communities: Community assets as key determinants. *Rural Society, 19*(4), 326–339.

McManus, P., Walmsley, J., Argent, N., Baum, S., Bourke, L., Martin, J., Pritchard, B., & Sorensen, T. (2012). Rural community and rural resilience: What is important to farmers in keeping their country towns alive? *Journal of Rural Studies, 28*, 20–29.

Moore, W. (1998). *Peninsular Malaysia and Singapore*. Singapore: Periplus Adventure Guides.

Mowbray, C. T., Woolley, M. E., Grogan-Kaylor, A., Gant, L. M., Gilster, M. E., & Shanks, T. R. (2007). Neighborhood research from a spatially oriented strengths perspective. *Journal of Community Psychology, 35*(5), 667–680.

Mujis, D. (2004). *Doing quantitative research in education with SPSS*. London: SagePublications.

Ngah, T.F. (2009). Penemuan batu bersurat. Retrieved from http://ww1.utusan.com.my/utusan/info.asp?y=2009&dt=0331&sec=Timur&pg=wt_01.htm.

Nik Mohd Kamil, N. F., Nasir, A., & Rashid, N. K. A. (2011). Factors that contribute to sustainable livelihood of the Orang Asli. *Journal of Sustainability Science and Management, 6*(2), 285–291.

Pendall, R., Foster, K. A., & Cowell, M. (2010). Resilience and regions: Building understanding of the metaphor. *Cambridge Journal of Regions, Economy and Society, 3*(1), 71–84.

Sekaran, U. (1992). *Research methods for business: A skill building approach* (2nd ed.). New York: Wiley.

Thulstrup, A. W. (2015). Livelihood resilience and adaptive capacity: Tracing changes in household access to capital in Central Vietnam. *World Development, 74*, 352–362.

Tran, V. Q. (2015). Household's coping strategies and recoveries from shocks in Vietnam. *The Quarterly Review of Economics and Finance, 56*, 15–29.

Walker, B., & Salt, D. (2006). *Resilience thinking: Sustaining ecosystems and people in a changing world*. Washington, DC: Island Press.

Walsh, F. (2007). Traumatic loss and major disasters: Strengthening family and community resilience. *Family Process, 46*(2), 207–227.

Wilson, G. A. (2012). Community resilience, globalization, and transitional pathway of decision-making. Geoforum. Retrieved from https://doi.org/10.1016/j.geoforum.2012.03.008.

Issues and Challenges in Domestic Water Services in Terengganu, Malaysia

Mahirah Kamaludin and Azlina Abd. Aziz

Abstract Efficient water supply services are crucial to people consumption as the water carries numerous function in people's daily life. In Terengganu, there are still deficiencies in access to services, which affect consumers in many areas. The inadequate water supply and poor quality of the services not only have undesirable influences on the health of the people but also affect the economy, environment, industrial and availability of water for various uses. Dealing the implications related to inefficient services can be a challenging undertaking. Therefore, this study was conducted in order to study the issues and challenges of domestic water services in Terengganu. Cheap water prices pressured heavily on water provider's side and it makes consumers take for granted, as they have no intention to conserve the resource. Presently, the state is ranked as the second lowest domestic water price in Malaysia and it was revised about 20 years ago. If the water provider is not charging the customer the amount that each additional unit of water costs to provide, then it leads to revenue instability which in the end it affect water services at consumer's side. In that case, inconsistency in services to cater high population growth because of rapid development of industrial and housing areas become challenges to water industry.

Keywords Domestic water service · Conservation · Terengganu

M. Kamaludin (✉)
School of Social and Economic Development, Universiti Malaysia Terengganu, Kuala Nerus, Terengganu, Malaysia

Institute of Oceanography and Environment, Universiti Malaysia, Terengganu, Kuala Nerus, Terengganu, Malaysia
e-mail: mahirah.k@umt.edu.my

A. Abd. Aziz
School of Social and Economic Development, Universiti Malaysia Terengganu, Kuala Nerus, Terengganu, Malaysia

M. T. Abdullah et al. (eds.), *Greater Kenyir Landscapes*,
https://doi.org/10.1007/978-3-319-92264-5_24

Introduction

The importance function of water to life cannot be denied, as water is the most valuable substance among all of our natural resources. Water has been a nucleus for its existence in every civilization. As such, there is no substitute for water and nothing can trade with this vital resource. There have been increases in water demand every year especially for household purposes. The household depends on a range of water sources with different characteristics such as quality, price, reliability and many more (Nauges and Whittington 2010). The uses of water vary such as for cooking, gardening, and it is also significance for industrial processes, recreational activities, and ecosystem support. Water management is becoming progressively more complicated because of high increases in population, increasing water consumption, water pollution, and land use conflicts (United Nations Population Information Network 1994).

Malaysia has easy availability of surface water from more than 150 rivers that contribute to 98% of the total national water use (Master Builders Association Malaysia 2008). Surface water consists of fresh water in lakes, streams, and rivers that collect and flows on the earth's surface (Tietenberg 2000). Malaysia has abundant surface water resources compared to its groundwater resources, but they are still not managed properly and many parts have issues on water stress (Chan 2009). Rapid economic and social developments are placing heavier demands on water resources. It is difficult for water companies to cater for the demand of large populations especially when they faced budgetary constraints to improve its services. Water supply coverage in Malaysia served to population is 96% in 2014 (Table 1).

Table 1 Urban and rural water supply coverage in 2013–2014 (% of population)

	2013			2014		
State	Urban	Rural	Average	Urban	Rural	Average
Johor	100	99.5	99.8	100.0	99.5	99.8
Kedah	100	96.5	98.3	100.0	96.5	98.3
Kelantan	59.5	63.4	61.5	61.3	65.1	63.2
Labuan	100	100	100	100.0	100	100
Melaka	100	100	100	100.0	100	100
Negeri Sembilan	100	99.8	99.9	100.0	99.8	99.9
Pulau Pinang	100	99.7	99.9	100.0	99.7	99.9
Pahang	100	96.0	98.0	100.0	96.0	98.0
Perak	100	99.2	99.6	100.0	98.2	99.6
Perlis	100	99.0	99.5	100.0	99.0	99.5
Sabah	99.8	73.1	86.0	99.8	73.1	86.5
Sarawak	99.7	76.0	83.8	99.8	76.6	94.0
Selangor	100	99.5	99.8	100.0	99.5	99.8
Terengganu	99.1	92.9	96.0	99.1	92.9	96.0
Malaysia	**97.0**	**92.5**	**95.1**	**97.1**	**92.6**	**95.3**

Source: Malaysia Water Association (MWA) 2015

The state with the lowest coverage of water supply in 2014 is Kelantan 61.3% for urban and 65.1% for rural areas.

Background of Study

Terengganu is one of the beautiful coastal states which include a number of marine parks such as Pulau Redang, Pulau Perhentian and Pulau Kapas. The state of Terengganu is divided into eight administrative districts of Setiu, Besut, Kuala Terengganu, Kuala Nerus, Hulu Terengganu, Marang, Dungun and Kemaman. This study portrays the performance and issues of water services in Terengganu. This study begins with the introduction of the study and a description of the case study. The second section is a discussion of water supply system issues in Terengganu, while the third section is an outline of management challenges and policy implications. The fourth section is a discussion of the study and sums up the conclusions for the issues of water supply services in the state.

Water Supply System Issues in Terengganu

Syarikat Air Terengganu Sdn. Bhd. (SATU) exclusively supplies water supply service in the state. Previously, Bekalan Air KIPC Sdn Bhd (BAKIPC) owned by Petrolium Nasional Berhad (PETRONAS) supplied treated water to the Kerteh Integrated Petrochemical Complex (KIPC) in Kerteh with capacity to produce up to 230 million litres of treated water per day. PETRONAS and Terengganu State Government signed a Share Sale Agreement (SSA) in May 2016 that results the ownership transfer of BAKIPC from PETRONAS to the State (Group Strategic Communications PETRONAS 2016). SATU incorporated under the Companies Act, 1965 on 18 November 1997 following the privatisation of water supply services in Terengganu (National Audit Department Malaysia 2011). SATU was established as a water corporation in order to increase efficiency and effectiveness of the services. The water corporation is a public entity and subsidiaries of the state government. Normally, state water department, state water corporation and private operator supplied water sector in Malaysia. Federal government is responsible to support financial assistance and state government is accountable to manage water supply and regulation.

SATU operates twelve water treatment plants in order to serve about 57.7% for domestic uses and 42.3% for non-domestic uses in 2014 in the state (MWA 2015). Figure 1 displays water treatment plants in Terengganu which is provided by solely water provider in the state, SATU. According to SATU, they used to have a water treatment plant in Kenyir that only operated for 3 years from 2004 to 2007. However, the operation was shut down because of the freshwater extraction in the area was not suitable for water supply.

Fig. 1 Water treatment plants in Terengganu. (Source: Syarikat Air Terengganu Sdn. Bhd. SATU 2015)

According to Ching (2012) water sector in the country still encountered many difficulties which can be concluded into four categories such as operational efficiency, ineffective governance and regulation, budget constraints and poor environmental performance. For instance, the operational inefficiency occurred because of two reasons, inability of water providers to decrease the high volume on non-revenue water (NRW) and presence of below cost tariff structure. NRW can be happened through **physical losses** from leaking and broken pipes, which are caused by poor maintenance and operations, the lack of active leakage control, and poor quality of underground assets. Some states demonstrated high NRW compared to NRW national average rate level (35.6%) which are Perlis (55.8%), Sabah (51.7%), and Kelantan (49.4%) in 2014 (Malaysia Water Association 2015). The decline of NRW is significant since less water loss means more water availability for consumers. In addition, some states in Malaysia have low water tariff that were recorded below than the national average price such as Pulau Pinang, Terengganu and Pahang (Ching 2012). Low water price limits their ability to generate revenues to maintain existing services and to broaden services in new areas.

Issues from Supply Side

According to SATU (Noriah, H., personal communication, 16 August, 2015), Terengganu may not have adequate financial sources to enhance water supply coverage. Water provider lacks the financial capability to cover the cost of operation and maintenance since the water price is just too low. Currently, Terengganu ranks second place lowest in Malaysia based on their domestic water tariff ranking in 2015, which is RM 0.52, applied on the first 35 m^3 and the last tariff revised in 1997 (MWA 2015). Normally, the water rates should be reviewed every 3–5 years but the water price in the state was last reviewed 20 years ago without any revision. Thus, the water company is operating the industry with very limited capital and they are very careful with expenses.

Below-cost tariff structure may threaten heavily on water provider with various services to implement in order to cater many costs, for example, abstraction, purification, water treatment, water distributions and many more. In the case of Terengganu, the government in the state regulates the water tariff. Presently, the lowest water tariff is in Pulau Pinang and Terengganu water tariff ranked second place lowest in 2015 for domestic user (Table 2).

Lower water prices make consumers take for granted on water sources and it is actually damaging the environment. By subsidizing domestic water tariff make them consume more water and it leads to water use inefficiently (Cassey 2005). Malaysians consume more water compared than other countries as Malaysians consume in average 212 litres per capita per day or about 141 bottles of 1.5 litres per person daily in 2014 (MWA 2015). Though, the consumption is far beyond the water consumption suggested by the World Health Organisation (WHO) that is 165 litres per day. Singapore, a water-stressed country consume about 165 litres per capita per day and they still making plans to decrease the usage to 147 l by 2020. A statistic by MWA (2015) states that Terengganu contribute 216 litres per capita per day which is above than national average consumption per capita per day. Asian Water Development Outlook (2007) stated that high water consumption generates

Table 2 Domestic water tariff ranking in 2014

States	Last tariff review	Average water tariff For first 35 m³ (RM/m³)
Pulau Pinang	2015	0.32
Terengganu	1997	0.52
Kelantan	2013	0.67
Sarawak[3]	1984	0.56
Perlis	1996	0.57
Pahang	1983	0.57
Sarawak[2]	1995	0.61
Sarawak[1]	1992	0.62
Kedah	2010	0.67
Negeri Sembilan	2015	0.68
Perak	2006	0.71
Melaka	2015	0.89
Selangor	2006	0.77
Labuan	1982	0.90
Sabah	2015	0.73
Johor	2010	1.05
National average		0.68

Source: Domestic Tariff (MWA) 2015
Sarawak[1] – Sibu, Kuching, Sri Aman, Limbang, Sarikei and Kapit
Sarawak[2] – Bintulu
Sarawak[3] – Other parts of Sarawak

pressure on the water supply infrastructure and water sources. It can degrade the environment through waste water disposal and the need to clear land to build water infrastructure in order to cater greater demand.

Access to clean, safe and affordable water supply are consumer's right in obtaining the best water services. A good form of tariff and subsidy should bring a bridge between the interests of all economic status of consumers including poor people and service provider. While, by subsidizing water tariffs oppose the concept of using pricing mechanisms to encourage efficient water consumption among consumers. Zetland (2011) points out that many countries were failed to execute a full cost recovery tariff structure which leads inefficient water consumption including Malaysia. The tariff structure makes the country is stated some of the lowest water price in the world. In Indonesia, only small minority of population has access to piped drinking water. Water quality is unreliable and must be boiled before consumption. The limitation by providing water supply to the population is the water price. Rietveld et al. (2000) said that governments want to keep low price in order to make the price affordable to households. Generally, inexpensive water price sounds politically and economically, however it generates low revenues and poor investment for water providers. Inexpensive water price encourage high level of access to water, including poorer sections of society (Biswas and Tortajada 2010).

Issues from Demand Side

Frequent interruptions can be occurred because of unexpected system failures in water services. Consumer concerns about the failure of infrastructure, which may affect their drinking water, since it may threaten the safety, quality, and health of the people (Tanellari et al. 2009). Contamination of water supply at homes can happen naturally because of human error or intentionally. For instance, the release of compost or other farming products can pollute the water. Moreover, water distribution can be affected by the amount of water available, quality of water and condition of physical components in distribution systems. In Terengganu, salt-water intrusion is one of the problems to some people since there is freshwater extraction station is located in the estuary. The maximum salinity intrusion was observed at 9.97 km inland from the river mouth and during the spring tide, higher tidal manages to penetrate saline water further inland (Lee et al. 2016).

Figure 2 and Table 3 show water services complaints as reported in 2014, in percentage and numbers, respectively. According to the official statistics received by the operator's Customer Services Department, MWA (2015), there are five major complaints were recorded. First, a problem related with pipe burst/breakages or leak, second; water quality, third; water pressure, fourth; water supply interruption and finally billing and meter. There are also other complaints related with illegal connections, asset vandalism, and many more.

Most states in Malaysia experience high complaints with regards to the problem related with pipe burst/breakages or leak. As of 2014, the highest percentage of the water services complaints comes from Selangor (472,785), followed by Kedah (413,275) and ranking number 3 of the 14 states is Johor (134,285). In Terengganu, about 57.84% of the complaints were reported to have a problem with the pipe burst/breakages or leak.

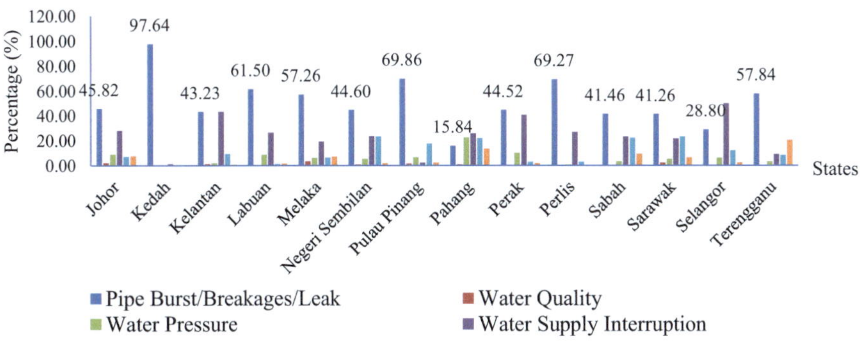

Fig. 2 Water services complaints in 2014 (%)

Table 3 Water services complaints in 2014

States	Pipe burst/ breakages/ leak	Water quality	Water pressure	Water supply interruption	Billing and meter	Others	Total
Johor	61,525	2960	12,015	37,737	9730	10,318	134,285
Kedah	403,527	734	2685	6059	50	220	413,275
Kelantan	16,257	526	743	16,257	3530	291	37,604
Labuan	893	4	128	385	19	23	1452
Melaka	24,712	1498	2715	8313	2785	3134	43,157
Negeri Sembilan	29,180	622	3539	15,425	15,338	1328	65,432
Pulau Pinang	43,517	977	4068	1413	10,903	1413	62,291
Pahang	2471	140	3480	3993	3433	2083	15,600
Perak	39,322	124	8923	35,765	2557	1626	88,317
Perlis	7456	8	88	2898	314	0	10,764
Sabah	18,050	227	1470	10,069	9631	4090	43,537
Sarawak	16,769	859	2162	8800	9447	2609	40,646
Selangor	136,163	2365	29,609	236,521	56,719	11,408	472,785
Terengganu	20,282	223	1172	3242	2958	7188	35,065
Total	820,124	11,267	72,797	386,877	127,414	45,731	1,464,210

Source: Malaysia Water Association (MWA) 2015

Management Challenges and Policy Implications

Terengganu ranked at second place in 2014 which demonstrates lowest water prices for domestic water services (RM0.52 for first 35 m^3) (Table 2). If the price is too low, the water company will be unable to maintain and sustain its operations and if the price of water is too cheap it leads to water wastage. Low pricing of water can cause a serious wastage of water, since demand is responsive to price. People believe that they can afford to pay the water bill thus they over use water every day and they do not take any positive actions to conserve water (Chan 2009). Positive changes in the price are a chance for consumers to enjoy clean and reliable water. Probably, water price increases will raise consumer awareness of water consumption and encourage water saving behaviour. Thus, by charging new and higher prices; it will induce lessening quantity in water demand (Nauges and Whittington 2010).

The state needs to increase their water tariff to cover the cost of replacing water pipes and maintaining water treatment plants, so that it can keep the water service quality at the required level. Besides, it can improve the water services delivery to consumers. Currently, water supply is at unacceptable level to cater consumer's demand. Consumers need an improvement in the services and water companies should know what needs to be done differently in the future. Thus, it requires major changes in planning management and investments from both sides.

Implication for Policy Makers

The government should regulate the market by controlling water prices in the state. The prices should be reliable in order to balance and protect both sides; producers and consumers. Water prices should be reflected the cost of water production in order to promote market efficiency. Even though the government fails to protect the consumers and sometimes it harms others, controlling the prices can avoid the pressure of operation cost increases. On the consumer's side, imposing higher prices will hurt them but it can encourage them to save the water, since there is no substitute for water. However, an increase in prices should not be a burden for the poor households considerably. Increment in prices should reflect the recoverable cost in the long term.

The government should facilitate the producer by giving subsidies to cover the high cost of production. The subsidies can at least help the producer to run the services smoothly in the best condition. When the government sets the ceiling prices and the water company is unable to cover the high cost of production, then the government should play its important role to help the industry by giving out the subsidies. Since water is considered as a 'public good', it should be supplied to people and meets socio-economic objectives. By granting subsidies to the water provider, the services provided by the company can be improved without having to increase higher tariff rate.

The policy makers should know how much investments are required, what should be taken to improve the services, and how one action taken will affect the others. Besides, the policy makers should take into consideration the socioeconomic characteristics, which somehow can affect the willingness to pay in order to improve water services. Sometimes socioeconomic characteristics give a big impact as it influences the significance of the model. In water demand studies, income as socioeconomic attribute plays a vital role. For example, higher levels of consumers, incomes make them able to use appliances using plenty of water in their lifestyles. According to Mahirah et al. (2013) and Vasquez et al. (2009), households with high level of income are willing to pay higher for water bill if there is an increase in water prices.

Implication for Water Companies

Water companies can propose options in allocating more investment or resources to upgrade the current domestic water services. The water companies can be customer-focused which can help them to improve their company's mission. There are numerous problems that should be highlighted in domestic water services in Terengganu. For instance, SATU should concentrate on some technical aspects which play a big role on the efficiency of the services. The old and rusty pipes should be replaced in order to improve quality of treated water supplied. Since the demand is increasing progressively, SATU should take actions to upgrade and build more new water treatment plants in order to capture more demands in the future. In addition, the water supply should be at a safe level with the aim to increase the quality of water and avoid the problems of murky and smelly water which affects consumers' household chores. Therefore, enhancing services to a high standard level may enhance the company's reputation.

Pricing of water services should precisely expose the true costs of providing high quality water and wastewater services to consumers in order to maintain infrastructure and plan for upcoming repairs, rehabilitation, and replacement of that infrastructure. The newly water price could encourage water companies to put consumers' priority at the right place with several changes which they can implement. If the water price is increased, it can lessen the financial burden of water companies by providing needed funds for further improvement in infrastructure, upgrade services and financial development. Revising prices can help the water companies to reduce cost somewhat but they still have to operate until the optimum production is achieved. The new designated prices would allow the consumers to manage their expenditures by controlling their water consumption. In addition, new water prices should be affordable to the people of the state, especially for poor households. The new price setting should be robust, fair and transparent to ensure long term sustainability of the water industry.

Good water management practices comprise successful maintenance based on water supply, water production, sewerage services, and many more. Thus, it necessitates a comprehensive plan in management for long-term benefits. By doing so, it will convince the consumers to contribute to the program, as they are satisfied and willing to pay for the maximum value for the services. However, water companies should balance the interest in improvement program and affordability of consumers to contribute in order to have an efficient policy. Water companies should learn from Perbadanan Bekalan Air Pulau Pinang Sdn. Bhd. (PBAPP) which has a good reputation in managing water supply services in Penang. For instance, PBAPP received the inaugural "Efficiency Award" in the water industry category in 2013, scoring a memorable "double" for Penang in the field of sustainable water supply management. PBAPP plays a vital role in good water supply management in Malaysia by meeting all consumers' water supply needs. The water company manages to meet the growing demand with new improved systems in their services by practising Certified Management System, NRW Management, Customer Care and new databases (iRMS and GIS). The water company is very committed in delivering the best services to bring satisfaction to their consumers and help to sustain the planet while still making profit.

Implication for Consumers

Consumers need to be educated about the specific value and benefits that they would get if the services were improved. What they would gain and what they would lose if the program of imposing higher water prices were implemented will affect their utility. Thus, by implementing an economic valuation studies in the state as in previous studies by Mahirah et al. (2013) in Kelantan, Anang (2013) in Johor and Rusli et al. (2011) in Selangor, many investment programs can be suggested for the benefits of the public.

By increasing the prices the public will be aware to conserve water and avoid water wastage. It will lead to economic effectiveness of water utilization among consumers. At the moment, consumers feel that the water is too cheap and this leads to water wastage and excessive water demand. Many alternative programs can be implemented to educate the public on water conservation such as seminar, poster, and workshop. Educating young generations about the significance of water in our life is required to maintain the benefits for future uses. Participating in awareness programs makes the consumers accountable for water conservation and appreciates the value of water, the most valuable among natural resources. By attending the programs, they will be aware of the need to use water efficiently and the programs will encourage positive attitude, motivation, and behaviour towards water management. Thus, education will indirectly help lessen the burden of water provider.

Punishment and fines can be implemented as an alternative for those water wasters who are not aware to conserve this precious sources. In some areas, which are having drought in their states such as Johannesburg, South Africa and Los Angeles, California, they were implemented water restrictions to users. Los Angeles has Water Conservation Plan, as the water wasters will face heavier fines and audits if they get caught watering the garden at the wrong time of day or using a hose to wash down a driveway. Johannesburg applies water restriction among users with heavy fines to be issued to non-complying residents as city's reservoirs are running dry because of water is being drawn out by consumers faster than it can be pumped in.

Conclusion

Water issues have been the focus of global and local concern. There are many challenges that still remain in water industry such as low water pressure, frequent interruption, murky and brackish water supply, impact of climate change on water supply and many more. The disruptions in water services affect household, business and even commercial production, which may result in huge losses. The function of water is vital since its quality may influence community health and people's lives itself rely on it. Therefore, the water supply program for domestic services needs to be continually improved to cater to the large growth in demand. Newly established programs, which bring to new upgraded facilities and infrastructures can help to improve capital, operation and maintenance costs.

In many circumstances, there is ineffective management of water resources. Since 1992, United Nations Conference on Environment and Development (UNCED) highlighted the need for water sector reforms throughout the world. The rapidly increasing demand for water over the past decades due to rapid growth in population, higher consumption levels in agriculture and industry sectors, household, urbanization and climate change requires the authorities and water providers to think of on new strategies to attain efficient water services. Poor pricing policies limit implementation of infrastructure projects to upgrade facilities and it implies low value of water, though the sources are valuable and insufficient. Water price should be able to cover both operations and capital expenditure in order to sustain in the long run. Efficient water price is very important to the country's future development, as it cannot be too cheap or pricey. Efficient water prices will secure stalk holders who invest in water services with sufficient revenues in order to manage, sustain, improve and expand the industry. Then it can provide better allocation of resources by increasing cost and operations efficiency and transparency in dealings of operators. Efficient allocation of resource will achieve when consumer's willingness to pay is equivalent to marginal cost of water supplying for consumers. Establishing water price to marginal cost signifies consumers pay a price equal to the increasing costs they impose on the system for providing the services. Accordingly, the consumers manage their consumption by changing behaviour to use water wisely.

Water is a fundamental and critical for sustainable and socio-economic growth. Water is also aiding as the critical link between the adaptations to climate change, strengthening the resilience of human well-being and the environmental systems. Since water not only has a crucial role in the dimensions of sustainable growth but also has a linkage to numerous key global challenges, therefore holistic approach is essential in water resource management for sustainable development.

Acknowledgements We sincerely acknowledge the financial support from the Ministry of Higher Education, Malaysia (MOHE) through the project Fundamental Research Grant Scheme (FRGS) 5934 vote no.: 8. The authors are jointly responsible for any inadvertent errors.

References

Anang, Z. (2013). *Assessing the effective demand for improved water supply service in malaysia: Focusing on Johor water company*. Unpublished doctoral thesis. Newcastle University, UK.

Asian Water Development Outlook. (2007). *Achieving water security for Asia*. Metro Manila: Asian Development Bank.

Biswas, A. K., & Tortajada, C. (2010). Future water governance: Problems and perspectives. *International Journal of Water Resources Management, 26*(2), 129–139.

Cassey, L. (2005). *Water tariff and development: The case of Malaysia, Working Paper Series No.117*. Manchester: Centre on Regulation and Competition.

Chan, W. N. (2009). Issues and challenges in water governance in Malaysia. *Iran Journal Environment Health Science Engineering, 6*, 143–152.

Ching, T.K. (2012). *Malaysian water sector reform: Policy and performance* (Unpublished doctoral thesis). Wageningen University, Netherlands.

Group Strategic Communications PETRONAS. (2016). PETRONAS hands over water treatment plant to Terengganu. Retrieved from http://www.petronas.com.my/media-relations/media-releases/Pages/article/PETRONAS-hands-over-water-treatment-plant-to-Terengganu.aspx.

Lee, H. L., Tanggang, F., Hamid, M. R., Benson, Y., & Razali, M. R. (2016). Modelling the influence of river flow and salt water intrusion in the Terengganu Estuary, Malaysia, presented at the Soft Soil Engineering International Conference 2015, Langkawi, Malaysia, 27–29 October 2015. Bristol: IOP Publishing.

Mahirah, K., Radam, A., Rahim, K. A., & Yacob, M. R. (2013). Consumer willingness to pay for domestic water services in Kelantan. *Pertanika Journal Social Science & Humanities, 21*(S), 1–12.

Malaysia Water Association (MWA). (2015). *Malaysia water industry guide 2012*. Kuala Lumpur: Malaysia Water Association.

Master Builders Association Malaysia. (2008). Safe and sustainable groundwater exploration. Retrieved from http://www.mbam.org.my/mbam/images/MBJ1Q08/CS1%20Safe%20&%20Sustainable%20(12-14).pdf.

National Audit Department Malaysia (2011). *Auditor General's Report 2010 on the Activities of the Terengganu State Government Departments and Agencies and the Management of State Government Companies*. Putrajaya, Malaysia: National Audit Department Malaysia.

National Water Services Commission. (2015). Water consumption. Retrieved from http://www.span.gov.my/index.php/en/statistic/water-statistic/water-rates.

Nauges, C., & Whittington, D. (2010). Estimation of water demand in developing countries: An overview. *The World Bank Research Observer, 25*(2), 263–294.

Rietveld, P., Rouwendal, J., & Zwart, B. (2000). Block rate pricing of water in Indonesia: An analysis of welfare effects. *Bulletin of Indonesian Economic Studies, 36*(3), 73–92.

Rusli, M., Radam, A., & Samdin, Z. (2011). Willingness to pay for domestic water service improvements in Selangor, Malaysia: A choice modeling approach. *International Business and Management, 2*(2), 30–39.

Syarikat Air Terengganu Sdn Bhd. (2015). *Statistik Loji Rawatan Air di Terengganu (Report)*. Kuala Terengganu: Syarikat Air Terengganu.

Tanellari, E., Bosch, D., & Mykerezi, E. (2009). *On consumers' attitudes and willingness to pay for improved drinking water quality and infrastructure*. Presented at the Agricultural & Applied Economics Association (AAEA & ACCI) 2009 Joint Annual Meeting, Milwaukee, Wisconsin, 26–28 July 2009. Milwaukee: AAEA.

Tietenberg, T. (2000). *Environmental and natural resource economics* (5th ed.). Boston: Addison Wesley Longman.

United Nations Population Information Network. (1994). Population and water resources. Retrieved from http://www.un.org/popin/fao/water.html.

Vasquez, W. F., Mozumder, P., Hernandez-Arce, J., & Berrens, R. P. (2009). Willingness to pay for safe drinking water: Evidence from Parral, Mexico. *Journal of Environment Management, 90*, 3931–3400.

Zetland, D. (2011). *The end of abundance: Economic solutions to water scarcity*. Amsterdam: Aguanomics Press.

Author's Profile

Aqilah Mohammad

Senior Lecturer (Mycology)
School of Marine & Environmental Sciences
aqilahmohammad@umt.edu.my

M. T. Abdullah et al. (eds.), *Greater Kenyir Landscapes*,
https://doi.org/10.1007/978-3-319-92264-5

Azlina Abdul Aziz

Associate Professor (Resource and Environmental Economics, Energy Economics)
School of Social and Economic Development
aqlina@umt.edu.my

Chong Ju Lian

Senior Lecturer (Genetic)
School of Marine & Environmental Sciences
julian@umt.edu.my

Edmand Bedurus

Postgraduate Student
Institute of Oceanography and Environment
edmand_andrew@yahoo.com

Hazlina Ahamad Zakeri

Senior Lecturer (Protein Biochemistry)
School of Fundamental Science
hazlina@umt.edu.my

Hazman Samsudin

Senior Lecturer (Financial Economy)
School of Social and Economic Development
hazman.s@umt.edu.my

Lee Gaik Ee

Senior Lecturer (Bryology)
School of Marine & Environmental Sciences
gaik.ee@umt.edu.my

Mahirah Kamaludin

Senior Lecturer (Environmental Economics)
School of Social and Economic Development
mahirah.k@umt.edu.my

Marzuki Ismail

Associate Professor (Air Pollution Management and Modelling)
School of Marine & Environmental Sciences
marzuki@umt.edu.my

Mohd Nasir Nawawi

Senior Lecturer (Environmental Economics)
School of Social and Economic Development
nasir@umt.edu.my

Mohd Tajuddin Abdullah

Professor (Zoology)
School of Marine & Environmental Sciences
mohd.tajuddin@umt.edu.my

Muhamad Safiih Lola

Associate Professor (Modelling Econometrics)
School of Informatics and Applied Mathematics
safiihmd@umt.edu.my

Muhammad Fais Fadzil

Postgraduate Student
Institute of Oceanography and Environment
a_bigthinker@yahoo.com

Noorhaslinda Kulub Abd Rashid

Senior Lecturer (Economy Development)
School of Social and Economic Development
lindarashid@umt.edu.my

Norhayati Mohd Tahir

Professor (Environmental Chemistry)
School of Marine & Environmental Sciences
hayati@umt.edu.my

Shahrudin Rohani

Senior Lecturer (Plant Ecology)
School of Marine & Environmental Sciences
rohanishahrudin@umt.edu.my

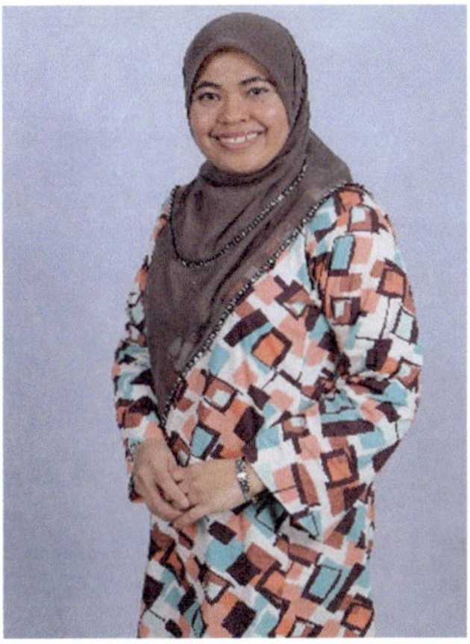

Suhal Kusairi

Senior Lecturer (Financial Economy)
School of Social and Economic Development
suhal@umt.edu.my

Suhaimi Suratman

Associate Professor (Environmental Chemistry)
Institute of Oceanography and Environment
miman@umt.edu.my

Suriyani Muhamad

Associate Professor (Social Economy)
School of Social and Economic Development
Suriyani_md@umt.edu.my

Suzana Misbah

Senior Lecturer (Mycology/Virology)
School of Fundamental Science
suzana_m@umt.edu.my

Tan Hock Seng

Senior Science Officer
Institute of Oceanography and Environment
hockseng@umt.edu.my

Thilahgavani Nagappan

Senior Lecturer (Advancement in Biodiversity)
School of Marine & Environmental Sciences
thilah.vani@umt.edu.my

Wahizatul Afzan Azmi

Associate Professor (Entomology)
School of Marine & Environmental Sciences
wahizatul@umt.edu.my

Wan Bayani Wan Omar

Senior Lecturer (Molecular Biology)
School of Marine & Environmental Sciences
bayani@umt.edu.my

Mohamed Nor Zalipah

Senior Lecturer (Vertebrate Ecology)
School of Marine & Environmental Sciences
zalipah@umt.edu.my

Glossary

A aquatic; antimicrobial; amphibians; aves; accommodation
B biogenic sources; biomass burning; birds; bats; bryophytes
C cytotoxicity; cover percentage; colony behaviour; composition; conservation;
D diversity; dynamic Model; development; domestic water service; Pielou evenness Index;
E etlingera; estimation; economic; environment; eco-tourism; E. Coli; E. megalocheilos; ecosystem; El-Nino
F flower morphometric; fruit-feeding butterflies; forest phenology; fruit-eating bats; fish richness; flora and fauna
G genetic diversity; gunung Tebu; Geniotrigona thoracica; gentamicin
H host specificity; herkogamy; habitat;
I Infestation; inflorescence; indigenous; inaugural;
K Kedah; Kuala Nerus; Kruskal-Wallis test;
L liverworts; livelihood; lagoon; Lasiocampidae; laksa asam;
M meteorological parameter; monsoon seasons; mosses; macrofungi; mangrove melastomatacea; mangrove; morphological characteristics; moth; Malaysia; mammals; multi-criteria analysis; Margalef Index;
N national water quality standard; nest architecture; nymphalidae; non-Volant small mammals; nutritious
O ornithology; oyster; ornaments; orchards
P physico-chemical properties; particulate matter; pollination; pteropodids; pollinators; Pulau Langkawi; population; principal component analysis
Q Quantification; questionnaires
R reptiles; richness; rats; River community; resilience; rhizophoraceae
S sustainable water; Setiu Wetland;; soils; sediment; sustainable management; ;;; seed; storage protein; species diversity; seed dispersal; squirrels; social; sustainability; Shannon Index; sampling survey;

M. T. Abdullah et al. (eds.), *Greater Kenyir Landscapes*,
https://doi.org/10.1007/978-3-319-92264-5

T Tasik Kenyir, Terengganu; Taman Negara Terengganu; trade-off analysis;
U UPEN; urbanisation; Unweighted Pair-Group Method
V vertical stratification; vulnerable; Vespertilionidae; visual encounter surveys
W willingness to pay; wetland; wellbeing; water quality index; water resources
X excreta
Y Youth Association
Z Zingiberaceae; zooplankton; Zingiberoideae

Index

M. T. Abdullah et al. (eds.), *Greater Kenyir Landscapes*,
https://doi.org/10.1007/978-3-319-92264-5

MIX
Papier aus verantwortungsvollen Quellen
Paper from responsible sources
FSC® C105338

If you have any concerns about our products,
you can contact us on
ProductSafety@springernature.com

In case Publisher is established outside the EU,
the EU authorized representative is:
Springer Nature Customer Service Center GmbH
Europaplatz 3, 69115 Heidelberg, Germany

Printed by Libri Plureos GmbH
in Hamburg, Germany